Collins Advanced Modular Sciences

People, Population and Behaviour

Peter Murray and Nick Owens

Series Editor: Mike Coles

Collins Educational

An Imprint of HarperCollins*Publishers*

Northern
Modular Science Scheme

Published by Collins Educational
An imprint of HarperCollins*Publishers*
77–85 Fulham Palace Road
Hammersmith
London
W6 8JB

© Peter Murray and Nick Owens 1997

Peter Murray and Nick Owens assert the moral right to be identified as the authors of this work.

First published 1997

ISBN 0 00 322410 4

All rights reserved. No part of this publication may be reproduced, stored in a retrieval system, or transmitted in any form or by any means, electronic, mechanical, photocopying, recording or otherwise, without the prior permission of the publisher.

Series design by Ewing Paddock at PearTree Design
Layout and composition by Newton Harris

Edited by Penelope Lyons

Picture research by Caroline Thompson

Illustrations by Barking Dog Art, Jerry Fowler, Fraser Williams, Hardlines, Peter Harper, Martin Shovel

Printed and bound by Scotprint, Musselburgh

Contents

Acknowledgements

Text and diagrams reproduced by kind permission of:
Blackwell Scientific Publications; Cambridge University Press; Hodder & Stoughton; Little, Brown & Co.; Oxford University Press; Southern Examination Group (SEG).

Every effort has been made to contact the holders of copyright material, but if any have been inadvertently overlooked the publishers will be pleased to make the necessary arrangements at the first opportunity.

The publishers would like to thank the following for permission to reproduce photographs (T = Top, B = Bottom, C = Centre, L = Left, R = Right):

Advertising Archive Ltd 76;
Allsport/C Cole 31, D Rogers 48, C Brunskill 54CL, M Powell 54C, M Hewitt 108TR;
Heather Angel 36R, 41, 45CR, 80B;
S A Thompson/Animal Photography 64CR;
Mr D Banerjee 27;
BBC Natural History Unit/J Sparks 70BR, C O'Reilly 70T, B Walton 96CL;
Biofotos/ J Thomas 72CL;
John Birdsall Photography 44, 45CL, 84R;
National Gallery, London/Index/Bridgeman Art Library, London 37R;
J C K Archive, London/Bridgeman Art Library, London 34;
British School of Motoring Ltd 35;
Bruce Coleman Ltd 96R/C & S Hood 9, J Burton 12TL&T, C Varndell 12C, A Purcell 12CL, 84C, H Reinhard 64CL, 65, 73, 92L&R, 93, G McCarthy 66, D Green 72CR, G Cubitt 74, J Cancalosi 81L, 104, P Davey

81R, Dr S Prato 86, I Arndt 92B, G Ziesler 100;
Bob Campbell 118;
Ron Austing, Cincinnati Zoo & Botanical Garden, Ohio 98L;
Cleveland Museum of Natural History, Ohio 115;
Tony Waltham/Geophotos 130;
The Ronald Grant Archive 102, 126;
Sally & Richard Greenhill 119CL, 125R, 131, 162TL,TR,CL,BR&BL, 163, 164;
Michael Holford 119B, 120;
Hulton Getty Collection 19T&L;
Hutchison Library/J Horner 52, E Lawrie 134, A Tully 143L, C Macarthy 153;
Inspire Foundation 6;
Emily King 175;
Carolco (courtesy Kobal) 22;
Andrew Lambert 108TL;
Don Lambert Photography 36TL&CL, 45B;
David Lawson 94;
Greg Lyons 38, 40, 127;
Magnum Photos/M Nichols 123, T Hoepker 177;
Metropolitan Police Service 77;
Peter Murray 98R, 99, 149, 150, 167, 171, 185;
NHPA/E A Janes 19B, W Paton 64T, M Lane 71, S Dalton 78T, B Beehler 84L, K Schafer 96TL, G Lacz 101, A Bannister 110,

145CR, C Ratier 147;
Nature Photographers Ltd/H van Lawick 82, 119TR, A Cleave 87L, P Sterry 90, J Hall 143R, E A Janes 143C;
Network Photographers/M Mayer 108CL, B Lewis 182C;
Orbis International 23, 26;
Nick Owens 87R;
Panos Pictures/E Marquez 18, N Cooper 37B, J Miles 125CL&T, S Sprague 125C, 184, J Hartley 142, 174, J-L Dugast 142, B Press 145TR, W G Feng 169;
M Hutson/Redferns 83;
Dr J M Round 54BL,BR&CR;
Science Photo Library 8, 14, 21, 28, 29, 32, 50, 54T, 59, 60, 78C 137, 154, 156, 173;
SHOUT Pictures 162CR;
Southern Water plc 182B;
Still Pictures/M Milligan 108CR, M Edwards 133, J Etchart 170;
Tony Stone Images title page, 17, 37CL;
Dr M Stroud 57T, 62;
Fiennes/Stroud/Howell/Sygma 47, 49, 57B;
C & S Thompson 138;
Roger Tidman 80T;
Dr G C Whitelam 39;
D Lawson/WWF-UK 106.

Cover photograph supplied by Tony Stone Images.

This book includes references to fictitious characters in fictitious case studies. For educational purposes only, photographs have been used to accompany these case studies. The juxtaposition of photographs and case studies is not intended to identify the individual in the photograph with the character in the case study. The publishers cannot accept any responsibility for any consequences resulting from this use of photographs and case studies, except as expressly provided by law.

To the student

This book aims to make your study of advanced science successful and interesting. The authors have made sure that the ideas you need to understand are covered in a clear and straightforward way. The book is designed to be a study of scientific ideas as well as a reference text when needed. Science is constantly evolving and, wherever possible, modern issues and problems have been used to make your study interesting and to encourage you to continue studying science after your current course is complete.

Working on your own

Studying on your own is often difficult and sometimes textbooks give you the impression that you have to be an expert in the subject before you can read the book. I hope you find that this book is not like that. The authors have carefully built up ideas, so that when you are working on your own there is less chance of you becoming lost in the text and frustrated with the subject.

Don't try to achieve too much in one reading session. Science is complex and some demanding ideas need to be supported with a lot of facts. Trying to take in too much at one time can make you lose sight of the most important ideas – all you see is a mass of information. Use the learning objectives to select one idea to study in a particular session.

Chapter design

Each chapter starts by showing how the science you will learn is applied somewhere in the world. Next come learning objectives which tell you exactly what you should learn as you read the chapter. These are written in a way which spells out what you will be able to do with your new knowledge, rather like a checklist – they could be very helpful when you revise your work. At certain points in the chapters you will find key ideas listed. These are checks for you to use, to make sure that you have grasped these ideas. Words written in **bold type** appear in the glossary at the end of the book. If you don't know the meaning of one of these words check it out immediately – don't persevere, hoping all will become clear.

The questions in the text are there for you to check you have understood what is being explained. These are all short – longer questions are included in a support pack which goes with this book. The questions are straightforward in style – there are no trick questions. Don't be tempted to pass over these questions, they will give you new insights into the work which you may not have seen. Answers to questions are given in the back of the book.

Good luck with your studies. I hope you find the book an interesting read.

Mike Coles, Series Editor
University of London Institute of Education, June 1996

Getting a grip

In 1981, Roger Fenn fell and broke his neck. The accident damaged his spinal cord above the point where nerves branch off to the arms. Nerve tissue, unlike bone, cannot repair itself easily, so Roger's arms and legs will be permanently paralysed. However, he can still move the muscles in his shoulders normally and in 1995 Roger had an operation to give him a 'bionic grip' in his right arm. He can now use his shoulder muscles to operate the electronic grip system (Fig. 1). The system gives Roger two sorts of grip: the power grip uses the whole hand to hold an object; the key grip uses the thumb and first two fingers to hold and manipulate an object. After 14 years of living without the use of his hand, Roger can now feed himself, clean his teeth, comb his hair, and let himself in and out of his house using an ordinary door key.

Designers of the grip system needed a thorough understanding of the structure of the nervous system. The electronics mimic the natural signals that pass along nerve cells to operate the muscles.

Roger Fenn uses the key grip to hold his pen for writing.

Fig. 1 The Neurocontrol Freehand System

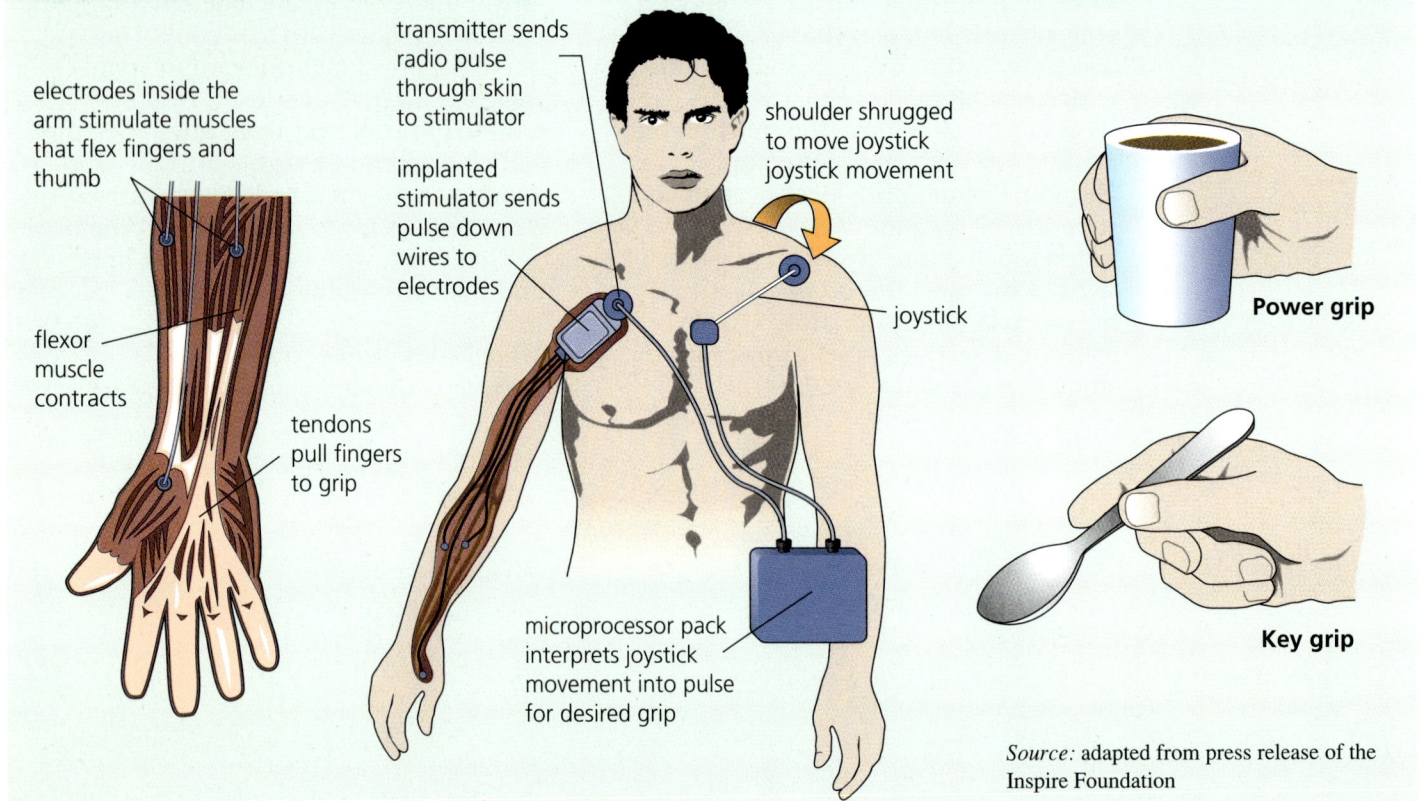

electrodes inside the arm stimulate muscles that flex fingers and thumb

flexor muscle contracts

tendons pull fingers to grip

transmitter sends radio pulse through skin to stimulator

implanted stimulator sends pulse down wires to electrodes

shoulder shrugged to move joystick joystick movement

joystick

microprocessor pack interprets joystick movement into pulse for desired grip

Power grip

Key grip

Source: adapted from press release of the Inspire Foundation

1.1 Learning objectives

After working through this chapter, you should be able to:

- **describe** the structure of a neurone and a nerve;

- **recall** the structure of the myelin sheath and explain its function;

- **explain** how nerve cells create a resting potential;

- **explain** how an action potential starts and travels down an axon;

- **list** the different types of sensory receptors;

- **explain** how a stimulus creates action potentials in receptor cells;

- **interpret** an electron micrograph of a synapse;

- **explain** how a message passes across a synapse;

- **recall** the processes of temporal and spatial summation, and adaptation;

- **predict** how specific drugs act at a synapse.

1.2 Component parts

The nervous system is composed of many nerve cells or **neurones**. They are more varied in shape and size than any other type of mammalian cell, but all have certain features in common (Fig. 2). The spinal cord and brain together make up the **central nervous system** (**CNS**). The human brain is possibly the most complex structure in the universe and contains roughly 10^{12} neurones. Many **nerves** branch out from

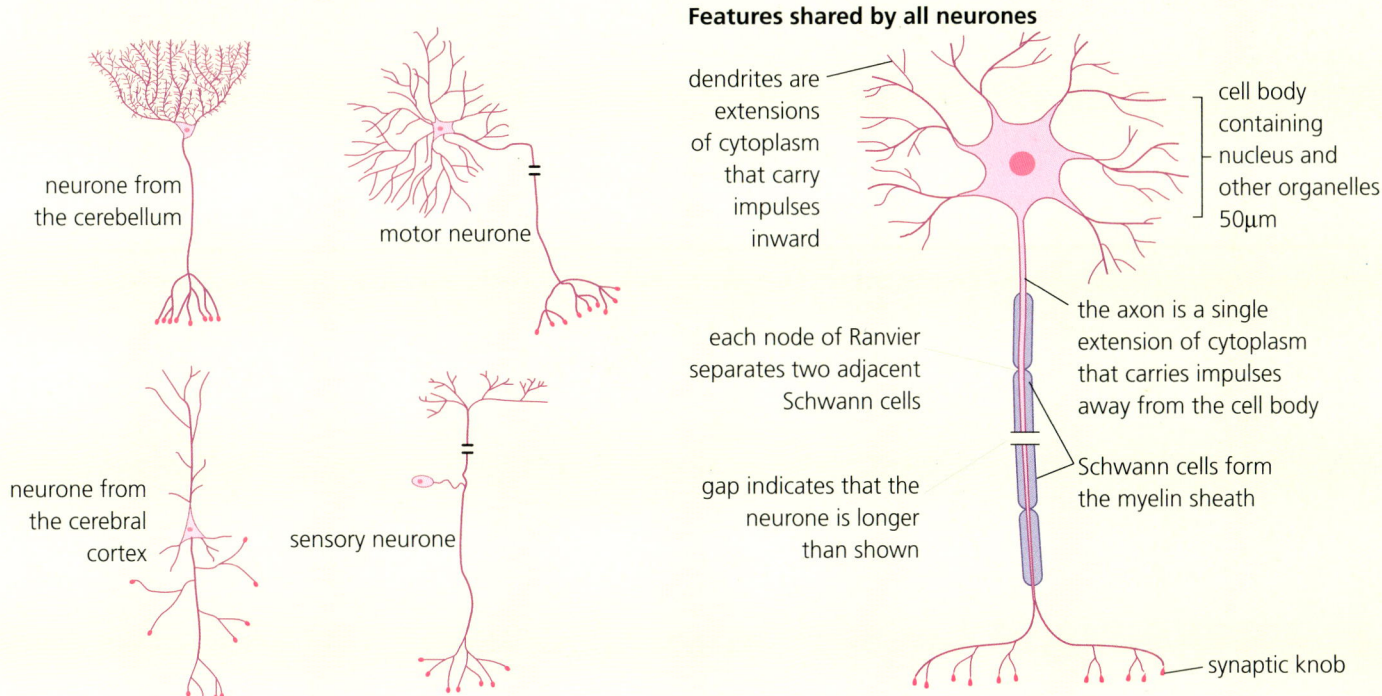

Fig. 2 A variety of neurones

neurone from the cerebellum

motor neurone

Features shared by all neurones

dendrites are extensions of cytoplasm that carry impulses inward

cell body containing nucleus and other organelles 50μm

neurone from the cerebral cortex

sensory neurone

each node of Ranvier separates two adjacent Schwann cells

gap indicates that the neurone is longer than shown

the axon is a single extension of cytoplasm that carries impulses away from the cell body

Schwann cells form the myelin sheath

synaptic knob

Table 1 Nerves and neurones

Type of nerve	Direction of impulses	Type of neurones within nerve	Effects of damage
motor nerve	from CNS to muscles	motor neurone	paralysis of muscles
sensory nerve	from sensory receptors to CNS	sensory neurone	loss of sensation

the CNS, carrying impulses to and from organs (Table 1). Each nerve is a bundle of neurones.

Nervous communication involves the initiation of a nerve impulse and its conduction from a **receptor** (a neurone that responds to a **stimulus** such as heat or pressure) to an **effector** (a muscle cell) via a coordinator in the CNS.

Neurones link together at junctions called **synapses**, which occur mainly in the CNS. One neurone can have as many as 10 000 synapses with other neurones. Impulses only pass from neurone to neurone through the synapses. All major axons are surrounded by **Schwann cells** which form a **myelin sheath** (Fig. 3). The main function of the myelin sheath is to increase the speed of conduction; it also electrically insulates axons from each other.

1 a What is the difference between a nerve and a neurone?
b Why does damage to a sensory nerve lead to loss of sensation?

Fig. 3 Schwann cells and myelination

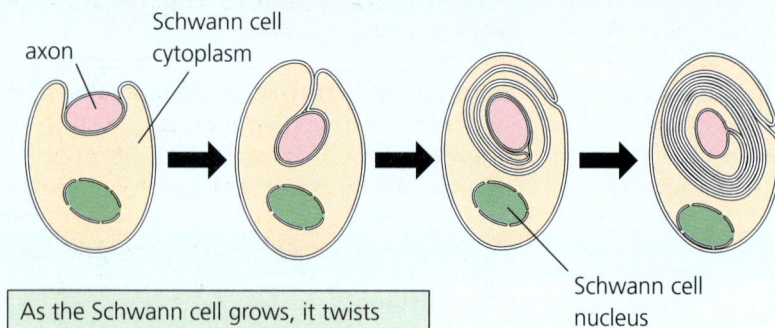

axon
Schwann cell cytoplasm
Schwann cell nucleus

As the Schwann cell grows, it twists around the axon many times; myelin is formed from many layers of Schwann cell membrane pressed together.

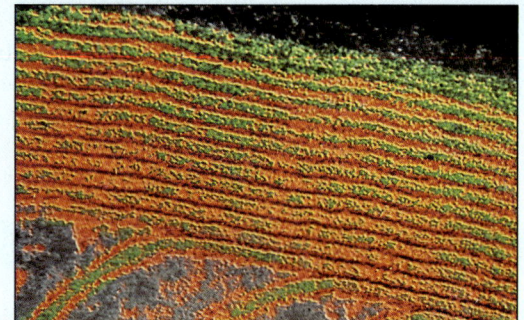

False-colour transmission electron micrograph of part of the myelin sheath (orange and green layers) of the human auditory nerve.

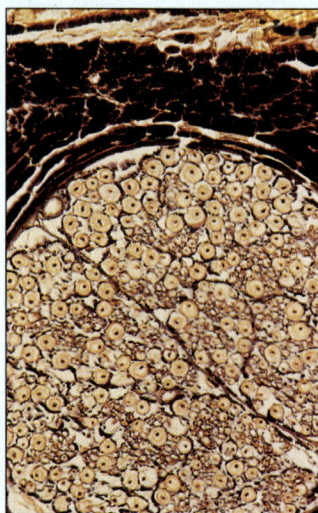

Light micrograph of a cross section through the human sciatic nerve showing myelinated nerve fibres of different sizes (yellow circles). The nerve is surrounded by loose connective tissue (dark brown).

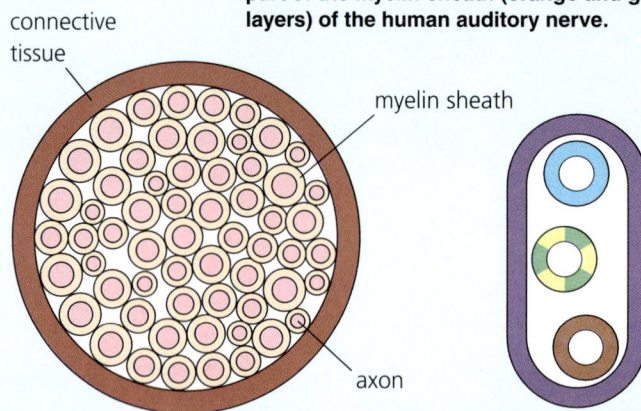

connective tissue
myelin sheath
axon

In a nerve, the axons are separated by myelin sheaths and the bundle is held together by connective tissue.

In an electric cable, the coloured plastic sheaths insulate the wires and the bundle is held together by an outer coat of plastic.

1.3 Waiting for action

When a neurone is *not* transmitting an impulse, the outside of the axon is positively charged compared with the inside. This potential difference is called the **resting potential**. A thin electrode can be placed inside an isolated squid axon to measure the resting potential; it is about -60 mV. A membrane with a potential difference across it is said to be **polarised**. When a membrane loses its potential difference, it has no charge and is described as **depolarised**.

If cyanide is added to a squid axon, the resting potential disappears. We know that cyanide stops respiration, so no ATP can be made in the presence of cyanide. Adding ATP to the squid axon restores the resting potential. So, ATP must be needed to *maintain* the resting potential. In fact, ATP is used to pump sodium ions (Na^+) and potassium ions (K^+) in opposite directions across the axon membrane. The **Na^+/K^+ pump** is an active transport mechanism that maintains a surplus of K^+ ions *inside* the axon, and a surplus of Na^+ ions *outside* the axon.

But what *establishes* the resting potential in an axon membrane? Permanently open diffusion channels in the membrane allow limited diffusion of Na^+ ions into the axon

Axons from mammals are very thin and difficult to study, but the axons of the squid are up to 1 mm wide. Although they lack a myelin sheath, squid axons are believed to work in the same way as mammalian axons.

and more rapid diffusion of K^+ ions out of the axon. This results in an excess of positive charge, in the form of K^+ ions, outside the axon. Without active transport, the diffusion of Na^+ ions and K^+ ions down their concentration gradients would lead to a gradual loss of the resting potential (Fig. 4).

2 Why is ATP needed to maintain the resting potential in an axon?

Fig. 4 Resting potential across an axon membrane

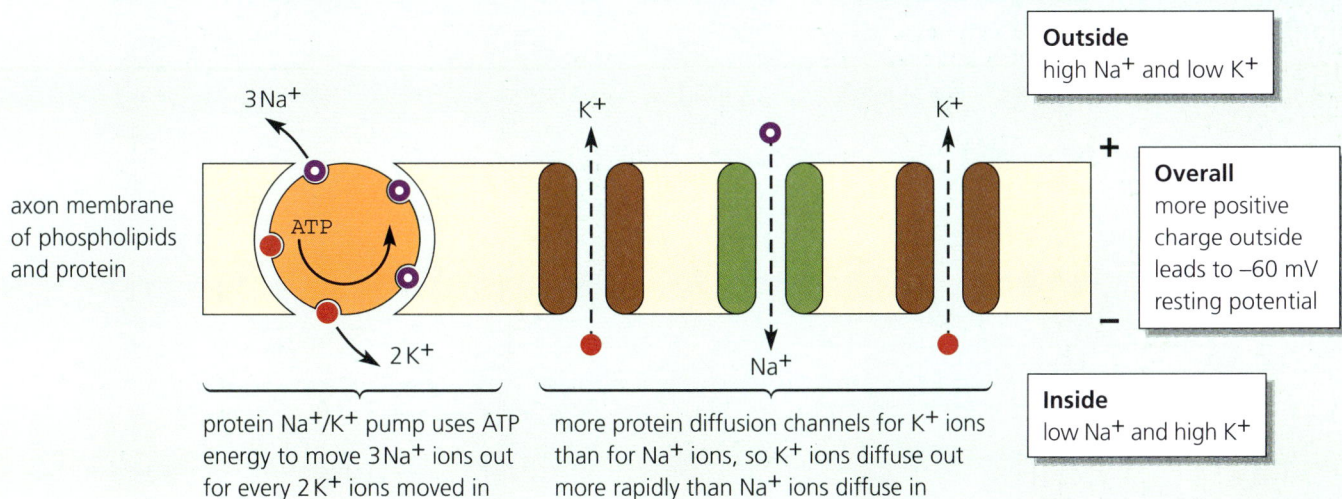

axon membrane of phospholipids and protein

ATP

3 Na⁺

2 K⁺

protein Na^+/K^+ pump uses ATP energy to move 3 Na^+ ions out for every 2 K^+ ions moved in

K^+

K^+

Na^+

more protein diffusion channels for K^+ ions than for Na^+ ions, so K^+ ions diffuse out more rapidly than Na^+ ions diffuse in

Outside
high Na^+ and low K^+

$+$

Overall
more positive charge outside leads to -60 mV resting potential

$-$

Inside
low Na^+ and high K^+

Key ideas

- The nervous system contains thousands of nerve cells (neurones) that vary greatly in shape and size; they link together at synapses. Neurones do not regenerate well.

- The brain and spinal cord make up the central nervous system, and contain most of the junctions or synapses.

- Nerves are bundles of axons; they transmit nerve impulses to and from the central nervous system.

- Neurones are electrically insulated from each other by Schwann cells.

- The small potential difference across the neurone membrane when it is not carrying an impulse is called the resting potential.

- The resting potential is maintained by active transport of Na^+ ions and K^+ ions across the membrane. Different diffusion rates of K^+ ions and Na^+ ions across the membrane establish the resting potential.

1.4 All-or-nothing

The resting potential is a store of potential energy in the axon. The neurone uses this energy to produce an **action potential**; a nerve impulse is the passage of an action potential along an axon (Fig. 5).

Fig. 5 Recording an action potential

The CRO sweeps a beam of electrons across a fluorescent screen; this is seen as a horizontal line. As an action potential passes, the change in voltage deflects the beam vertically up then down; this causes the line to form a wave.

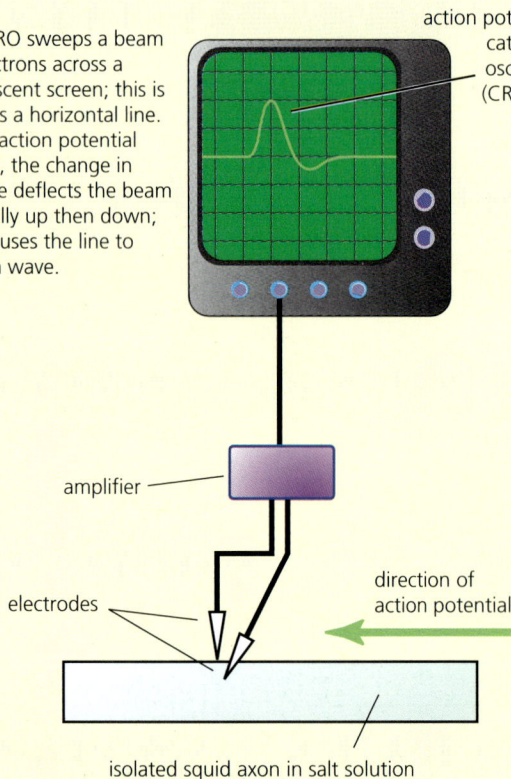

action potential on cathode ray oscilloscope (CRO) screen

amplifier

electrodes

direction of action potential

isolated squid axon in salt solution

First, the membrane is depolarised as the potential difference across it drops to zero. Then the potential momentarily reverses, as the membrane becomes relatively positive on the inside. The reverse polarisation peaks at about +40 mV, before returning towards the resting potential of −60 mV. There is a slight overshoot (hyperpolarisation) to −70 mV before the membrane settles down. The whole event takes about 3 ms.

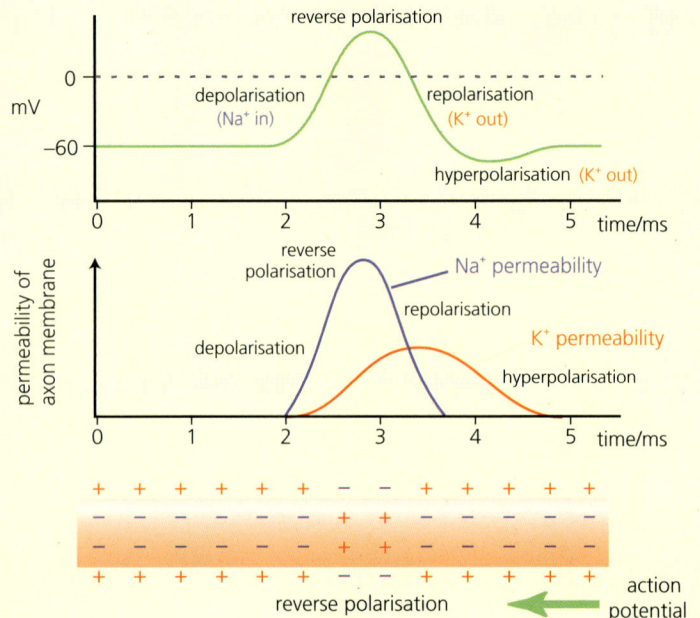

To understand the action potential, we must look in more detail at the different types of ion diffusion channels in the axon membrane. Besides the permanently open diffusion channels, there are also **voltage-gated channels** that open when the voltage across the membrane falls. Even a slight drop in potential difference (that is, even a slight depolarisation) triggers the chain of ion movements that make up the action potential (Fig. 6).

When a part of the membrane depolarises, Na^+ ions diffuse in and some positive ions are pushed along the axon to the next part of the membrane so it becomes slightly depolarised. Sodium diffusion channels there start to open and that part of the membrane fully depolarises. This pushes Na^+ ions ahead to the next section, and so on. In this way, the action potential passes along the axon as a **wave of depolarisation** (Fig. 7).

3 Na^+ ions enter the axon by diffusion, but leave the axon by an active pump mechanism. Explain why a different mechanism is used for Na^+ ions entering and leaving the axon.

4 Outline the changes in permeability of the axon membrane to Na^+ ions during one action potential.

Stimulation

Nerve impulses need a push or stimulus to start them off. A stimulus is defined as *a change in the environment*. In other words, some form energy is released in the animal's surroundings. Animals detect stimuli using sensory receptors. The bigger the stimulus, the greater the number of action potentials generated; the smaller the stimulus the fewer the number of action potentials. It is the number of action potentials in a given time that is variable, not their size.

Fig. 6 Permeability changes during an action potential

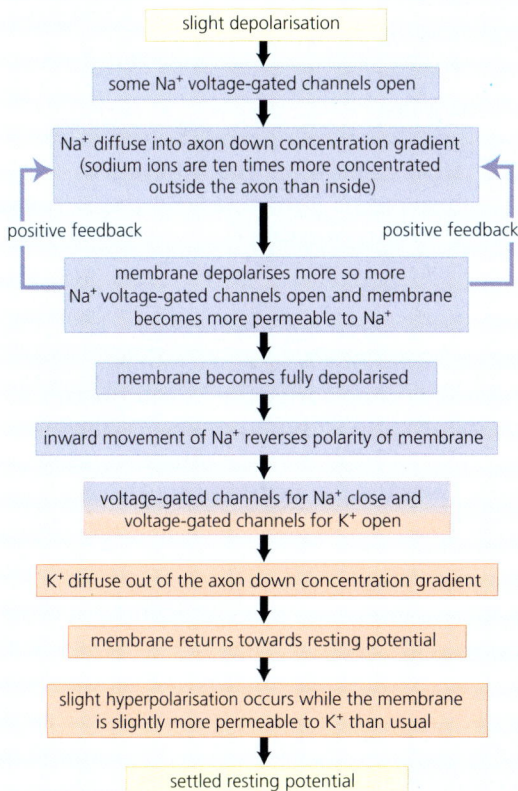

slight depolarisation

↓

some Na^+ voltage-gated channels open

↓

Na^+ diffuse into axon down concentration gradient (sodium ions are ten times more concentrated outside the axon than inside)

positive feedback   *positive feedback*

↓

membrane depolarises more so more Na^+ voltage-gated channels open and membrane becomes more permeable to Na^+

↓

membrane becomes fully depolarised

↓

inward movement of Na^+ reverses polarity of membrane

↓

voltage-gated channels for Na^+ close and voltage-gated channels for K^+ open

↓

K^+ diffuse out of the axon down concentration gradient

↓

membrane returns towards resting potential

↓

slight hyperpolarisation occurs while the membrane is slightly more permeable to K^+ than usual

↓

settled resting potential

Fig. 7 How an action potential travels along a membrane

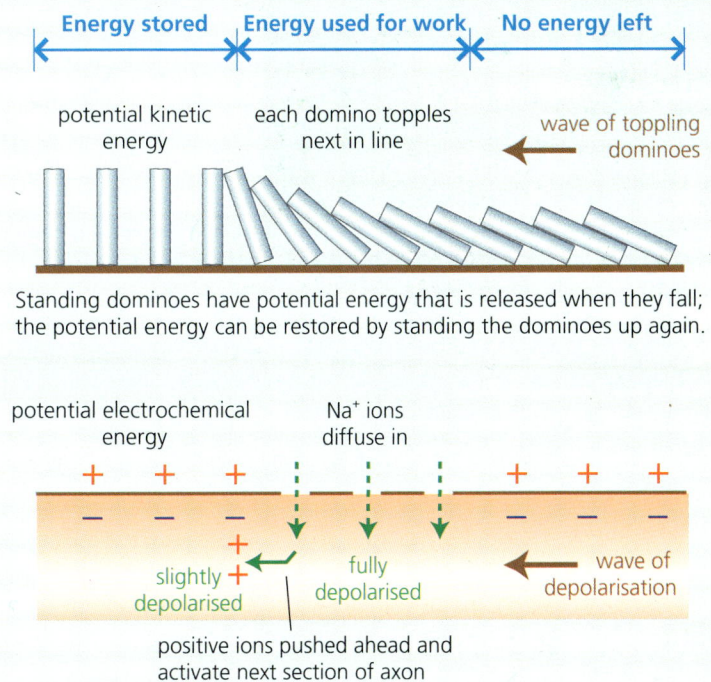

Energy stored   **Energy used for work**   **No energy left**

potential kinetic energy   each domino topples next in line   wave of toppling dominoes

Standing dominoes have potential energy that is released when they fall; the potential energy can be restored by standing the dominoes up again.

potential electrochemical energy Na^+ ions diffuse in

slightly depolarised   fully depolarised   wave of depolarisation

positive ions pushed ahead and activate next section of axon

The resting potential is a store of potential energy used as the action potential spreads rapidly along the axon, each section triggering the next; the resting potential must be restored before another action potential can occur.

Worms respond to light, grasshoppers to sound, snakes to warmth and moths to chemical substances because in each case the animal has special receptors with gated channels for sodium that are opened by the particular stimulus.

Table 2 Stimuli for gated channels

Stimulus	Type of receptor	Examples
light	photoreceptor	• light receptors in worms • rods and cones in eye
movement	mechanoreceptor	• balance and sound receptors in inner ear • touch and pressure receptors on skin • stretch receptors in muscle • sound receptors in grasshopper legs
temperature change	thermoreceptor	• heat and cold receptors in skin • heat receptors in snakes and fleas
chemical substance	chemoreceptor	• taste buds in tongue • smell receptors in nose • smell receptors in moth antennae

Each type of receptor has a different type of gated channel for sodium. In the eyes, gated channels are opened by light energy, in the ear by movement of fluid, and so on (Table 2). In all cases, the stimulus makes receptor cells more permeable to Na^+ ions which then diffuse in through the receptor cell membrane, depolarising it slightly. If this depolarisation is great enough, a series of action potentials is triggered.

Depolarisation in a sensory receptor is called the **receptor potential**. If the receptor potential exceeds a particular level, called the threshold level, action potentials occur. The stronger the stimulus, the greater the depolarisation. The greater the depolarisation, the greater the number of action potentials passing per second. So, a loud noise creates many action potentials per second in the auditory nerve, whereas a soft noise creates fewer (Fig. 8).

The *size* of each action potential is not affected by the strength of a stimulus. The resting potential is always about –60 mV and the peak of the action potential is always about +40 mV, a difference of about 100 mV. Action potentials always rise from about –60 mV to about +40 mV, or they don't happen at all. This is the **all-or-nothing law**.

Q 5 How can we tell the difference between a light touch on the skin and a strong pressure?

During the 3 milliseconds or so that it takes for a nerve impulse to pass a point on the axon, no new impulse can pass. For a few milliseconds after the action potential, the membrane is hyperpolarised. During this time, it takes a bigger stimulus than usual to trigger an action potential. So, there is usually a slight pause between one action potential and the next. Most axons conduct up to 250 potentials per second, though wide axons can conduct up to 1000 action potentials per second.

The speed of conduction is fixed for any particular axon. Wide axons conduct faster than narrow ones, and myelinated axons conduct much faster than non-myelinated (Fig. 9).

Fig. 8 Stimulus strength and action potential frequency

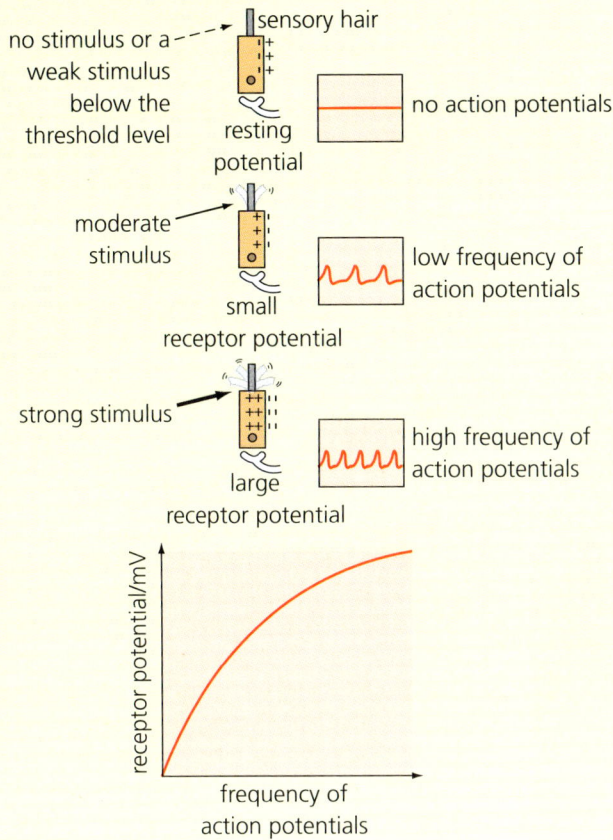

Fig. 9 Myelin and conduction speed

Unmyelinated axon

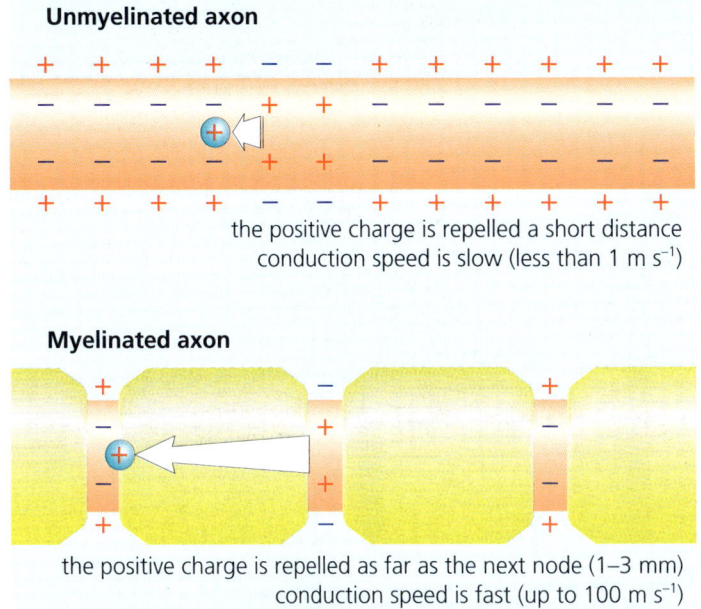

the positive charge is repelled a short distance
conduction speed is slow (less than 1 m s^{-1})

Myelinated axon

the positive charge is repelled as far as the next node (1–3 mm)
conduction speed is fast (up to 100 m s^{-1})

The myelin of the Schwann cells is tightly pressed to the axon membrane and prevents any ion exchange except at the small gaps between Schwann cells called nodes of Ranvier. As sodium ions diffuse in at a node, they repel other positive ions along the axon ahead of them. Positive ions repelled as far as the next node depolarise the membrane enough to trigger an action potential at that node. This 'jumping' is rapid compared with conduction by ion diffusion at the nodes.

Key ideas

- Resting potential is a measure of energy stored in neurone membranes. Energy is needed to transmit nerve impulses. A nerve impulse consists of many action potentials.

- An action potential is a wave of depolarisation that passes along the axon membrane.

- As an action potential passes a point on the axon, the polarity briefly reverses, as the membrane becomes positive inside. This change is caused by movements of Na$^+$ ions and K$^+$ ions.

- The axon membrane contains voltage-gated ion channels. These channels change their permeability to Na$^+$ ions and K$^+$ ions, depending on the membrane potential.

- The action potential is conducted by a chain reaction. Each section of the axon triggers the section ahead of it by the movement of positive charge ahead of the wave.

- Action potentials are started by a stimulus causing depolarisation in a receptor cell. Different types of receptor cell respond to different stimuli but the result is always depolarisation caused by increased permeability to Na$^+$ ions.

- A larger stimulus produces more action potentials in a given time, but does not increase their size.

- The myelin sheath increases the speed of conduction and insulates axons from each other.

1.5 Transmitters

It took Roger Fenn several weeks after his operation to learn how to use the electronic grip. At first he found it difficult to coordinate his new hand. But after a year, it became second nature. He can quickly make just the right shoulder movements for the electronics to interpret and send electrical impulses down the wires in his arm to produce the kind of grip he wants. He can even grip with enough force to open a bottle of wine using a bottle opener!

When we make hand movements, hundreds of motor neurones carry impulses to muscles. These muscles must contract in the correct sequence and with the appropriate strength. The electronic grip is limited because it contains only eight channels to convey impulses to the arm muscles.

The brain must constantly change the information it sends along motor neurones to move the hand. For example, when we write, our fingers make many changes of direction and force. Movements are adjusted as we see what we write. If we don't look at what we are writing, the letters are likely to wander off the line and change size and shape. The brain takes account of what we see, and makes the necessary adjustments.

Synapses

Such control would not be possible without synapses. Synapses can pass signals onward, or they can block signals. Synapses behave rather like traffic police, controlling the pathways that nerve impulses travel along. Synapses between neurones might direct an impulses to just one other neurone, or spread it out to several. Synapses also occur at junctions between nerve and muscle, where they are called **neuromuscular junctions**.

Besides their role in controlling deliberate movement such as writing or lifting a cup, synapses are just as important in regulating involuntary movement such as heartbeat. Movement of this sort is under the control of the **autonomic nervous system**.

Cells do not join at a synapse; there is a small gap of about 20 nm (1 nanometre is 1 millionth of a millimetre). Impulses cross this gap as chemical **transmitters**. Transmitters are stored at nerve endings in tiny bags of membrane called **synaptic vesicles**. Over 40 transmitters have now been identified (Table 3). Synapses where the transmitter is **acetylcholine** are called **cholinergic** synapses; those using **noradrenaline** are called **adrenergic** synapses. Transmitters can be either **excitatory** (they increase the activity of the next cell) or **inhibitory** (they decrease the activity of the next cell).

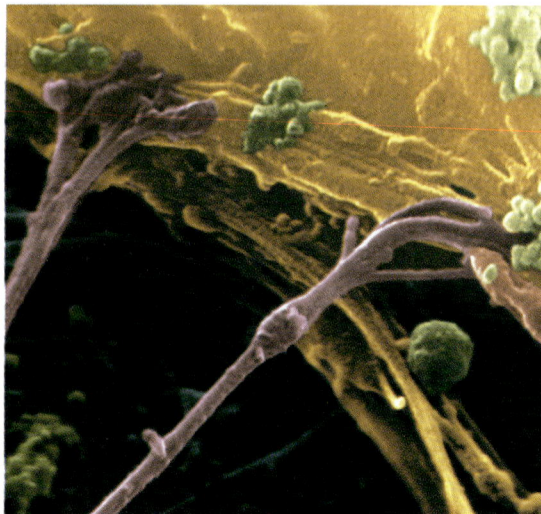

False-colour scanning electron micrograph of the junction sites (synapses) between nerve fibres (purple) and a neurone cell body (yellow).

Table 3 Some chemical transmitters

Transmitter	Where found	Action
acetylcholine	sensory and motor nerves	excitatory
	central nervous system	excitatory
	neuromuscular junctions	excitatory
	autonomic nervous system	inhibitory
noradrenaline	autonomic nervous system	excitatory
glutamic acid	brain	excitatory
γ-amino butyric acid (GABA)	brain	inhibitory
glycine	spinal cord	inhibitory
dopamine	brain	excitatory or inhibitory depending on the type of synapse
endorphin	brain	excitatory

There is a delay of about half a millisecond while an impulse crosses a synapse. This is the time needed for the transmitter to diffuse across the gap, bind to a receptor protein and so activate or inhibit the next cell (Fig. 10).

6 **What is the difference between the electronic grip and the natural system in (a) the source of energy and (b) the way information travels to the muscles?**

Transmitters are quickly removed from synapses. If they were not, neurones would keep on firing uncontrollably. The transmitter acetylcholine is used in many motor and sensory neurones. At cholinergic synapses, a very fast-acting enzyme called **acetylcholinesterase** breaks down acetylcholine into acetic acid and choline. These substances are taken back into the synaptic knob and ATP energy from mitochondria is used to resynthesise the acetylcholine which is returned to the vesicles.

7 **Eserine is a substance that inhibits acetylcholinesterase. What is the effect of eserine at (a) a cholinergic synapse and (b) an adrenergic synapse?**

Fig. 10 Transmission across a synapse

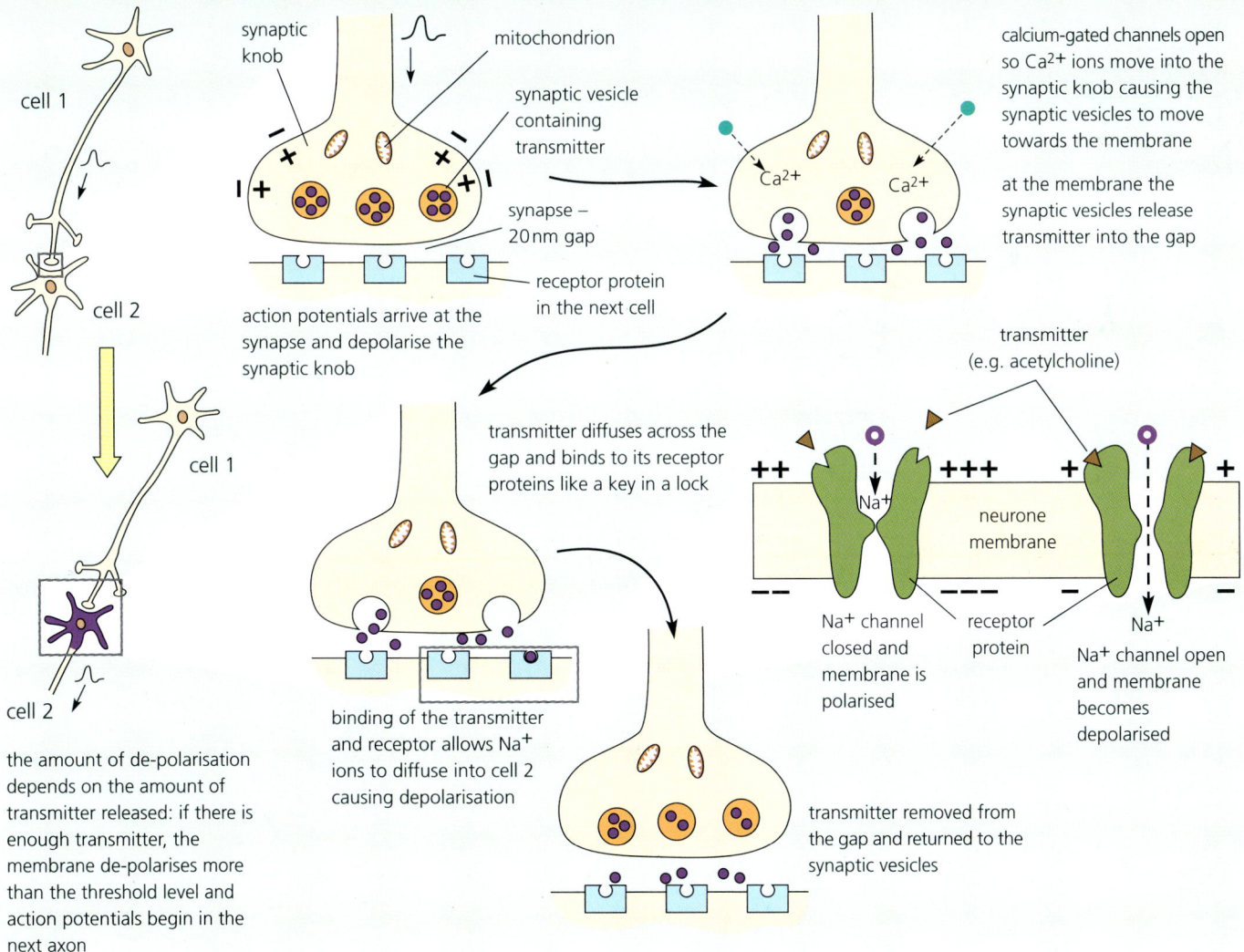

cell 1

cell 2

the amount of de-polarisation depends on the amount of transmitter released: if there is enough transmitter, the membrane de-polarises more than the threshold level and action potentials begin in the next axon

synaptic knob

mitochondrion

synaptic vesicle containing transmitter

synapse – 20 nm gap

receptor protein in the next cell

action potentials arrive at the synapse and depolarise the synaptic knob

calcium-gated channels open so Ca^{2+} ions move into the synaptic knob causing the synaptic vesicles to move towards the membrane

at the membrane the synaptic vesicles release transmitter into the gap

transmitter diffuses across the gap and binds to its receptor proteins like a key in a lock

binding of the transmitter and receptor allows Na$^+$ ions to diffuse into cell 2 causing depolarisation

transmitter removed from the gap and returned to the synaptic vesicles

transmitter (e.g. acetylcholine)

neurone membrane

Na$^+$ channel closed and membrane is polarised

receptor protein

Na$^+$ channel open and membrane becomes depolarised

15

Since the transmitter is removed almost as quickly as it is released, several action potentials might be needed in quick succession to produce enough transmitter to trigger the next cell. The need for several action potentials to be added together is called **temporal summation**. In other cases, the transmitter from two or more neurones acting side-by-side is needed. This kind of adding together is called **spatial summation** (Fig. 11).

Controlling behaviour

Summation is important for healthy brain functioning. For example, it might be better to ignore a weak stimulus, such as mild pressure on the skin, but a stronger stimulus might cause injury and so require a response. A strong stimulus sends a high frequency of nerve impulses to the brain.

When temporal summation has released enough transmitter to cross the synapse, the brain registers pain and brings about avoiding action.

Spatial summation controls behaviour such as drinking. For example, a deer drinks when three conditions are met:
• the deer must be thirsty;
• the stimulus of water must be present;
• there must be no predators or other danger nearby.

Put simply, this could be controlled by three neurones converging on a 'drinking' neurone. When the three neurones are all firing together, enough transmitter builds up to fire the drinking neurone.

Synapses can become tired. This happens when so many action potentials arrive in so short a time that the cell runs out of transmitter (or its components). The

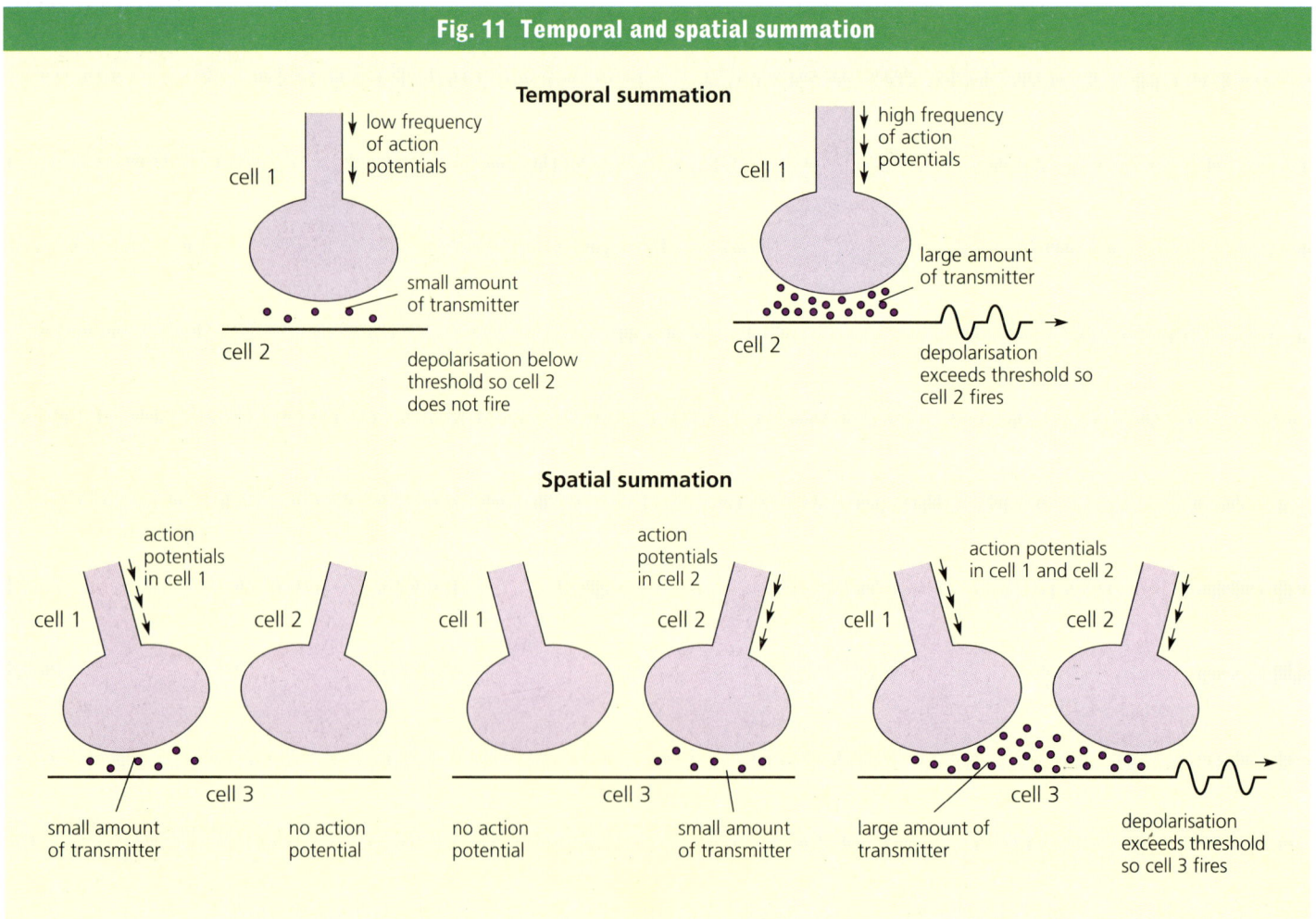

Fig. 11 Temporal and spatial summation

Temporal summation

cell 1 — low frequency of action potentials
cell 2 — small amount of transmitter — depolarisation below threshold so cell 2 does not fire

cell 1 — high frequency of action potentials
cell 2 — large amount of transmitter — depolarisation exceeds threshold so cell 2 fires

Spatial summation

action potentials in cell 1
cell 1 — cell 2
cell 3 — small amount of transmitter — no action potential

action potentials in cell 2
cell 1 — cell 2
cell 3 — no action potential — small amount of transmitter

action potentials in cell 1 and cell 2
cell 1 — cell 2
cell 3 — large amount of transmitter — depolarisation exceeds threshold so cell 3 fires

We are likely to ignore stimuli such as the hum of traffic noise when we are used to it.

impulse can no longer cross the synapse. This loss of response at a synapse is known as **adaptation** and it means that animals ignore stimuli that go on for a long time. For example, we do not notice stimuli such as the tick of a clock or the feel of our clothes brushing against our skin, once we are used to them. Such stimuli are irrelevant to our survival. Sudden changes in our environment, like the unexpected approach of a car in a quiet lane, are much more important.

8 **Why does a shortage of ATP reduce the amount of acetylcholine in the vesicles of an active synapse?**

Learning and development

Some transmitters have a more gradual effect on synapses. The receptors they bind to send **secondary messengers** inside the cell body. These can change the activity of enzymes or switch genes on or off. In this way, transmitters can, for instance, promote the synthesis of more ion channels or receptor proteins. This would alter the future response of the synapse. Such changes are important in development and learning. For example, a baby gradually begins to smile when it sees the faces of its parents. The response becomes stronger and stronger as the baby becomes more familiar with their faces. This could be due to the synthesis of more receptor proteins in the synapses linking the visual areas in the baby's brain to a 'smiling control' area.

So, the nervous system is not just a mass of fixed circuits, like a computer. By changing the activity of synapses, the brain can learn and adapt – almost as if it were rewiring itself. Patterns of electrical activity in the brain change and develop all the time. But unlike a computer, the brain needs sleep to reorganise and adjust itself. Recently, scientists have developed computers which use 'neural networks' made of artificial silicon neurones. Like real neurones, these artificial neurones show summation and also increase their rate of 'firing' action potentials when the strength of the 'stimulus' increases. Electronic engineers are working with anatomists and physiologists to make these systems more like the brain. It is hoped that in the future such networks can be designed to recognise faces and understand speech, but we still have a long way to go.

9 **How can some transmitters affect learning?**

More than you think

In the introduction to his book *The Astonishing Hypothesis*, Francis Crick writes: *The Astonishing Hypothesis is that 'You', your joys and your sorrows, your memories and your ambitions, your sense of personal identity and free will, are in fact no more than the behaviour of a vast assembly of nerve cells.* One of the most difficult aspects of the nervous system to explain is: how can ion movements in neurones give us consciousness and self-awareness? Yet we accept that certain other complex structures have qualities that we would not predict by looking at their component parts. For example, the components of a motor car do very little until they are fitted together – but then the whole assembly can be driven down the motorway at 70 mph.

10 **Can we understand aspects of the brain such as consciousness and behaviour by studying individual brain cells?**

Key ideas

- Nerve impulses cross synapses using chemical transmitters. Transmitters are released from vesicles and diffuse across the small gap that is the synapse.

- Transmitters fit protein receptors on the target cell like a key in a lock. When the transmitter binds to the protein, the membrane of the target neurone is depolarised.

- Several action potentials, sometimes from more than one neurone, produce sufficient transmitter to activate the target neurone. This is summation.

- If the supply of transmitter runs out, the impulse can no longer cross the synapse. This is adaptation. Summation and adaptation help to control and adjust the activity of the brain.

- Synapses are essential for interpreting information from sensory receptors and for learning.

1.6 Potent chemicals

Transmitters are released in tiny amounts, only 500–1000 molecules from each synaptic knob. So, drugs that affect transmitters or their binding sites can have powerful effects when given in fairly small doses. Some chemicals, many of them from plants, have a dramatic effect on the nervous system.

Preventing transmitter uptake

Amphetamines and cocaine prevent the uptake of noradrenaline from the synaptic gap. So, noradrenaline remains in the gap and the neurone keeps on firing. Synapses that use noradrenaline have the effect of speeding up the heart, breathing rate and brain activity. This is similar to the effect of the hormone adrenaline, which prepares the body for emergencies. The result is that amphetamines make a person feel energetic and carefree, which is why they are often known as speed. Before the harmful effects were known, amphetamines were given to pupils with poor attention spans, to help them concentrate.

Erythroxylon coca is a South American shrub. Cocaine paste is made from the leaves after they have dried in the sun. Coca plantations employ thousands of farmers in South America, and the crops are worth millions of pounds. The plants mature in 6 months and the leaves can be picked six times a year. It is an excellent cash crop for poor countries. Cocaine taken illegally is usually snorted up the nose; it can also be made into a more powerful solid called crack, which is smoked.

During World War II, amphetamines were given to British, German and Japanese troops to boost morale and fight fatigue.

Hitler is said to have been injected with a form of amphetamine up to five times a day.

Binding to transmitter sites

The nicotine molecule is a similar shape to acetylcholine, so it can bind with acetylcholine receptors and open sodium channels (Fig. 12). This is how nicotine stimulates the nervous system. But the nervous system gradually adjusts to nicotine by reducing the number of acetylcholine receptors. Once this has happened, the nervous system is less active than normal unless nicotine is present. This is the reason why nicotine is addictive.

11 a Caffeine has a similar shape to an excitatory brain transmitter called adenosine. How does caffeine increase alertness?

b How could someone become addicted to caffeine?

Atropine also binds to acetylcholine receptors, but does not open the sodium channels. It blocks the receptors, and prevents acetylcholine binding to them (Fig. 12). This happens in motor neurones and causes muscle paralysis.

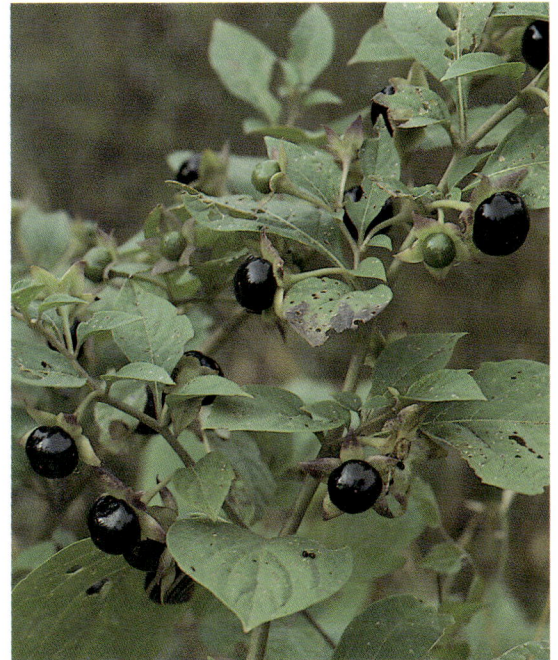

Deadly nightshade (*Atropa belladonna*) is very poisonous. It contains the alkaloids atropine, solanine and hyosyanine. It was called 'belladonna' because sixteenth-century Italian women used it to dilate their pupils – which they thought made them more attractive.

Fig. 12 Drugs and transmitter binding sites

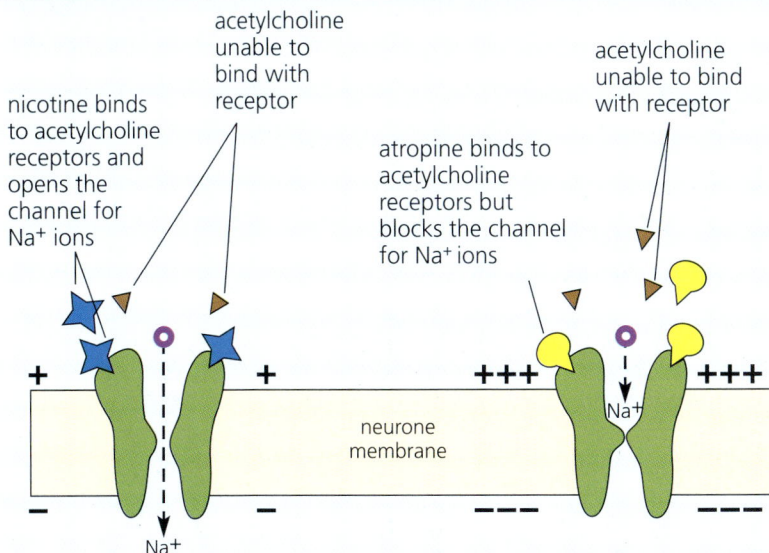

nicotine binds to acetylcholine receptors and opens the channel for Na$^+$ ions

acetylcholine unable to bind with receptor

atropine binds to acetylcholine receptors but blocks the channel for Na$^+$ ions

acetylcholine unable to bind with receptor

neurone membrane

Na$^+$

Curare has a similar effect to atropine. It is obtained from the bark of the tree *Strychnos toxifera* and is used on the tips of arrows by South American Indians because it paralyses any animal hit by a treated arrow. Fortunately, curare is inactive when taken by mouth, so the meat can safely be eaten. Curare used to be used as a muscle relaxant in surgical operations.

12 Digitalin comes from foxgloves and stimulates the nerves controlling the heart. How could a large dose of digitalin could cause death?

Phencyclidine comes from 'magic mushrooms'; it excites brain receptors for amino acids and causes neurones in the visual or auditory areas to become overactive. This causes hallucinations: seeing or hearing things that aren't there. Hallucinations can also occur in some mental disorders such as schizophrenia – a disease that affects about 1% of the population. The drugs used to control schizophrenia block the receptors for a brain transmitter called dopamine. So, it is thought that an abnormality of dopamine receptors might be the cause of this disease.

13 Why do you think so many plants have developed substances that are harmful to animals?

Tranquillisers are drugs that reduce tension. Benzodiazepine tranquillisers, such as valium, do this by increasing the binding of inhibitory transmitters in the brain. Inhibitory transmitters hyperpolarise rather than depolarise the membrane of the next neurone. This makes the next neurone less excitable. Valium reduces stress and anxiety, but it can be addictive.

Key ideas

- Some drugs interfere with transmitter release or with the removal of a transmitter from synapses.

- If a drug's method of action is known, its effect at a synapse can be predicted.

- Some drugs have the same or nearly the same shape as a transmitter. These drugs can bind to the transmitter receptors; some of them have the same effect as the transmitter, some of them block the effect of the transmitter.

1.7 Making it twitch

Because communication in the nervous system uses charged ions, it is not surprising that a severe electric shock can harm the system. Electricity affects the junctions between nerves and muscles in a way that makes the muscles contract violently and stay contracted, so it is impossible to let go when holding a live mains wire.

The first evidence that muscles could be stimulated electrically came in 1762 when an Italian called Galvani touched the muscles of dead frogs with pairs of different metals. The metals produced electric currents which made the frog muscles twitch. The same principle is applied in the electronic grip that restored Roger Fenn's hand movements. The greater the electrical stimulation, the more the muscles contract.

In the neuromuscular junction, there is a chemical synapse between the motor neurone and the muscle fibres. This synapse works in a similar way to neurone–neurone synapses. Synapses give more accurate muscle control than is possible by direct electrical stimulation, because synapses can control small parts of each muscle separately.

The transmitter at neuromuscular junctions in skeletal muscle is acetylcholine. An action potential in the motor neurone releases acetylcholine that diffuses across the neuromuscular junction, binds to receptors in the muscle fibre membrane and produces an action potential. The muscle fibre responds by contracting. There is no summation at neuromuscular junctions; one action potential causes one **muscle twitch** (Fig. 13).

14 What are the main similarities and dissimilarities between neurone–neurone synapses and neuromuscular junctions?

Many people lose parts of their limbs entirely. Road accidents, anti-personnel mines, industrial injuries, disease, and birth defects are responsible for many such disabilities. In the past, artificial limbs were simple hooks or pegs. Much more sophisticated artificial limbs are now being made, but even with modern electronics, they do not match the real thing.

In one promising line of research, neurones are grown on silicon chips. The growth of the cells is aided by **growth factors** that encourage neurones to develop in cultures outside the body. When the neurones are grown in set geometrical patterns on silicon, impulses can be passed between the chip and the neurones. The hope is that it may be possible to link non-living silicon chips with the living body.

Fig. 13 Neuromuscular junction

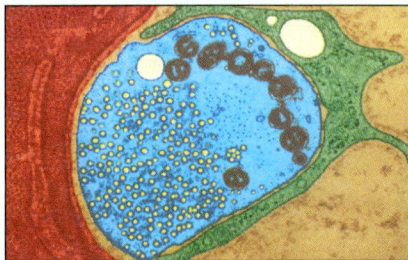

False-colour transmission electron micrograph of a neuromuscular junction. The end of the motor axon is coloured blue and the muscle fibre is red. The Schwann cell forming the end of the myelin sheath is green.

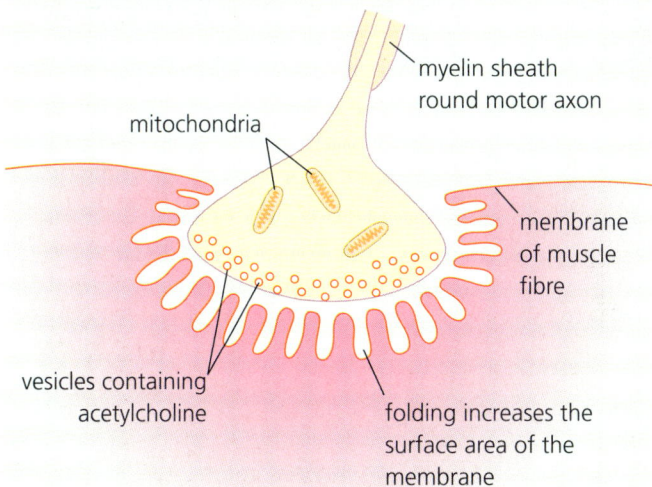

mitochondria

myelin sheath round motor axon

membrane of muscle fibre

vesicles containing acetylcholine

folding increases the surface area of the membrane

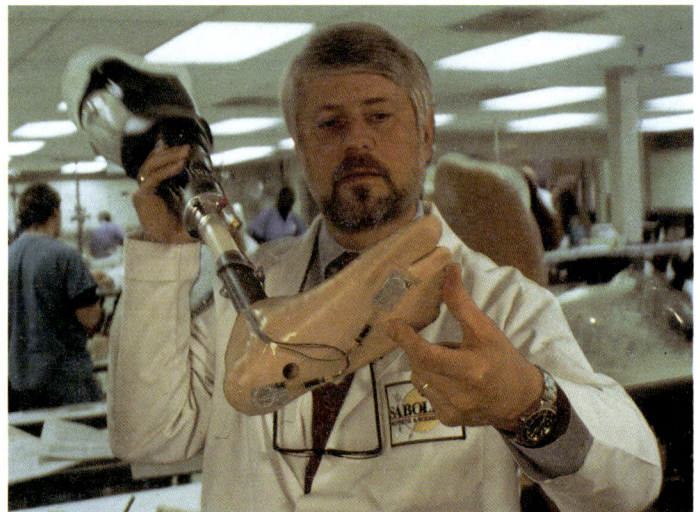

John Sabolich designed this artificial leg. He has put pressure sensors in the sole of the foot so when it it used for walking, the patient gets a tingling sensation as the foot touches the ground. This helps the patient to learn to walk in a natural way.

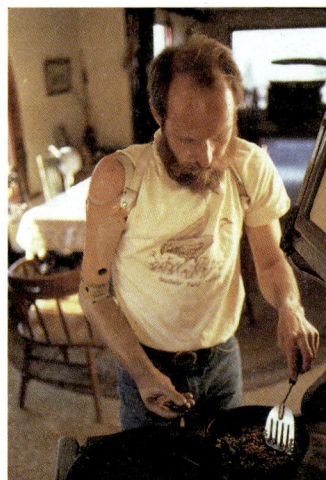

Electrodes on the skin of the upper arm pick up electronic activity in the remaining arm muscle. The signal is electronically amplified and used to move the artificial arm.

Work is also under way on chemical fibres that contract when the pH changes. So, in the future we might see chip–neurone hybrids controlling artificial muscles with abilities as good as natural muscles. But this will not be for many years; muscles are very complex and difficult to imitate (Chapter 3).

Although it is possible to help people like Roger Fenn, operations like his are very expensive and are not available through the Nation Health Service even though hundreds of people would benefit. Two other people besides Roger had the operation in 1995 and nine more operations are planned in an on-going 3-year programme.

There's a long way to go before reality catches up with the Terminator.

Key ideas

- Impulses pass from nerve to muscle by chemical transmission, using acetylcholine, at neuromuscular junctions.

- One action potential arriving at a muscle fibre causes it to make one twitch. There is no summation.

- There is a great need for artificial limbs, and some progress is being made in their development.

Eye in the sky

There are 42 million blind people in the world and about 80% of them live in poor countries. The majority of these people are unnecessarily blind. With medical care, their sight could be restored or blindness prevented. In the UK and the USA there is one eye specialist for every 30 000 people, but in Africa there is only one eye specialist for every million people.

Every 3 weeks, at the invitation of local doctors, Orbis, a flying eye hospital, flies into a new country. Orbis's aim is to share and transfer knowledge between eye surgeons, nurses, anaesthetists and biomedical engineers around the world. Commercial pilots give up their holidays to fly Orbis to its next destination. Everybody in the crew helps to unpack and set up the operating theatre and recovery area – and everyone, irrespective of rank, receives the same salary.

Orbis's recent work in the Sudan has been followed by Lucy, a pupil in Stamford, who won a national competition to film a report on Orbis for BBC's *Newsround*. She describes her experience:

It's difficult to imagine an operating theatre on a DC10 plane. It has a classroom in the front where the local doctors watch the operations live on closed-circuit TV. I was impressed by the perfect teamwork shown by the Orbis medical staff. The 9-year-old girl Haiam, whose operation we followed, was amazed when she first realised she could see again. Her smile was heart-warming to watch; she had never seen her family before or her house.

Orbis on its way.

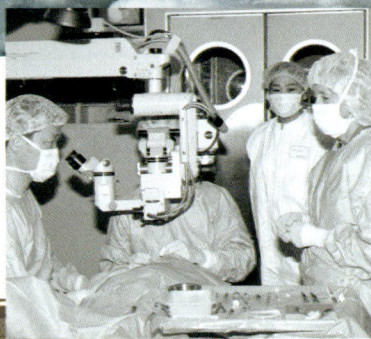

An operation in progress.

Doctors watching the operation.

2.1 Learning objectives

After working through this chapter, you should be able to:

- **recall** that detecting light begins with the absorption of light by a pigment;

- **explain** how a mammalian eye focuses an image on the retina and controls the amount of light that enters the eye;

- **explain** how the retina converts light energy into nerve impulses;

- **explain** how the arrangement of rod and cone cells in the retina gives rise to differences in sensitivity and acuity;

- **explain** colour vision in terms of the trichromatic theory;

- **understand** that impulses in the optic nerve are analysed and interpreted by particular parts of the forebrain.

2.2 Creating a good image

Living organisms can detect light by various mechanisms that all begin with the absorption of light by a pigment (Fig. 1).

Eyes to see with

Many blind people can distinguish between light and dark, and perhaps the direction of a light source. However, they cannot focus light rays to produce a clear image. For normal vision we need a system of lenses and apertures that can focus light from objects at different distances, and can adapt to changes in light intensity (Fig. 2).

Fig. 1 Responses to light

Euglena lashes its flagellum to move towards the light source when the stigma casts a shadow on the photoreceptor.

Earthworms withdraw into their burrows when light falls on photoreceptors in the skin at the head and tail.

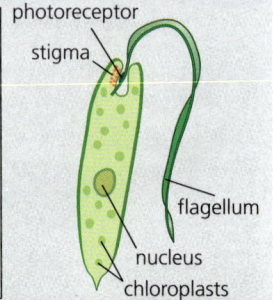

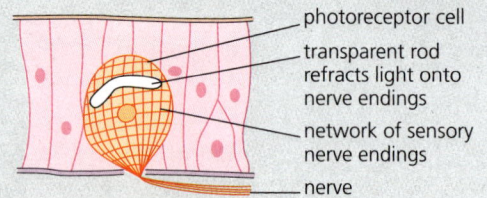

photoreceptor
stigma
flagellum
nucleus
chloroplasts

photoreceptor cell
transparent rod refracts light onto nerve endings
network of sensory nerve endings
nerve

Source: adapted from Vines and Rees, *Plant and Animal Biology* (Vol. 1), Pitman, 1964

Fig. 2 Structure of the human eye

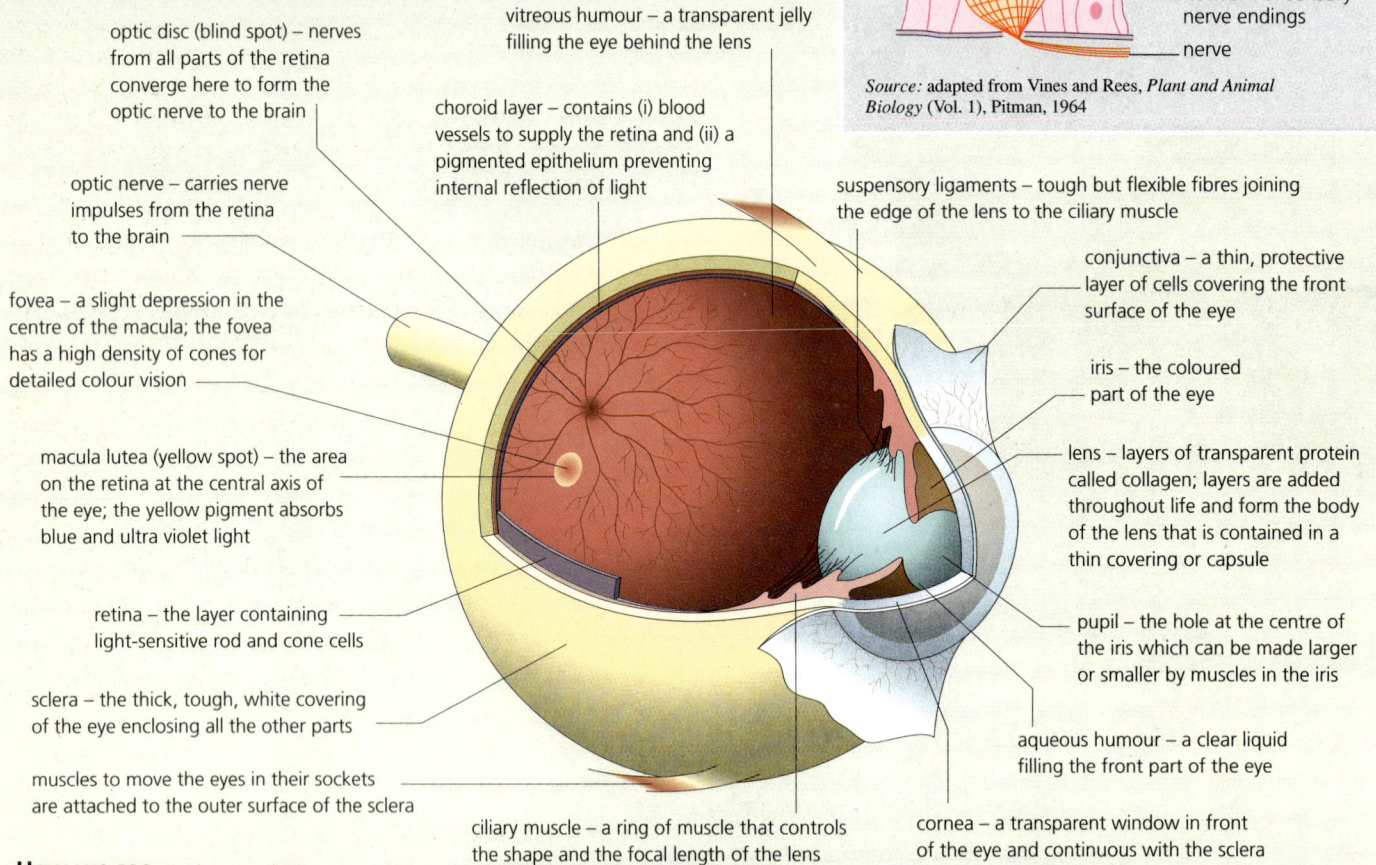

optic disc (blind spot) – nerves from all parts of the retina converge here to form the optic nerve to the brain

vitreous humour – a transparent jelly filling the eye behind the lens

choroid layer – contains (i) blood vessels to supply the retina and (ii) a pigmented epithelium preventing internal reflection of light

optic nerve – carries nerve impulses from the retina to the brain

suspensory ligaments – tough but flexible fibres joining the edge of the lens to the ciliary muscle

conjunctiva – a thin, protective layer of cells covering the front surface of the eye

fovea – a slight depression in the centre of the macula; the fovea has a high density of cones for detailed colour vision

iris – the coloured part of the eye

macula lutea (yellow spot) – the area on the retina at the central axis of the eye; the yellow pigment absorbs blue and ultra violet light

lens – layers of transparent protein called collagen; layers are added throughout life and form the body of the lens that is contained in a thin covering or capsule

retina – the layer containing light-sensitive rod and cone cells

pupil – the hole at the centre of the iris which can be made larger or smaller by muscles in the iris

sclera – the thick, tough, white covering of the eye enclosing all the other parts

aqueous humour – a clear liquid filling the front part of the eye

muscles to move the eyes in their sockets are attached to the outer surface of the sclera

ciliary muscle – a ring of muscle that controls the shape and the focal length of the lens

cornea – a transparent window in front of the eye and continuous with the sclera

How we see

light enters cornea → rays converge towards pupil → pupil (controlled by iris muscles) regulates amount of light passing through lens → lens focuses light onto retina → retina converts light energy to nerve impulses → nerve impulses travel to brain in optic nerve → brain interprets nerve impulses to give vision

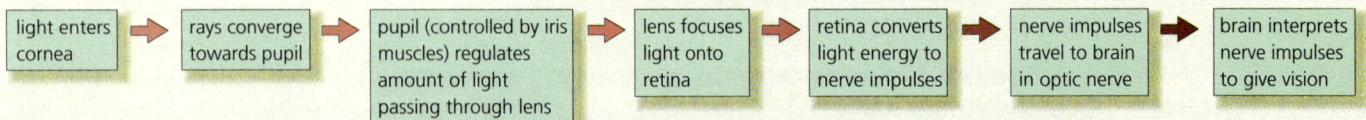

Source: drawing adapted from Mueller and Rudolf, 'Light and Vision', *Time Life*, 1969

The human eye is a sphere about 2.5 cm in diameter. It is divided into two chambers by a lozenge-shaped lens that is more convex at the back than the front. The chamber in front of the lens is filled with **aqueous humour** and the rear chamber is filled with **vitreous humour**. The aqueous humour is secreted and drained away continuously, keeping the inside the eye slightly above atmospheric pressure. The pressure is the same in both chambers of the eye and creates a smooth, curved internal surface for the **retina**, where the image is focused.

Light is focused onto the retina by the **cornea** and **lens** which together refract (bend) the light rays entering the eye. Most of the refraction is done by the cornea. However, it is a fixed shape and cannot make adjustments for near and distant vision. Adjusting the focal length of the eye for different distances is called **accommodation** and is done by changes in the shape of the lens. The lens enables us to focus on an object only 10 cm away or on a distant star, merely by a reflex altering of its thickness (Fig. 3).

1a Why is close eye work, such as reading, more tiring for the eyes than looking into the distance?

b Study Figure 3. How does the internal pressure of the eye affect the shape of the lens during accommodation?

Lenses and cataracts

The problem for many who seek help from the Orbis team is a cataract – cloudiness of the lens. Cataracts cause more than 50% of the blindness in many countries; around the world 17 million people suffer from this eye disease. Usually, cataracts occur in older people, but they can also occur in babies (perhaps as a result of rubella in pregnancy) and can be caused by injury or infection.

To improve the sight of patients with a cataract, the whole lens is replaced with an artificial one. The anaesthetised eye is held open, and a small flap of conjunctiva folded back. A cut is then made along the edge of the cornea, and the lens is very carefully removed, first through the dilated pupil, and then via the cut in the cornea. The most modern technique is to remove the

Fig. 3 How we focus light

Front view

lens | suspensory ligaments | ciliary muscle

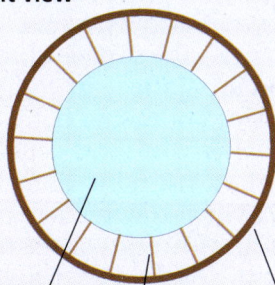

- ciliary muscle relaxed
- ciliary ring opened by pressure inside eye
- suspensory ligaments taut
- lens flat and thin

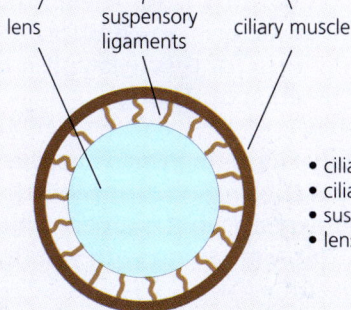

- ciliary muscle contracted
- ciliary ring closed
- suspensory ligaments slack
- lens rounded and plump

In cross section

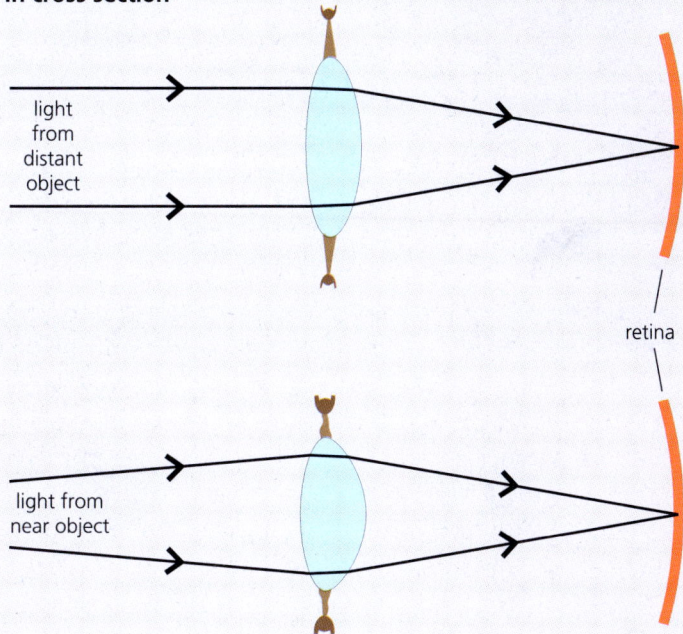

light from distant object

light from near object

retina

central part of the lens, leaving a section of the lens capsule, like an open bag, to hold the implant. The lens implant has two curved projections to hold it in place inside the lens capsule (Fig. 4).

Haiam was selected for treatment because the surgery made an excellent teaching case. One eye was treated by the Orbis surgeons while 50 Sudanese surgeons watched the operation in the Orbis classroom. One of them later operated on Haiam's second eye. The operations were very successful. The transfer of skills and knowledge will benefit many other Sudanese people.

Folding lenses have recently been developed for use in some patients. These can be inserted into a very small cut in the eye, and spring open when put in position. In the future, lens cells grown by tissue culture might be used as implants, but animal trials have found that when cultured lenses are used, the lens capsule becomes opaque. This problem must be overcome before human trials can begin.

Lens replacement surgery gives the patient much clearer vision, and the world will seem more vividly coloured. However, a replacement lens lacks the slight yellow colouring of the natural lens, so colour vision is not restored to its exact natural state.

2 a How does a cataract affect vision?
b Give two reasons why an artificial lens is not as good as a healthy real one.

The eye accurately controls the amount of light falling on the retina. Too much light would damage the light-sensitive cells in the retina, but there must be enough light to stimulate these cells. Light entering the eye passes first through the cornea and then through the **pupil** on its way to the lens and eventually the retina. The iris muscles control the size of the pupil (Fig. 5).

The retina

Impulses from the light-sensitive cells in the retina carry visual information to the brain via the **optic nerve**. Narrow nerves pass to the optic nerve across the surface of the retina, partly obscuring the light coming in. In general, this does not matter but it would do at the **fovea**, where the most detailed vision occurs. In fact, the nerve fibres skirt round the fovea, which is why there is a slight depression in the retina at the fovea. The point where the optic nerve leaves the eye is called the **optic disc**. There is no retina at the optic disc, so light focused on it cannot be seen; for this reason it is often called the **blind spot** (Fig. 6).

3 a Study Figure 6. What is happening when the cross disappears? How do you explain what happens when you do the same thing with your left eye closed?
b Why are we not usually aware of the blind spot?

Fig. 4 Lens implant

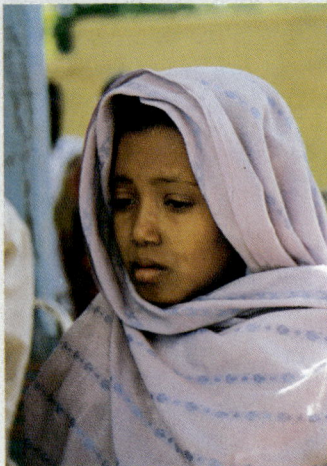

This is Haiam, age 9, before her operation. Both her eyes were affected by cataracts, so she couldn't see. Apart from cloudy lenses, her eyes were healthy. After the operation, Haiam needed glasses because lens implants cannot change shape to accommodate objects at differing distances.

Lens implants like this enabled Haiam to see again. This implant is drawn actual size. Implants are made from an inert material called polymethyl methacrylate (PMMA) and are made in one piece so the hairs can't fall off.

Fig. 5 The iris and light

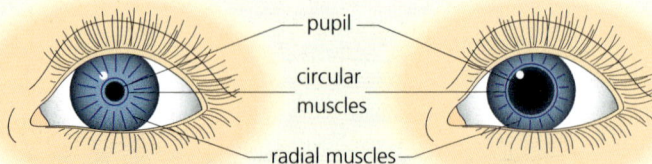

pupil
circular muscles
radial muscles

In bright light
The circular muscles contract and reduce the size of the pupil.

In dim light
The pupil is opened as the radial muscles contract and pull the edges of the iris away from the centre.

Fig. 6 The blind spot

Hold the page at arm's length, close your right eye and stare at the dot with your left eye. Bring the page gradually closer until the cross disappears. Now try it again with your left eye closed.

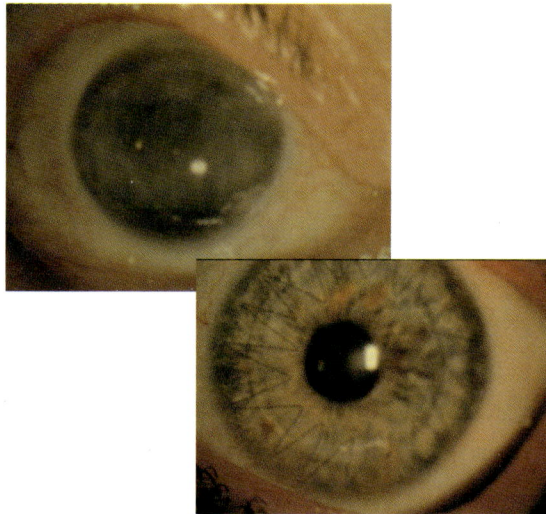

Damage to the cornea makes it go opaque so light cannot enter the eye, but a healthy cornea can be transplanted and stitched into place using very fine nylon thread. Corneal transplants were first performed in the late nineteenth century.

Blood vessels enter the eye at the optic disc and spread out into the **choroid layer** to supply the retina. In a disease called glaucoma, the aqueous humour does not drain away and pressure builds up inside the eye. This stops the blood flow to the retina and the optic nerve. If not treated, the condition eventually leads to blindness.

All convex lenses bend blue light more than red light (other colours of light are bent by an in-between amount depending on their position in the spectrum). This means that when red light is focused on the retina, blue light is focused just in front of the retina, so blue parts of an image are slightly out of focus. This problem is partly solved by the slight yellowness of the lens which filters out some of the blue from light entering the eye. The **yellow spot** surrounding the fovea also helps.

The cornea

In contrast to the retina, the cornea has a very poor blood supply, as blood vessels crossing the cornea would obscure vision. Any serious damage to the cornea is not repaired by the body and can lead to permanent blindness. Fortunately, the lack of blood supply also means that corneas can be transplanted with little risk of rejection.

The Orbis surgeons use a round cutting disc for corneal transplants. The surgeon places the disc on the surface of the eye and twists to make a circular cut in the damaged cornea. A round piece of cornea is then lifted out and replaced by a donor cornea that has been trimmed to the exact size in advance. Given the role of the cornea, it is important that no scar tissue forms and that the corneal surface does not become distorted.

4 Why is it important not to distort the cornea during a corneal transplant?

Key ideas

- Most of the refraction of light is done by the cornea. The lens makes the fine adjustments.

- Ciliary muscles contract to enable the lens to focus light from near objects on the retina. When the ciliary muscles relax, pressure within the eye, not another muscle, expands the ciliary ring and thus stretches the lens to allow for distant vision.

- The lens is elastic and thickens automatically when no force is acting to stretch it. It is expanded for near vision and thinned for distant vision.

- The iris controls the size of the pupil by reflex action. The pupil expands in dim light and constricts in bright light.

- The retina has a rich blood supply via vessels in the choroid layer; the cornea has very little blood supply.

2.3 Seeing the light

Colour vision is found in humans, monkeys and apes, insects, fish, and birds. Animals with colour vision are able to detect the wavelength of light. There are two types of light-sensitive cells in the human retina: rods and cones (Fig. 7).

Cone cells give colour vision and also allow us to see in detail (they give good **acuity**). There is maximum acuity at the fovea because here there are many densely packed cones and each **bipolar cell** is stimulated by a single cone cell; elsewhere

Fig. 7 Structure of the retina

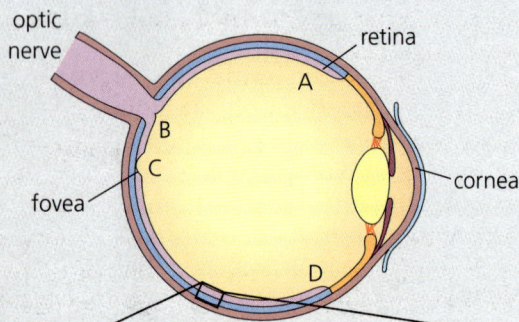

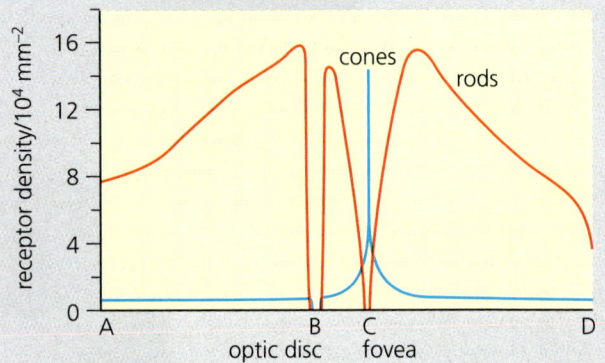

1: internal limiting membrane
2: nerve fibre layer
3: ganglion cell layer
4: inner plexiform layer
5: inner nuclear layer
6: outer plexiform layer
7: outer nuclear layer
8: layer of rods and cones
9: pigment epithelium
10: external limiting membrane
11: choroid layer

bipolar cell

ganglion cell
synapse
amacrine cell
bipolar cell
horizontal cell
synapse
rod nucleus
cone nucleus
rod
cone
extension of pigment cell
pigment layer
pigment
epithelium cell
blood vessel
pigment

This graph shows the relative density of cones and rods at different distances from the fovea; most cones are at the fovea, and there are no receptor cells at all at the optic disc (blind spot).

Source: adapted from Aidley, *The Physiology of Excitable Cells*, Cambridge University Press, 1975

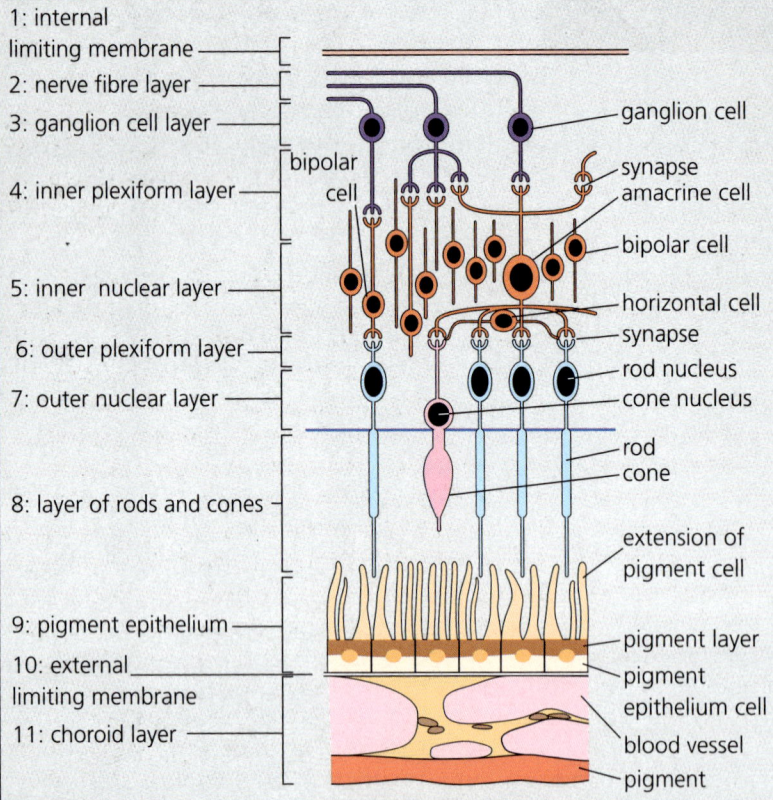

Light micrograph of a section through a normal human retina and choroid layer.

Source: adapted from Freeman and Bracegirdle, *An Atlas of Histology*, Heinemann Educational Books, 1991

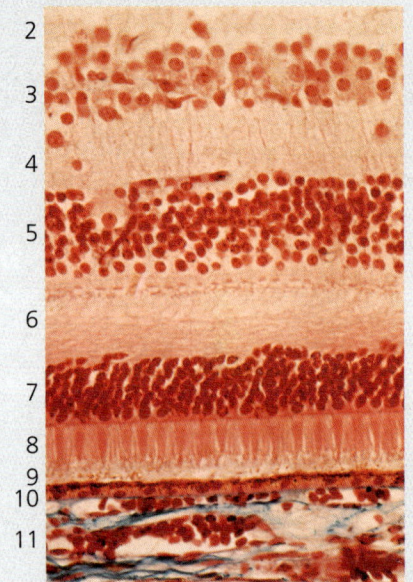

in the retina there are fewer cones and each bipolar cell is stimulated by several cones. Bright light is needed for cones to work well. On the other hand, rods are more effective in dim light, because up to 45 rods are linked to each bipolar cell (Table 1). The rods pool the information they send to the brain. This gives less acuity, but better **sensitivity**. When we look directly at something in good light, light from it is focused on the fovea, and we see it clearly. But when there is not much light, as when we look at a faint star, it is more easily seen if we look slightly to one side of it. This is because light is then focused to one side of the fovea where there are more rods, rather than on the fovea itself where there are more cones.

5 a **Why can't we see colours in dim light?**
  b **Why do rods give us greater sensitivity in dim light?**
  c **Why do cones gives us greater acuity in bright light?**

Visual pigments

Many pigments are bleached by light. This is why curtains in a sunny window slowly fade. A similar change happens to the pigments in rods and cones, but in these cases the bleaching is both rapid and reversible.

Each rod is packed with about 10^8 molecules of the light-sensitive pigment **rhodopsin** arranged in discs called lamellae. The rhodopsin molecule breaks down, in the presence of light, into **retinal** and the protein **opsin**.

But when we enter a dark room, the light-bleached pigment is restored to its unbleached form, rhodopsin. This is called **dark adaptation** and is why we are gradually able to see more as as we 'get used to' the lack of light. In the dark, it takes about 30 minutes to remake all the rhodopsin from the retinal and opsin.

Each rod produces about three new discs per hour; worn out discs of rhodopsin are broken down by epithelium cells. In complete darkness, rhodopsin is very stable, so there is little chance of any false signals being given. If rhodopsin were not so stable, rods might send signals to the brain when no light was falling on them.

6 **Why can't we see very well when we first enter a poorly lit room from a brightly lit area?**

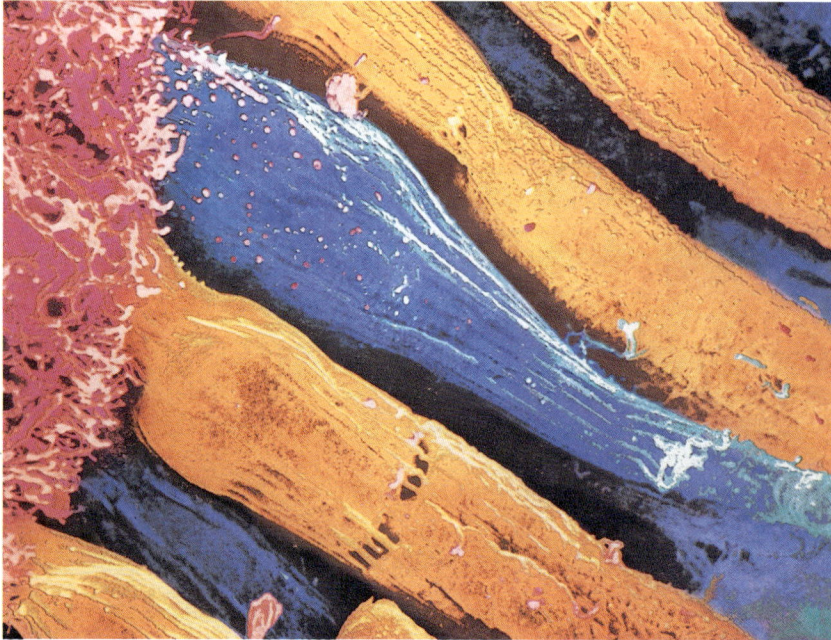

False colour scanning electron micrograph of rod cells (orange) and cone cells (blue). At dusk, we cannot see in as much detail or in such strong colours because we are using rods rather than cones.

Table 1 Rods and cones		
	Rods	Cones
Number per eye/millions	120	6
Size/µm	50 × 3	60 × 10
Absorption peak wavelength/nm	500 (green)	420 (blue)
		531 (green)
		558 (red)
Retinal convergence	15–45 rods to 1 bipolar cell	in fovea, 1 cone to 1 bipolar cell
Sensitivity	good: 1 photon gives response	poor: several hundred photons needed
Acuity	poor	good

The **photochemical change** of rhodopsin to retinal and opsin leads to a **cascade amplification** and causes an increase in the resting potential (hyperpolarisation) of the rod cell membrane (Fig. 8). Most receptor cells are *depolarised* by a stimulus rather than *hyperpolarised* (Chapter 1). However, in rod cells, hyperpolarisation leads indirectly to the generation of nerve impulses in the optic nerve.

Fig. 8 Rod cells, rhodopsin and light

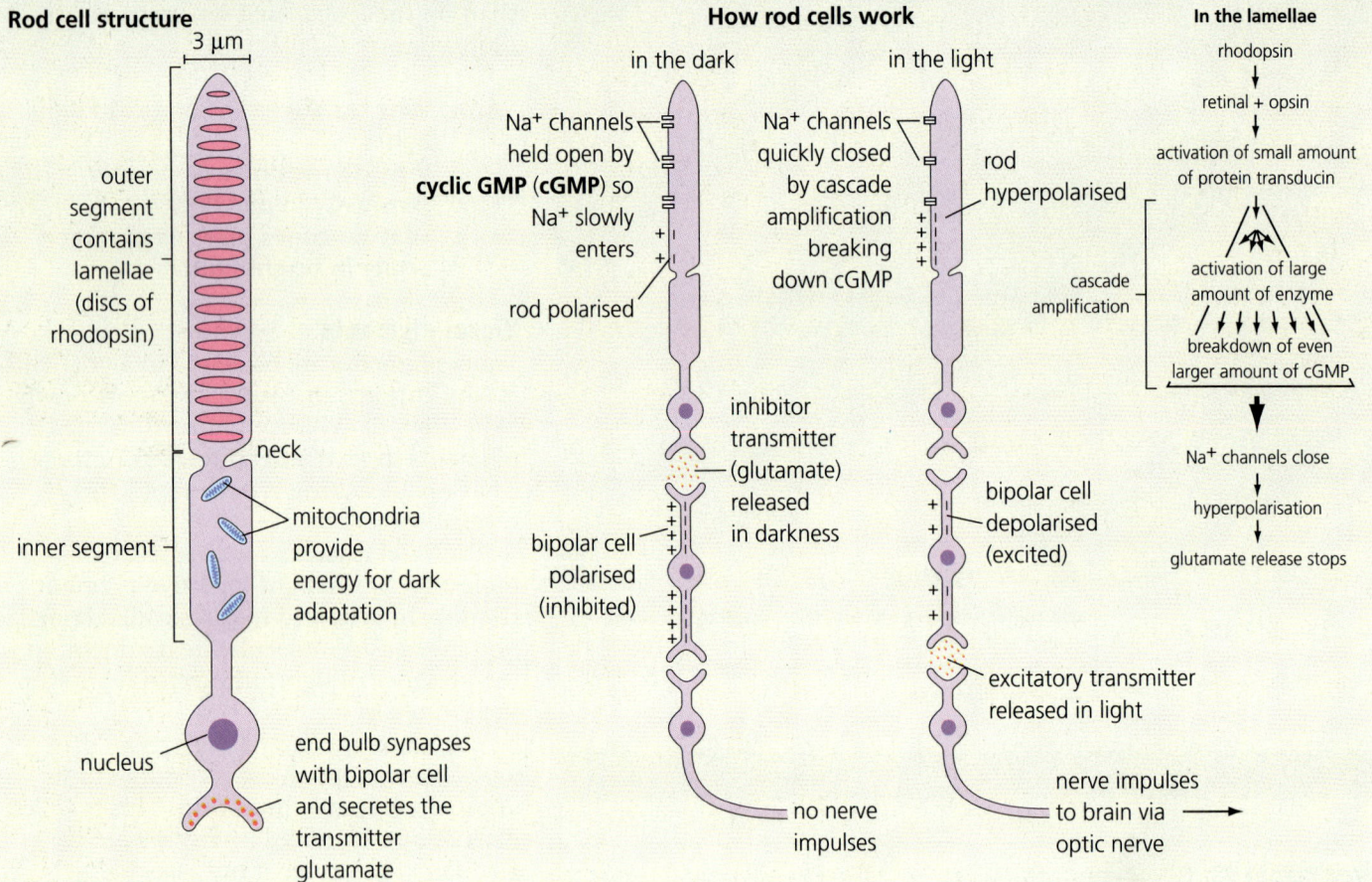

Rod cell structure

3 μm

outer segment contains lamellae (discs of rhodopsin)

neck

mitochondria provide energy for dark adaptation

inner segment

nucleus

end bulb synapses with bipolar cell and secretes the transmitter glutamate

How rod cells work

in the dark

Na+ channels held open by **cyclic GMP (cGMP)** so Na+ slowly enters

rod polarised

bipolar cell polarised (inhibited)

inhibitor transmitter (glutamate) released in darkness

no nerve impulses

in the light

Na+ channels quickly closed by cascade amplification breaking down cGMP

rod hyperpolarised

cascade amplification

bipolar cell depolarised (excited)

excitatory transmitter released in light

nerve impulses to brain via optic nerve

In the lamellae

rhodopsin

↓

retinal + opsin

↓

activation of small amount of protein transducin

↓

activation of large amount of enzyme

↓

breakdown of even larger amount of cGMP

↓

Na+ channels close

↓

hyperpolarisation

↓

glutamate release stops

What happens to rhodopsin

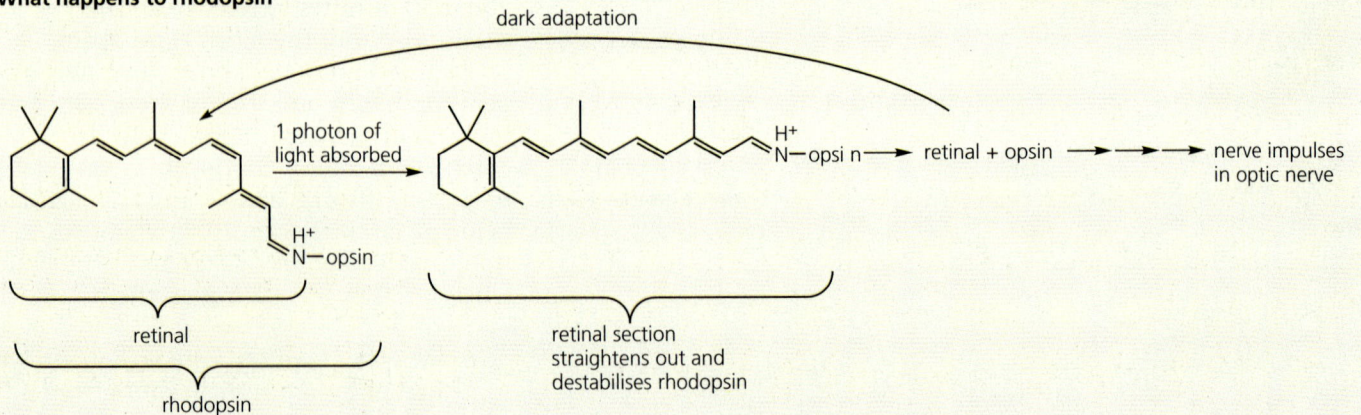

dark adaptation

1 photon of light absorbed

H^+ N—opsin

retinal

rhodopsin

H^+ N—opsin → retinal + opsin → nerve impulses in optic nerve

retinal section straightens out and destabilises rhodopsin

Violent blows to the head, as are sometimes delivered in boxing, can detach the retina. Frank Bruno has now retired because of the risk of blindness.

Some causes of blindness

Vitamin A is needed for the synthesis of both rhodopsin and the cone pigments. Lack of vitamin A in the diet is a common cause of blindness in developing countries. In its early stages, vitamin A deficiency leads to poor night vision, but later it causes complete and permanent blindness.

The retina can become detached from its junction with the pigment epithelium. This can occur by sudden jarring and sometimes occurred after the earlier forms of lens replacement surgery. The retina can be rejoined using lasers. Sometimes fluid must by removed from under the detached area, by making an incision through the sclera to drain it away. Orbis surgeons demonstrate these techniques.

7 Light from the centre of our field of vision is focused on the cones at the fovea. What evidence is there, from what we see, that we have cones in other parts of the retina?

2.4 Colouring in

Rods respond most strongly to bluish-green light. Cones are divided into three types, each of which has a different sensitivity to light, but the ranges of sensitivity overlap and most wavelengths of light stimulate at least two types of cone (Fig. 9).

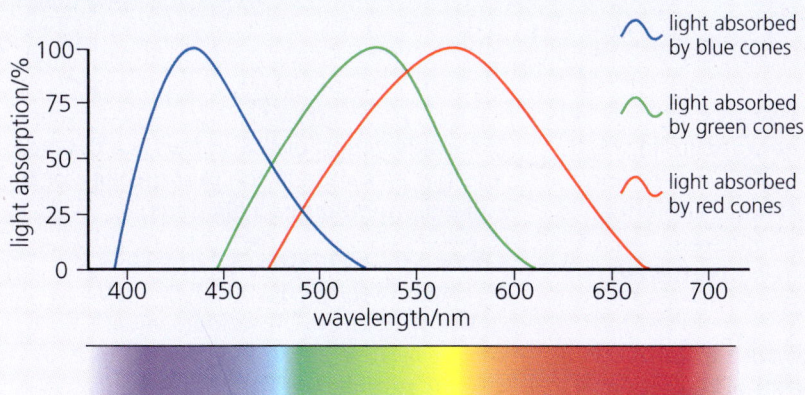

Fig. 9 Cones and light absorption

The discovery of three types of cone supports the **trichromatic theory** of colour vision. This theory states that we see all the colours of the visible spectrum by a mixing of the three primary colours – blue, green and red. A white wall reflects all the colours of the spectrum back to the eye; the white parts of a colour computer monitor or TV screen are made up of red, green and blue dots. In both cases, the three types of cone are stimulated equally. The brain interprets any combination of messages from the three types of cone as a particular colour.

8 a Which cones are stimulated by light of wavelength 600 nm?
 b Why do you see an orange colour when light of 600 nm stimulates the cones?
 c Which cones are stimulated by the white paper and black letters of this page?

Fig. 10 After-images

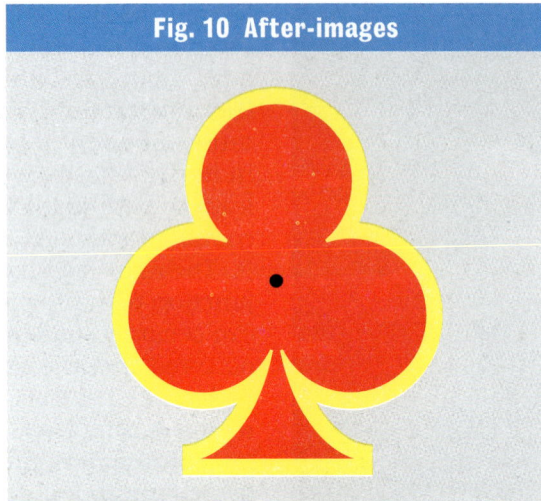

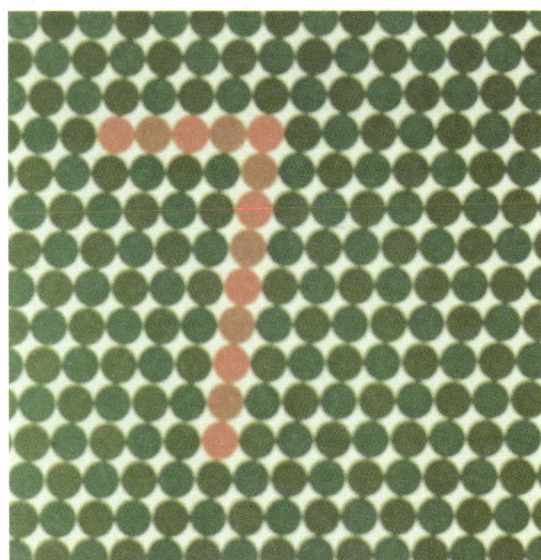

People with full colour vision see the number 7 in this pattern of coloured dots, those with red–green colour-blindness do not.

Colour-blindness

About 8% of men and less than 1% of women have faulty colour vision. The genes for making the cone receptor pigments (cone opsins) are on the X chromosome. Because men have only one X chromosome and women two, a defective gene for cone pigment is more likely to be expressed in men. Many colour-blind men lack either the red cones or the green ones and so confuse red, blue-green and grey. For example, bright red roses can look the same colour as the leaves, and scarlet clothes may look dark grey. Complete colour blindness in which all cone pigments are absent is very rare.

The genes for the cone opsins have now been cloned. The base pair sequences show that the three cone opsins differ only slightly from each other. Since the cone opsins are also similar to rhodopsin, it seems likely that colour vision evolved by slight mutations to the rhodopsin gene.

As we get older, the number of cone cells declines, so brighter and brighter light is needed for good acuity. This is why older people need very bright light for reading.

9 **Why is colour-blindness more common in men than in women?**

10 **Stare at Figure 10 for 20 seconds, then look at a white surface. What colours do you see? This is called an after-image. How does the trichromatic theory explain after-images?**

Key ideas

- There are two types of receptor cells in the retina: rods and cones. They have different functions and properties. These differences occur because they contain different pigments and are connected to other cells in the retina in a different way.

- Detecting light starts with light absorption by rhodopsin in the rod cells of the retina and involves (i) bleaching of rhodopsin, (ii) amplification of the signal by a cascade, and (iii) a change in the permeability of the rod cell membrane to Na^+ ions.

- There are three types of cone. Each is sensitive to different, but overlapping, parts of the spectrum. The colours we see depend on the relative amount that each type of cone is stimulated. This is the trichromatic theory of colour vision.

2.5 Making sense of information

In the retina

The retina processes information from the receptors before it reaches the brain. This is carried out by **horizontal cells**, which join up with several rods, cones and bipolar cells, and also by **amacrine cells**, which link to bipolar and ganglion cells.

The horizontal and amacrine cells emphasise changes in contrast (brightness). They do this by a combination of stimulation and inhibition of nearby cells. This processing means that the brain is alerted to the outlines of objects and to movements because both involve changes in contrast. So, we quickly notice any changes in our surroundings, such as a slight movement or a flashing light like a car indicator, but continuous stimuli are soon ignored. This sort of information is especially important in the recognition of food, prey, or danger.

Colour signals are also processed. The ganglion cells show responses to *four* colours: blue, green, red *and yellow*. No cone responds specifically to yellow light, so this response of the ganglion cells must be a result of the influence of nearby retina cells on each other.

In the brain: colour and movement

Further processing occurs in the brain. Electrical recordings have been made from monkey and cat brain cells while the animals look at images on a computer screen.

The visual cortex is the region of grey matter where the information from the optic nerves is transformed into vision (Fig. 11).

The brain is able to put this 'electrical storm' of information together and perceive objects. The way in which this happens is still not fully understood. We believe that information about colour, movement and form are analysed separately by the brain, and finally combined. Added to that, we can call up visual information and relate it to past experience – for example, we recognise a familiar face when we see it.

What we see is not simply a pattern of dots, one from each receptor. Both the retina and the brain analyse and modify the visual stimuli they receive. Unimportant messages are screened out, and relevant details, such as changes in the surroundings, or contours, are enhanced. The brain also uses 'common sense' to

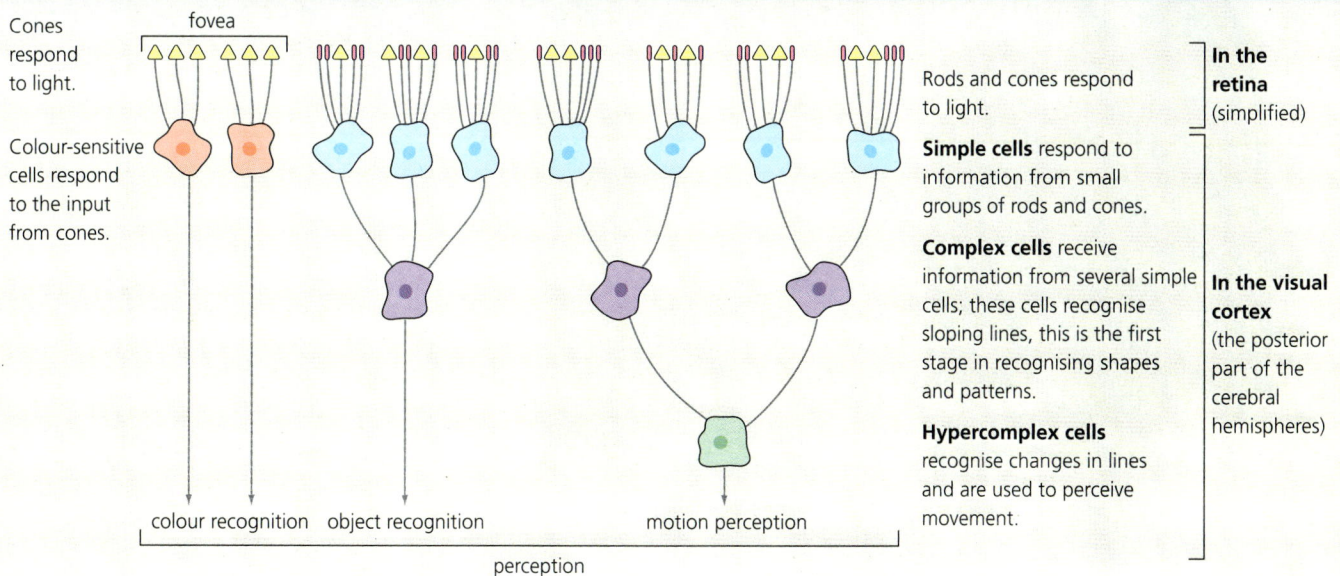

Fig. 11 Visual cortex and perception

Cones respond to light.

fovea

Colour-sensitive cells respond to the input from cones.

Rods and cones respond to light.

Simple cells respond to information from small groups of rods and cones.

Complex cells receive information from several simple cells; these cells recognise sloping lines, this is the first stage in recognising shapes and patterns.

Hypercomplex cells recognise changes in lines and are used to perceive movement.

In the retina (simplified)

In the visual cortex (the posterior part of the cerebral hemispheres)

colour recognition   object recognition   motion perception

perception

interpret visual images. For example, the image cast on the retina is upside down, yet we see the world the right way up. Similarly, we may interpret the colour of our car as 'blue', even when we see it under an orange street light which changes its colour.

In the brain: distance

The two eyes do not act independently. For example, both are needed to judge distance accurately. We can sometimes tell how far away a familiar object is by its apparent size; the further away an object is, the smaller it seems, but the brain can be confused over this (Fig. 12).

11 Study Figure 12. Why does the upper horizontal line look longer than the lower one at first sight?

The brain can judge distances much more accurately by comparing the images from each eye (Fig. 13). The images are different because the central axes of the eyes are about 64 mm apart. The difference between the images is greater for near objects than it is for distant objects. The brain uses this information to compute distance.

Many puzzles and even some interior decoration schemes depend on our ability to confuse ourselves about what we see. This is a painted wall, not a balcony. If you look carefully, you can see a door cut into the centre section.

Humans are remarkably skilled at making these judgements. It is likely that such abilities evolved in our tree-living ancestors, enabling them to jump from branch to branch in the canopy. It has also been suggested that colour vision in primates had advantages such as enabling individuals to judge the ripeness of fruit, and so avoid the risks of climbing up to sample it.

Fig. 12 Deceiving the brain

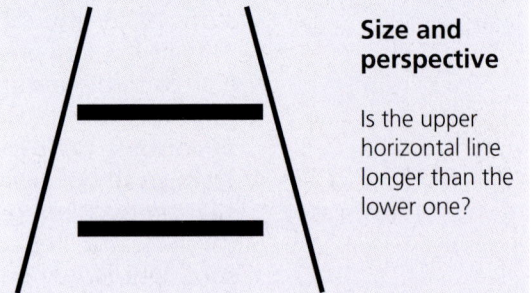

Size and perspective

Is the upper horizontal line longer than the lower one?

Fig. 13 Judging distance

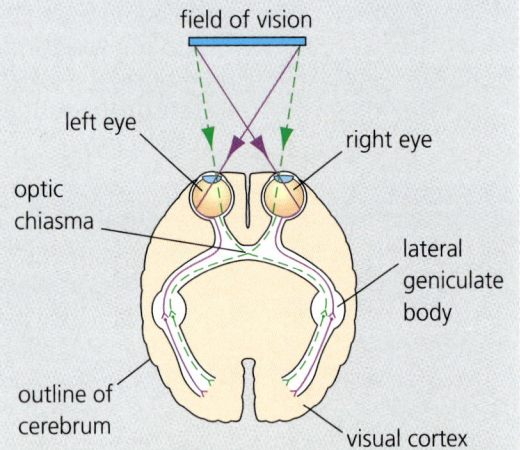

The green and purple lines show the pathways of light from each side of the field of vision to each side of the brain. Half of the nerve fibres carrying information from each eye cross over at the optic chiasma, so *all* the information from any particular point goes to either the right or left side of the brain. Each side of the brain has two slightly different images of the opposite side of the visual field and judges distance by comparing them.

Looking to the future

The level of understanding that we now have about the ways in which the eyes and the brain function together is very sophisticated. It is certainly in advance of the knowledge needed to help many who have visual problems. Who knows what the Orbis teams of the next century could be offering to help blind and partially sighted people round the world?

Judging distance is essential when driving a car. We can tell how far away another vehicle is, and also judge how quickly it is approaching.

Key ideas

- Processing of visual information begins in the retina by cells affecting the activity of their neighbours.

- Integration of the information reaching the visual cortex emphasises edges, boundaries and movements.

- The brain is able to use the different images from the two eyes to judge distance.

- The brain integrates the information from thousands of nerve cells about colour, shape and movement. From such information it produces the experience of seeing the world, and can store and retrieve visual memories.

..
3 Seeing it the plant's way

I manage a plant nursery that grows chrysanthemums, poinsettias, kalanchoes and begonias for the potted plant trade. Pot-grown plants are very popular these days. Lots of people want to buy them as gifts, and they are often used as decoration in hotels and shopping centres. This means that fresh flowering plants are wanted on a regular basis all year round.

Many plants naturally flower in the spring or autumn, but by giving them the correct length of light and darkness, I can produce plants in flower all through the year. I can predict almost to the day when a batch of plants will flower. This extends our growing season and means we can make efficient use of space in the greenhouses.

Growing plants is big business.

Much more is known about plants grown for the potted plant trade than about the Titan arum flowering at Kew Gardens in August 1996. It comes from open, evergreen forest in Sumatra, but nothing is known about what triggers it to produce this 2.05 m flowering structure.

3.1 Learning objectives

After working through this chapter, you should be able to:

- **explain** how day length affects the flowering season in plants;

- **recall** that plant species differ in their response to the seasons: some flower during spring and autumn, others flower in the summer;

- **explain** the role of phytochrome pigments in responding to day length;

- **recognise** the general features of tropisms in plants;

- **explain** how auxins control phototropism.

3.2 To flower or not?

In temperate regions, such as Britain, day length varies between about 20 hours in mid-summer to about 8 hours in mid-winter. Plants produce flowers when the **photoperiod** (the length of the daily light and dark periods) is suitable (Table 1). The growth and development of plants in response to photoperiod is called **photoperiodism**.

Table 1 Flowering conditions

Plant type	Conditions for flowering	Examples
short-day plants	short days and long nights (spring and autumn)	primulas, cowslip, strawberry, some orchids, chrysanthemums, coffee, tobacco, rice, sugar cane
long-day plants	long days and short nights (summer)	cereals and grasses, ice-plant (Sedum spectabile), clover
day-neutral plants	day length not important	sunflower, pea, cucumber

Being long-day plants, cereals flower in summer so the harvest is collected in late summer or early autumn; rice is a short-day plant that is planted early in the rainy season in Asia (June) and is harvested 100–180 days later.

Q 1 Why do some plants that flower in the spring, such as primulas, sometimes flower again in the autumn?

Not all plants are equally sensitive to photoperiod. Cocklebur is a very sensitive short-day American plant in which a single short-day cycle triggers flowering. Although it used to be thought that plants flowered in response to the day length, experiments with artificial light have shown that the *dark period* is much more important in causing flowering. Cocklebur has a critical dark period of 8.25 hours. If this is exceeded, flowering follows. So, short-day plants would be more sensibly called long-night plants. In these plants, the dark period must be uninterrupted by any light, otherwise flowering does not occur. In cocklebur, even a 1-minute flash of light during a 9-hour dark period prevents the flowering process. This is called the night break effect.

Despite the fact that the flowers 'follow the sun', day length is not important in initiating flowering in sunflowers.

Using the night break effect

In the nursery, the plants are first grown to a good size without developing flowers. The night break effect is used to control flowering. Chrysanthemums are short-day (long-night) plants, so they require over 8 hours of continuous darkness to start making flower buds. Flower formation is prevented during the winter by giving the plants a 4-hour light period in the middle of the night. The timing and length of the night break is computer-controlled.

After about two weeks, the plants are big and bushy. The flowering process is then started by giving the plants 11 hours of light and 13 hours darkness every day for about 10 weeks. Automated black screens are used to keep out the light when necessary. The artificial long nights trigger flower bud formation and, after 10 weeks, the plants are in full flower ready to sell (Fig. 1).

Florists need potted plants that are bushy, with all the flower buds opening at the same time.

2 What is the night break effect?

Tests using light of different colours show that red light has the best night break effect. However, if red light is quickly followed by far-red light, the night break effect does not happen (Table 2). In a fairly rapid sequence, it is the *last* light period that is important.

Fig. 1 Dark period and flowering

Flowering in soybeans

Soybean flowers are not produced unless the dark period is at least 10 hours. The length of the light period (4 hours or 16 hours) makes no difference to the length of dark period needed to trigger flowering.

Source: adapted from Wareing and Phillips, *The Control of Growth and Differentiation in Plants*, Pergamon, 1975

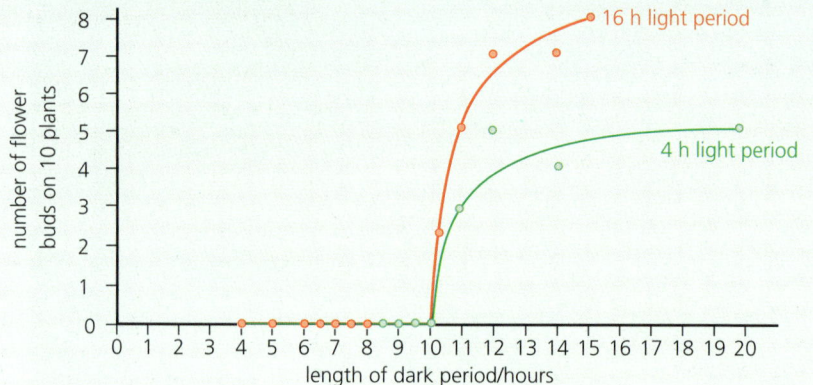

Graph: number of flower buds on 10 plants (y-axis, 0–8) vs length of dark period/hours (x-axis, 0–20). Curves for 16 h light period and 4 h light period.

Flowering in chrysanthemums

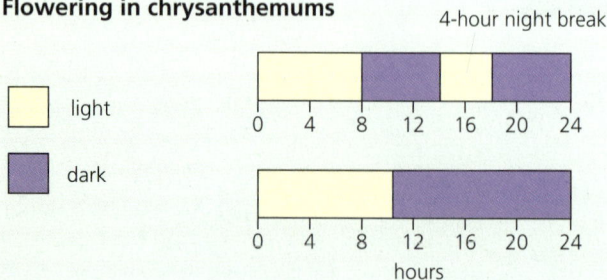

light
dark

4-hour night break

2 weeks of long days (days are 'long' because of the night break) — bushy plant no flower buds

10 weeks of short days (long nights) — flowering plant

Table 2 Effect of red and far-red light on flowering in short-day plants

Light exposure	Night break effect	Flowering response
red	✓	✗
red/far-red	✗	✓
red/far-red/red	✓	✗
red/far-red/red/far-red	✗	✓

However, in Cocklebur, the night break effect of red light is *not* prevented if more than 30 minutes passes before far-red light is shone on the plant.

3 a Would the rapid sequence far-red/red/far-red produce the night break effect in a short-day plant?

b Is flowering in a cocklebur prevented by the sequence far-red/red/45 minutes/far-red?

Detecting photoperiod

It is the leaves, not the shoots producing the flower buds, that are sensitive to the photoperiod, so photoperiodism must involve two stages:

- there must be a leaf pigment absorbing red and far-red light;
- a message must travel from the leaves to the growing shoots.

A pigment that can detect red and far-red light in plant leaves has been identified; it is **phytochrome**. Phytochrome occurs in minute quantities and exists in two forms, **Pr** and **Pfr** (Fig. 2). In this sense, phytochrome in plants is the equivalent of rhodopsin in the rods in mammalian eyes (Chapter 2): both receive light, the molecules change shape, and subsequently trigger a change in the organism.

4 How does the Pr form of phytochrome change to the Pfr form?

Fig. 2 Phytochrome and light

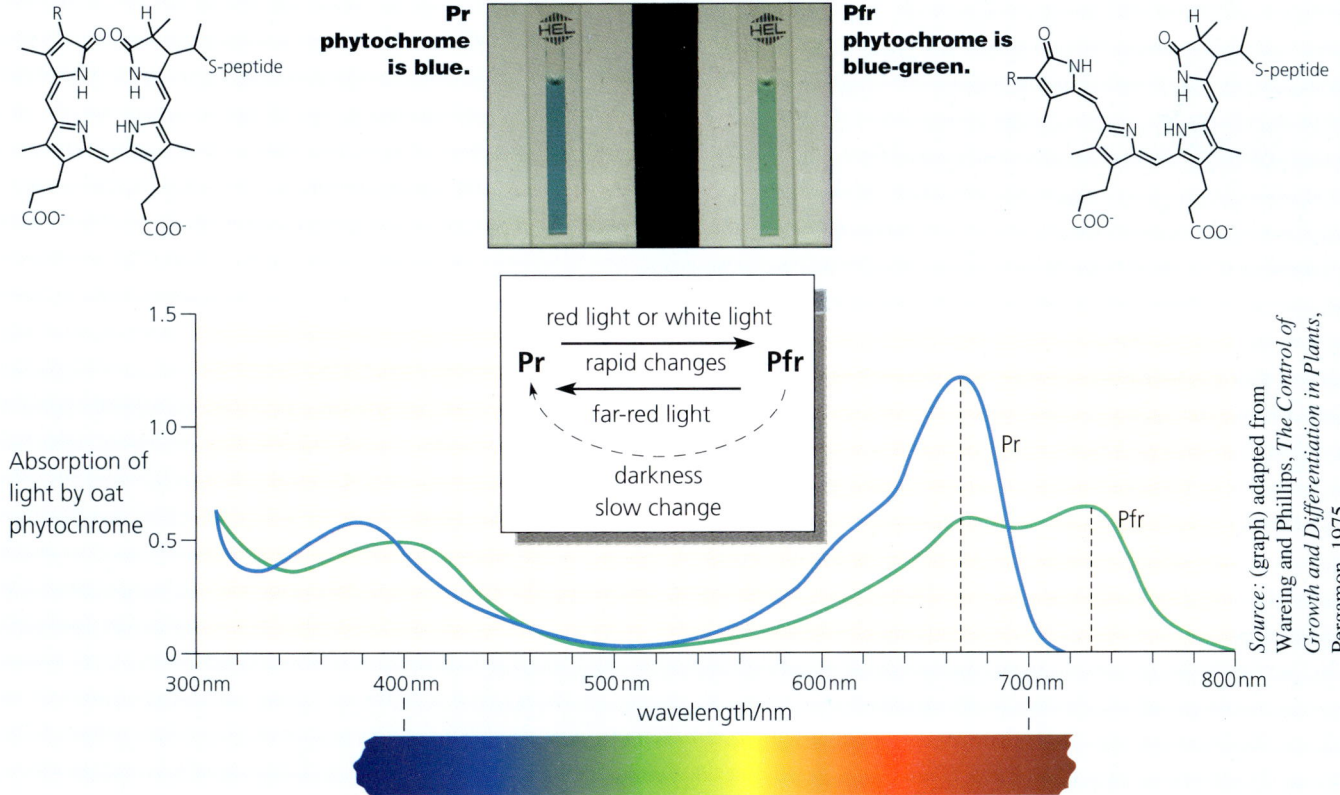

Pr phytochrome is blue.

Pfr phytochrome is blue-green.

red light or white light

Pr → rapid changes → Pfr

far-red light

darkness slow change

Absorption of light by oat phytochrome

Source: (graph) adapted from Wareing and Phillips, *The Control of Growth and Differentiation in Plants*, Pergamon, 1975

Short days with long nights lead to an excess of the Pr form of phytochrome. So, in short-day plants, a high Pr/Pfr ratio is believed to cause flowering. A period of light during the dark period acts as a night break and prevents flowering because it reduces the amount of Pr.

Long days and short nights lead to an excess of the Pfr form of phytochrome. In long-day plants, it is a high Pfr/Pr ratio which causes flowering. In this case, a period of light during the dark period increases the amount of Pfr, and so stimulates flowering, rather than prevents it.

Q 5 How can flowering be prevented in a short-day plant, such as a primrose, during the spring?

Phytochrome causes flowering by increasing the amount of a flower-promoting substance, **florigen**. Florigen has not yet been isolated or identified, despite much searching. It is presumed to travel from the leaves to the growing shoots, where it causes flower buds to form. Strong evidence for the existence of florigen comes from grafting experiments (Fig. 3).

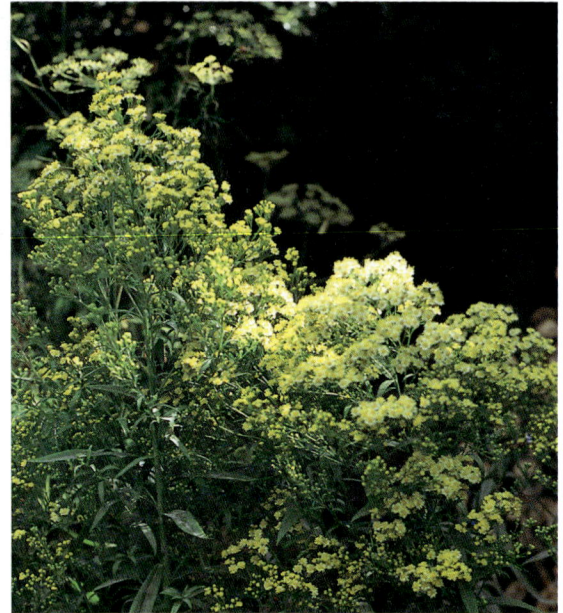

Even within a single species, plants may differ in their response to day length between different regions. For instance, the wide-ranging golden rod (*Solidago*) is a short-day plant in some areas, but a long-day plant in others.

Q 6 If florigen could be manufactured, what could it be used for commercially?

At least three different kinds of phytochrome are found in plants. Phytochromes not only affect flowering, they also influence germination, stem growth, and leaf fall in some plants. The responses of plants to day-length are complex and every species is slightly different.

Biological clocks

Plants show all kinds of daily rhythms. For example, bean plants (*Phaseolus vulgaris*) hold their leaves horizontally in the day but point them downward at night. This rhythmical change of leaf position continues even if the bean plants are kept in constant darkness, so it is an **endogenous** rhythm (it is produced inside the plant, not in response to changes in the environment). Such rhythms use **biological clocks**, which 'measure' a 24-hour period. Biological clocks are not yet fully understood, but they might work by using diffusion across membranes.

Fig. 3 Does florigen exist?

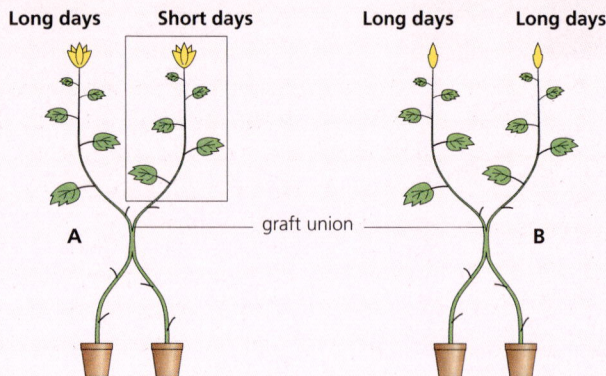

In experiment A, two cocklebur plants are grafted together. The top of the left-hand plant is exposed to long days, but the right-hand plant is exposed to short days. Both plants flower. In the control experiment, B, two cocklebur plants are similarly grafted and both plants exposed to long days. Neither flowers. The experiment shows that a message passes from the short-day side to the long-day side in experiment A, causing the long-day side to come into flower. The chemical messenger is called florigen.

Source: adapted from Wareing and Phillips, *The Control of Growth and Differentiation in Plants*, Pergamon, 1975

Photoperiodism must involve a biological clock, because plants are able to 'measure' the length of light and dark periods. Although change in the phytochrome from Pr to Pfr (or Pfr to Pr) could work as a clock by changing the ratio of Pr to Pfr over time, biological clocks do not respond to temperature, and conversion of Pr to Pfr *is* affected by the temperature. Also, the conversion of Pr to Pfr can happen in less than hour, so the changes couldn't accurately measure a 24-hour period. On the other hand, it appears that changes in the Pr/Pfr ratio happen fairly rapidly at dusk. Perhaps the changing Pr/Pfr ratio resets the biological clock to zero so it can then 'measure' the dark period and control the production of florigen and other substances.

Q 7 Why can't the biological clock involved in photoperiodism simply be the changing Pr/Pfr ratio?

Key ideas

- Plants are adapted to local conditions where they grow. They develop and come into flower in the appropriate season.

- The photoperiod controls the timing of the flowering season in many plants. The length of the uninterrupted dark period is the important factor.

- Photoperiod is detected by pigments called phytochromes in plant leaves.

- Phytochrome occurs in two forms, Pr and Pfr, which can change into one another. The ratio of Pr to Pfr is altered by changes in the photoperiod as the seasons change. The Pr/Pfr ratio controls flower formation.

- Phytochrome causes flowering through a flower-promoting substance called florigen. Florigen travels from the leaves to the shoots causing flower buds to form. Florigen has not yet been chemically identified.

3.3 Seeking the light

In the nursery, plants are grown using both artificial and natural light. One of the problems that must be overcome is the tendency of plants to bend or grow towards the light source. So, poinsettias grown close to glasshouse windows all lean slightly towards the south, and chrysanthemums grown at the edges of the trays lean towards the middle where the lamps are directed. The most up-to-date nurseries have installed high-pressure sodium lights that give a good spread of illumination and result in evenly growing plants.

These tulips have bent stalks because they are growing towards a uni-directional light source.

Table 3 Effects of two tropisms		
Tropism	Positive	Negative
phototropism	growth towards light	growth away from light (rare)
geotropism	growth towards gravity	growth away from gravity

The growth of plants towards or away from a stimulus is called a **tropism** (Table 3). A germinating seed first sends out a root to anchor the plant in the soil. The root grows downward in a positive response to gravity (**positive geotropism**). Next, a shoot grows upward through the soil in a negative response to gravity (**negative geotropism**). When the shoot reaches the surface, it grows towards sunlight (**positive phototropism**), perhaps through a gap in the surrounding plants, so gaining as much light as possible for photosynthesis.

In corn and grasses, the new shoot is covered in a layer of protective cells called a **coleoptile**. Coleoptiles are positively phototropic and grow towards the light. Phototropism is often investigated in schools using barley coleoptiles to show that although it is the coleoptile *tip* which perceives the light, it is the *middle region* that bends (Fig. 4). So, there must be a means of communication between the tip and middle of the coleoptile.

Q 8 a In experiments on tropisms, it is important to use a large sample of coleoptiles. Why?
b Describe how you could investigate the way coleoptiles respond to gravity.

Plant physiologists have concluded that three things are needed for positive phototropism in coleoptiles to occur:
- a pigment in the tip that acts as a receptor for light and somehow triggers the synthesis of a chemical messenger;
- a chemical messenger to carry the signal from the growing tip to the middle of the coleoptile;
- a response (to the chemical messenger) that causes the shaded side of the coleoptile to grow more than the side in the light, thus bending the coleoptile towards the light.

Q 9 Are the conclusions drawn from experiments with coleoptiles true for all plants?

Fig. 4 Phototropism in barley

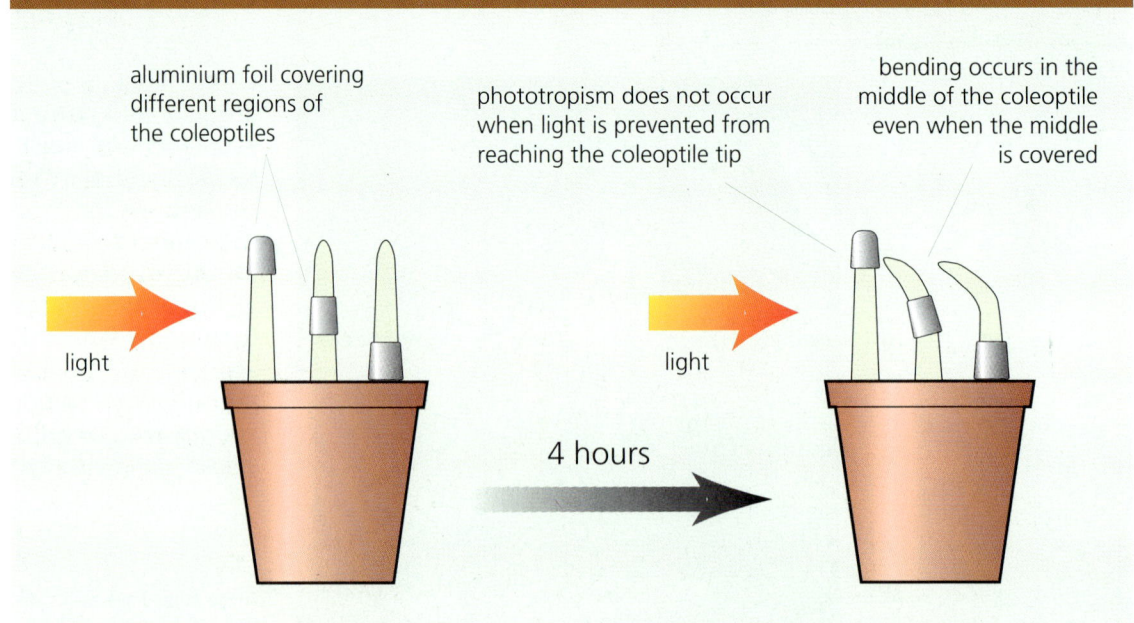

aluminium foil covering different regions of the coleoptiles

phototropism does not occur when light is prevented from reaching the coleoptile tip

bending occurs in the middle of the coleoptile even when the middle is covered

light

light

4 hours

Fig. 5 Light colour and phototropism

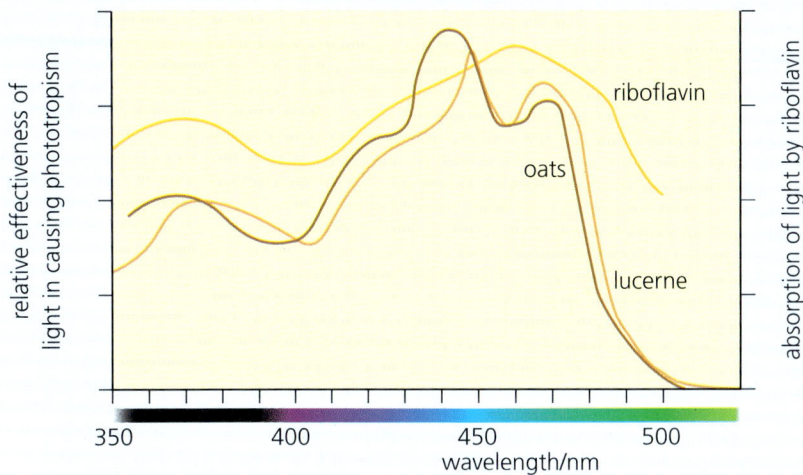

The **action spectra** of oats and lucerne show that blue light is most effective at causing phototropism in both species. The **absorption spectrum** of riboflavin has a similar shape. Riboflavin can join to a protein to make flavoprotein.

Source: adapted from Salisbury and Ross, *Plant Physiology*, Wadsworth, 1992

Fig. 6 Indoleacetic acid – an auxin

This is the structure of indoleacetic acid (IAA) one of the most common auxins in plants. There are at least three other similar auxins controlling growth in plants.

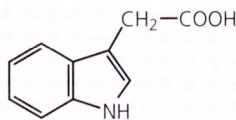

IAA and growth

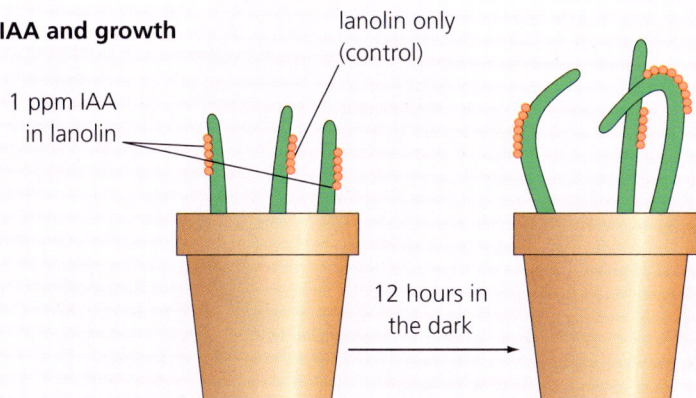

1 ppm IAA in lanolin

lanolin only (control)

12 hours in the dark

IAA causes the treated side of the coleoptile to grow faster than the untreated side, so the coleoptile bends.

The pigment

Blue light causes the strongest phototropism in many plants. This suggests that the receptor pigment is yellow, because yellow pigment absorbs blue light. So, the pigment may be a **flavoprotein** in the cell membrane (Fig. 5).

The chemical messenger

Many different **plant hormones** act as chemical messengers. **Auxins** are a group of hormones that can travel from cell to cell and can cause stems to grow faster (Fig. 6).

The response

Positive phototropism happens when the shaded side of the stem grows faster than the illuminated side. There are two possible explanations for this in plants that have coleoptiles:

• 1: the auxin is moved from the illuminated to the shaded side of the coleoptile at the tip;
• 2: the auxin is inactivated on the illuminated side of the coleoptile.

Support for explanation 1 comes from experiments examining the effect of light on auxin in the tips of maize coleoptiles, and also from experiments that show intact maize coleoptiles grow less on the illuminated side and more on the shaded side than control plants grown in even illumination.

However, researchers using oat coleoptiles and sunflowers (which do not have coleoptiles) could not get the same results as were obtained for maize coleoptiles. In fact, they could not find any difference in auxin concentration, but they did find a higher concentration of an **auxin inhibitor** on the illuminated side of the stem in sunflowers. This supports explanation 2, that auxin is inactivated on the side nearest to the light.

Figure 7 (overleaf) shows some of the experiments that support explanation 1.

Fig. 7 Experiments with maize coleoptile tips

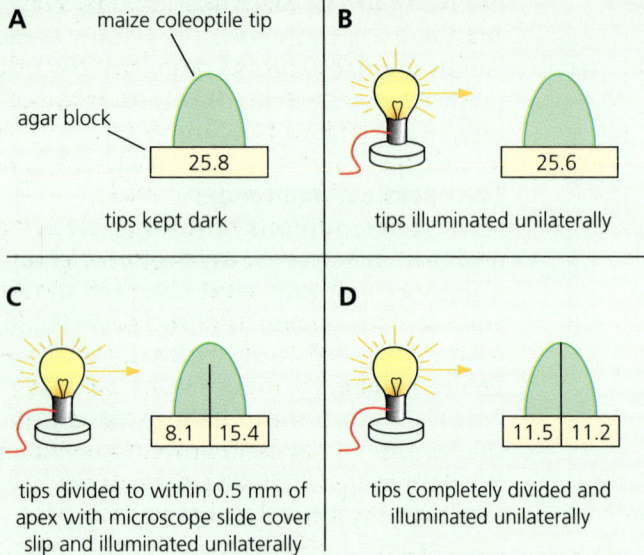

A

maize coleoptile tip

agar block

| 25.8 |

tips kept dark

B

| 25.6 |

tips illuminated unilaterally

C

| 8.1 | 15.4 |

tips divided to within 0.5 mm of apex with microscope slide cover slip and illuminated unilaterally

D

| 11.5 | 11.2 |

tips completely divided and illuminated unilaterally

The numbers represent the amount of auxin passing into the agar blocks from the coleoptile tips. Comparison of A and B shows that light does not cause the destruction of auxin. Auxin in the partly divided tip (C) is transported away from the illuminated side, above the dividing barrier, but in the completely divided tips (D) this was not possible.

Source: adapted from Salisbury and Ross, *Plant Physiology*, Wadsworth, 1992

Growth of maize coleoptiles in uniform light (control) and in light shining from one side. The shaded side of the coleoptiles grew more than the control; the illuminated side grew less. This could be explained by movement of auxin from the illuminated side to the shaded side.

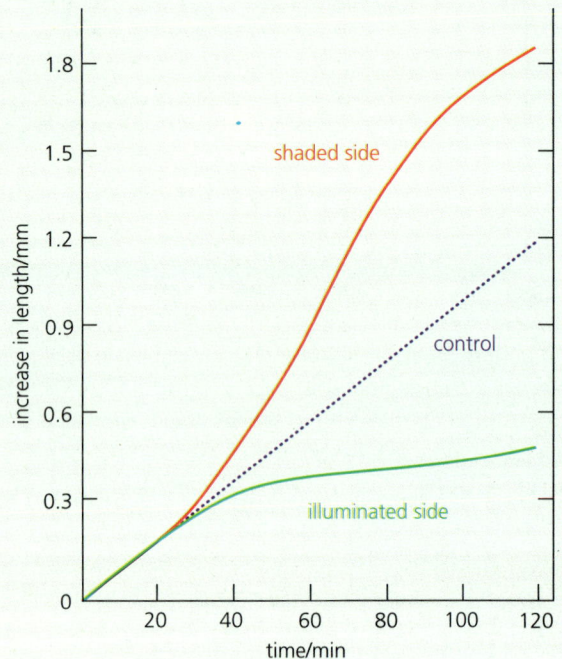

Q 10 How does Figure 7 support the idea that auxin has been moved rather than destroyed?

11 How does the difference in auxin concentration between the illuminated and the shaded sides of a coleoptile tip cause a phototropic response?

A problem for both possible explanations is that auxin is unlikely to move quickly enough from the tip to the region of bending to cause the rapid phototropism seen in some growing shoots. Coleoptiles begin bending towards light within 5 minutes. Although auxin uses ATP energy to move from cell to cell and travels about ten times faster than it could by diffusion, its speed is still only 1 cm per hour. This is not fast enough to move from the coleoptile tip to the region of bending in 5 minutes.

In fact, plant physiologists have not yet agreed about the exact mechanism of phototropism in any plant.

I sell a range of commercial plant growth regulators to people whose business is plants. Plant growth regulators are chemicals that mimic the action of plant growth hormones.

Some commercial applications

Farmers and horticulturists need a good understanding of the effects of many different plant hormones on growth and development. After the chemical structure of the auxin indoleacetic acid (IAA) was determined, researchers started trying to make artificial alternatives. There is now quite a range of commercially produced **artificial auxins** and some are even more effective than IAA. They have many different uses in the plant trade even though the detail of how they act is not always fully understood.

If a length of stem is cut from a parent plant and placed in water or moist soil, it often grows roots from the bottom of the stem. This is called 'taking a cutting' and growing new plants from cuttings is very common in the plant trade. Dipping the cut base of the stem into IAA improves the rate at which new roots are developed, but artificial auxins can be used instead. Some of the artificial auxins are easier to make, cheaper, and have a longer-lasting effect because they are not inactivated by the plant.

Research into plant hormone weedkillers started as long ago as 1940 – when it was considered to be a military secret! Weedkillers of this type are toxic to broad-leaved plants because the artificial hormones are absorbed through the leaves and then transported to the growing points. They also kill the roots so the plant does not grow back again after the stems and leaves have died away. Although they are thought to be harmless to animals at the concentrations in which they are applied, indiscriminate use of hormone weedkillers can cause serious ecological problems.

Some fruits can develop without the occurrence of pollination or fertilisation, so the fruits are seedless. For this to happen, IAA is required. Examples of seedless fruit are seedless grapes, bananas, and

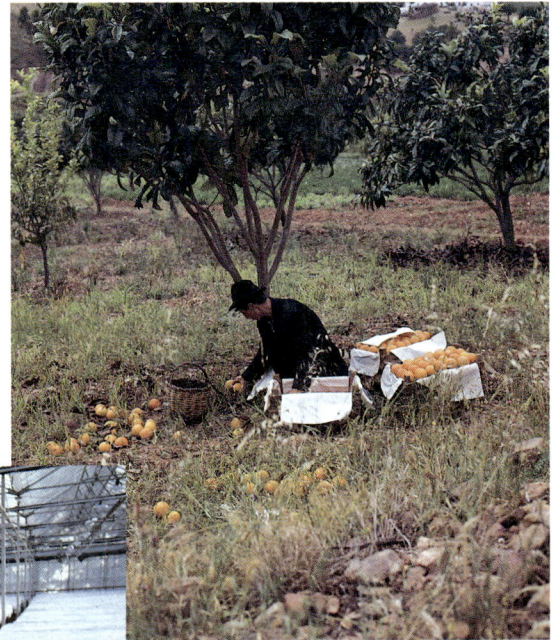

Auxin-based selective weedkillers are absorbed and transported more quickly by broad-leaved plants than by grass or corn. They are useful in keeping cornfields, lawns and bowling greens weed-free.

Rooting compounds are artificial auxins available in powder form.

Unlike this small-scale producer, large commercial growers of apples, pears, and citrus fruits use plant growth regulators developed from auxins to prevent ripe fruits from dropping off the tree so the fruit can be efficiently harvested.

pineapples. Artificial auxins are used commercially in this area to increase the size and number of the fruit produced.

Some plants grow taller than the grower would like. This happens in wheat and barley and causes the crop to fall over, which makes harvesting difficult. Pot-grown plants also need to be bushy because spindly ones are less attractive to customers. Besides the auxins, there are also gibberellin growth hormones that normally cause stems to elongate. These hormones can be inhibited by commercial **anti-gibberellins** so the plants remain stocky.

12 a What is IAA?
b What are plant growth regulators?

13 a List three uses for artificial auxins.
b What are anti-gibberellins used for?

Key ideas

- Plants change their direction of growth in response to light and gravity. These responses are called tropisms.

- Shoots usually grow towards light (positive phototropism) and away from gravity (negative geotropism). Roots usually grow towards gravity (positive geotropism).

- In many plants, the shoot tip detects light direction. The receptor pigment is most responsive to blue light and might, therefore, be a flavoprotein.

- The receptor pigment responds to light and somehow stimulates the synthesis of plant hormones called auxins.

- Auxins cause the shaded side of a stem to grow more than the illuminated side. Thus, the stem grows toward the light.

- Auxins are either redistributed by light towards the shaded side of the shoot, or inactivated on the lit side.

- Phototropism and geotropism help plants to grow their shoots toward the sun (for photosynthesis), and their roots into the earth (for anchorage).

- Anti-gibberellins inhibit gibberellin growth hormones.

- Plant growth regulators that mimic plant growth hormones and have widespread commercial use.

Mind and muscle

Dr Mike Stroud and Sir Ranulph Fiennes say goodbye to the plane and crew who brought them to the start of their long trek.

On 9 November 1992, Sir Ranulph Fiennes and Dr Mike Stroud began their attempt at the first unassisted crossing of the Antarctic continent. Ninety-five days later, more dead than alive, they achieved their aim. They had pulled their 220 kg sledges across 2300 km of ice and snow. The energy used was more than the equivalent of running a marathon every day. No pack dogs or food depots supported them: all was achieved by human muscle power.

The aim of the expedition was to test the human body's ability to withstand extreme conditions, and also to raise money for medical research. Feats of endurance can tell us a great deal about energy balance, temperature control, water balance, and muscle physiology. During the expedition Fiennes and Stroud took blood and urine samples, and their food intake was measured precisely. Before and after the trek, they tested their heart and lung functions and muscle strength. Small samples of muscle tissue were taken from their legs.

Dr Stroud and Sir Ranulph Fiennes took risks, but their chances of survival were greatly improved by their understanding of their bodies and what could be achieved. Every day, medical workers use their understanding of the human body in less dramatic but important situations to treat people with illnesses. How do scientists help to build this understanding?

Antarctica

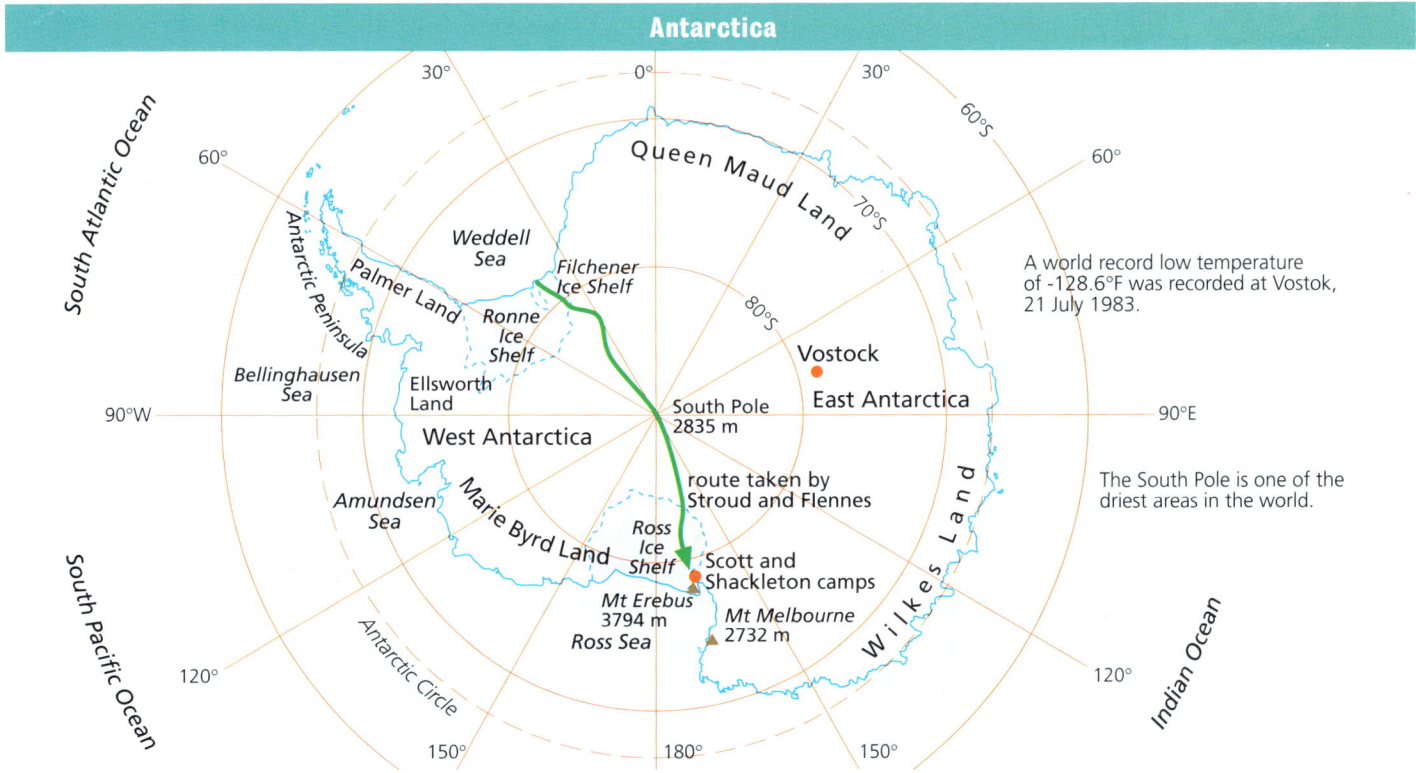

A world record low temperature of -128.6°F was recorded at Vostok, 21 July 1983.

The South Pole is one of the driest areas in the world.

4.1 Learning objectives

After working through this chapter, you should be able to:

- **explain** how antagonistic muscles are used to move a joint;

- **interpret** the structure of skeletal muscle seen with a light microscope and an electron microscope;

- **explain** the sliding filament hypothesis of muscle contraction;

- **describe** the functioning of a reflex action involving three neurones;

- **recall** the functions of sensory, motor and association areas in the cerebral hemispheres of the brain;

- **explain** the roles of the hypothalamus and the medulla in the brain;

- **understand** the structure and function of the autonomic nervous system and its division into sympathetic and parasympathetic components.

4.2 Muscle power

Stroud and Fiennes's polar trek made extreme demands on their muscles and other body systems. To understand how the trek was achieved, we need to know how:
- muscles are constructed and arranged;
- the brain controls the muscles;
- the brain controls heartbeat and fluid balance.

Like tug-of-war teams, muscles can only pull.

Muscles attached to bones are known as **skeletal muscles**. Muscles can pull but not push, so they are arranged in **antagonistic pairs**, one on each side of a joint (Fig. 1). Movement depends on muscles, bones and nerves. Nerves transmit impulses between muscles and the central nervous system, and the muscles pull on the bones to which they are attached by tendons.

1 **Study Figure 1. How do the gastrocnemius and tibialis anterior muscles work as an antagonistic pair to move the foot?**

Muscles are called effectors, because they create an effect, such as a movement. During walking, the muscles that are being stretched are relaxed, but not completely so. They provide some resistance so the legs do not move too suddenly. Other leg muscles (not shown in the figure) are used to keep the body balanced, as weight is moved from limb to limb. As muscles contract, tendons pull on the bones. Tendons do not stretch or break because of the presence of the fibrous protein, collagen. Collagen in bones makes the bones themselves slightly bendy and less brittle (Table 1).

Fig. 1 Skeletal muscles used in walking

To bend the knee, the biceps femoris muscle contracts while the quadriceps femoris relaxes. To extend the knee, the quadriceps femoris contracts while the biceps femoris relaxes.

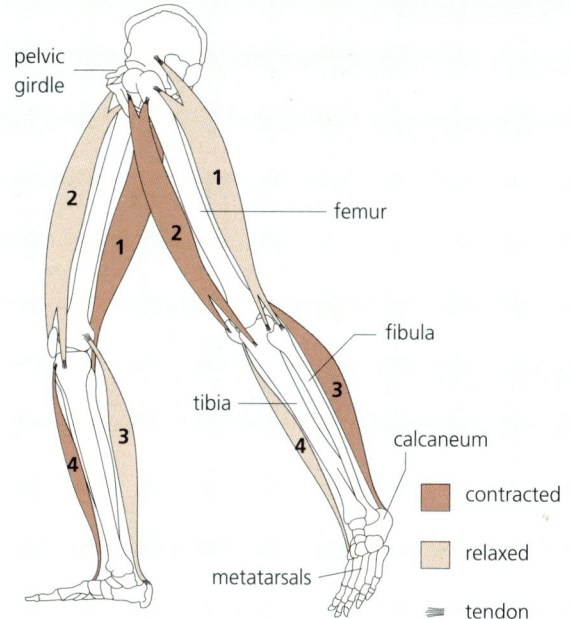

pelvic girdle

femur

fibula

tibia

calcaneum

metatarsals

contracted

relaxed

tendon

1 biceps femoris

2 quadriceps femoris

3 gastrocnemius

4 tibialis anterior

Skeletal muscles used in walking

Muscle	Origin (stationary bone)	Insertion (moving bone)	Action
biceps femoris	pelvic girdle & femur	tibia & fibula	bends knee (flexor)
quadriceps femoris	pelvic girdle & femur	patella & tibia	straightens knee (extensor)
gastrocnemius	femur	calcaneum (heel bone)	pulls heel up & toes down (extensor)
tibialis anterior	tibia	metatarsals (foot bones)	pulls top of foot up (flexor)

Table 1 Structures used in movement

Structure	Material	Properties	Function
tendons	collagen	non-elastic, tough	connect muscles to bones
ligaments	collagen and elastin	slightly elastic, tough	connect bones to bones
bones	tricalcium phosphate calcium carbonate collagen	rigid	levers for movement
muscles	actin, myosin	contractile, elastic	movement of bones

At the joints between bones are ligaments containing elastin. Elastin is a protein with branched fibres that can stretch to one-and-a-half times their original length and spring back again. Elastin in ligaments allows joints to move slightly, but prevents dislocation.

2 Elastin tends to split as we get older. How does this affect joints?

Key ideas

- Movement depends on muscles, bones, and nerves.

- Muscles cannot push. They are arranged in antagonistic pairs in order to move joints.

- Skeletal muscles are attached to bones by non-elastic but flexible tendons. Ligaments hold bones to bones and are more elastic than tendons.

4.3 Muscle movement

Muscle structure

Skeletal muscles consist of overlapping fibres, held together by connective tissue, with a tendon at each end. When the fibres contract, the muscle shortens, pulls the tendons and moves the bones.

With a light microscope, muscle fibres are seen to be striped across their length. An electron microscope shows more detail (Fig. 2).

3 Study Figure 2. Which protein is present in the H-zone?

Sliding filaments

Muscle consists largely of two proteins, **actin** and **myosin**, forming the thin and thick strands or **filaments.** Myosin filaments are about twice the thickness of actin filaments.

Fig. 2 Structure of skeletal muscle

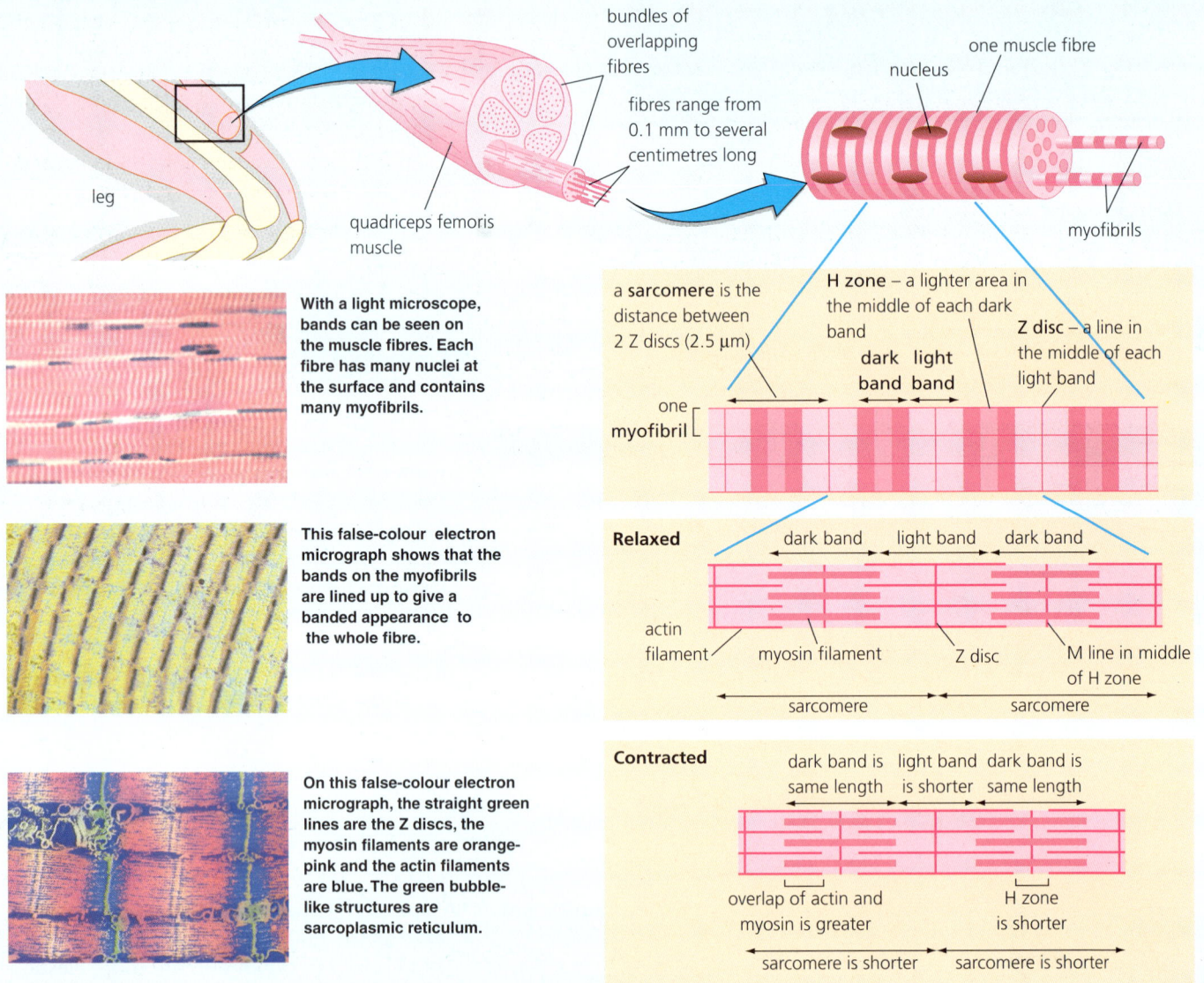

leg

quadriceps femoris muscle

bundles of overlapping fibres

fibres range from 0.1 mm to several centimetres long

nucleus

one muscle fibre

myofibrils

With a light microscope, bands can be seen on the muscle fibres. Each fibre has many nuclei at the surface and contains many myofibrils.

This false-colour electron micrograph shows that the bands on the myofibrils are lined up to give a banded appearance to the whole fibre.

On this false-colour electron micrograph, the straight green lines are the Z discs, the myosin filaments are orange-pink and the actin filaments are blue. The green bubble-like structures are sarcoplasmic reticulum.

a **sarcomere** is the distance between 2 Z discs (2.5 μm)

one myofibril

H zone – a lighter area in the middle of each dark band

dark light band band

Z disc – a line in the middle of each light band

Relaxed

dark band   light band   dark band

actin filament

myosin filament

Z disc

M line in middle of H zone

sarcomere

sarcomere

Contracted

dark band is same length

light band is shorter

dark band is same length

overlap of actin and myosin is greater

H zone is shorter

sarcomere is shorter

sarcomere is shorter

When a myofibril contracts:
- the sarcomeres become shorter;
- the light bands become shorter;
- the dark bands stay the same length.

In the 1950s, Jean Hanson and Hugh Huxley at London University put forward the **sliding filament hypothesis** which suggested that muscular contraction comes about by the actin filaments sliding between the myosin filaments, using ATP energy. This theory is still accepted. In the region of overlap, six actin filaments are arranged neatly around each myosin filament (Fig. 3).

4 a Do either the actin or the myosin filaments change their length when a sarcomere shortens?

b Why does the H-zone shorten when a sarcomere contracts?

How it works

Electron micrographs of the dark bands show **cross-bridges** between the myosin and actin filaments. These bridges are part of the myosin molecules, and push on the actin filaments to make the myofibril shorten. This is called the **ratchet mechanism** because the actin molecules are moved along one step at a time by the myosin heads. The mechanism is turned on and off by a **calcium switch** (Fig. 4).

Fig. 3 Sliding filament model

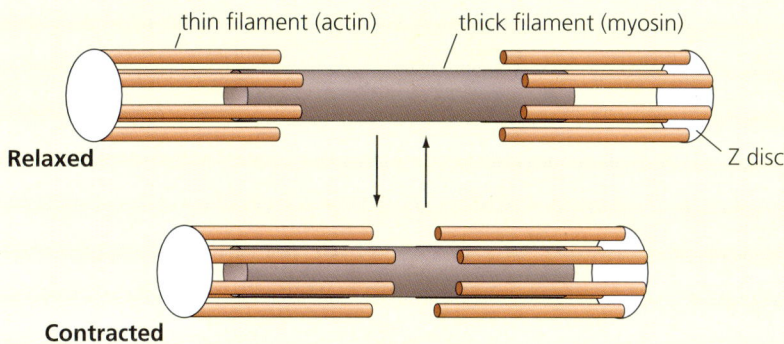

thin filament (actin)   thick filament (myosin)

Relaxed

Z disc

Contracted

Fig. 4 How the myosin cross-bridges work

actin   myosin

Z disc   Z disc

Transverse section

Myosin cross-bridges point in six directions, in a spiral.

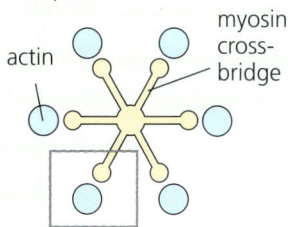

actin

myosin cross-bridge

Calcium switch

Ca²⁺ ions bind to the switch protein and move it away from the binding site so the myosin can bind to the actin and make a power stroke.

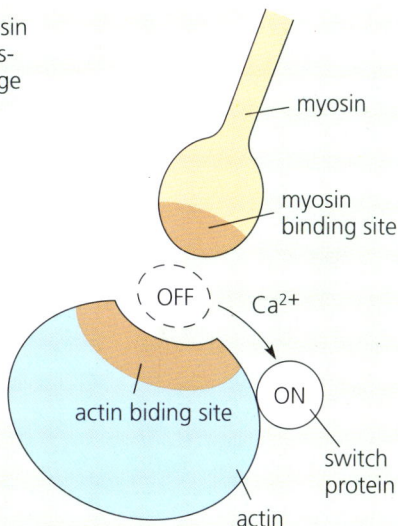

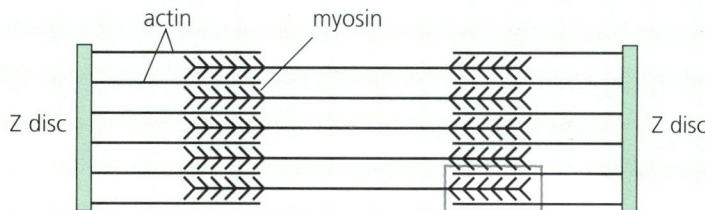

myosin

myosin binding site

OFF   Ca²⁺

ON

actin biding site

switch protein

actin

ATP enables cross-bridges to detach and swing forward

calcium ions enable cross-bridges to bind to actin and make power stroke, using ATP energy

sarcomere shortens and ADP and P are released

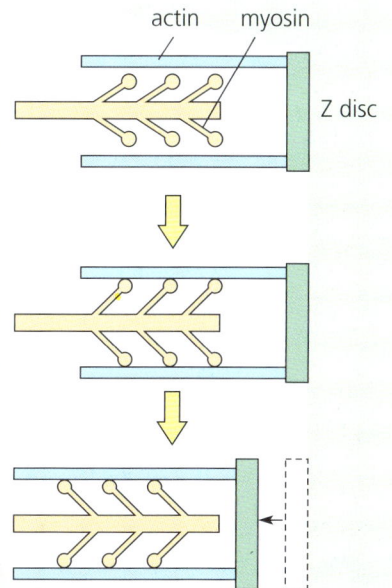

actin   myosin

Z disc

Each myosin cross-bridge can bind to, push against and release actin many times a second. Each swing of a cross-bridge uses the energy from one molecule of ATP. Many thousands of cross-bridges working together create the power of the muscle.

Q5 Why is ATP required when a myofibril shortens?

The calcium switch

When we are awake, muscles are used to maintain posture, even if we are keeping still. For example, the neck muscles hold the head in place when sitting or standing upright. All movement needs graded control; lifting a sheet of paper requires less muscular effort than lifting a book. In the Antarctic, Stroud and Fiennes needed almost all the muscular strength they could produce to start their 220 kg sledges moving.

To understand how the force produced by a muscle is controlled, we must look in more detail at the action of the calcium switch. In living muscles, calcium ions (Ca^{2+}) are stored in membrane sacs, called

During sleep, most of the body's skeletal muscles are relaxed.

the **sarcoplasmic reticulum**, that surround each myofibril. When the muscle is resting, Ca^{2+} ions are pumped into the sarcoplasmic reticulum by active transport.

The arrival of an action potential at a neuromuscular junction, sets off a chain of events leading to the contraction of myofibrils (Fig. 5).

Fig. 5 Cross section through part of a muscle fibre

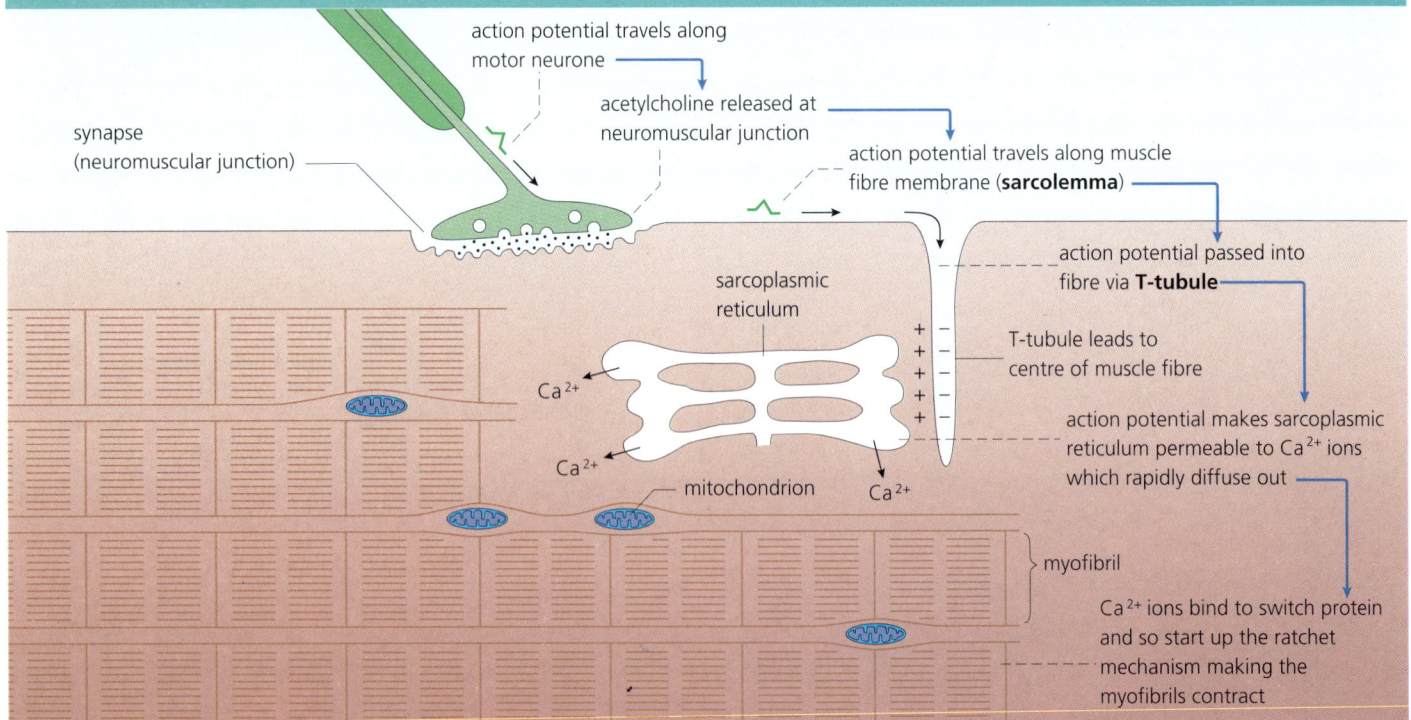

action potential travels along motor neurone

acetylcholine released at neuromuscular junction

synapse (neuromuscular junction)

action potential travels along muscle fibre membrane (**sarcolemma**)

action potential passed into fibre via **T-tubule**

sarcoplasmic reticulum

T-tubule leads to centre of muscle fibre

Ca^{2+}

action potential makes sarcoplasmic reticulum permeable to Ca^{2+} ions which rapidly diffuse out

Ca^{2+}

mitochondrion

Ca^{2+}

myofibril

Ca^{2+} ions bind to switch protein and so start up the ratchet mechanism making the myofibrils contract

6 Draw a flow chart showing how a nerve impulse arriving at a neuromuscular junction causes a muscle fibre to contract.

After death, bodies go rigid. This is called rigor mortis and is due to muscle contraction. Once death has occurred, ATP production stops and Ca^{2+} ions leak out of the sarcoplasmic reticulum.

7 How does the lack of ATP after death contribute to rigor mortis?

Strength of contraction

In order to control movement, the brain can control how strongly a muscle contracts. It does this by changing:
• how much each fibre contracts;
• how many fibres contract.

When one action potential arrives at a neuromuscular junction, it crosses the synapse and causes one brief contraction or **twitch** of the fibre (Chapter 1). If a second action potential arrives before the fibre has fully relaxed, the second twitch adds to the effect of the first, and a greater contraction occurs. A rapid sequence of action potentials causes a continuous strong contraction of the fibre called a **tetanus** (Fig. 6). So, the more nerve impulses there are per second in a motor nerve, the stronger the contraction of each muscle fibre will be.

Each motor neurone serves about 150 muscle fibres. One motor neurone and its associated muscle fibres are together called a **motor unit** (Fig. 7). Action potentials in a motor neurone cause all the muscle fibres in its motor unit to contract together. If a stronger contraction is needed, the brain recruits more motor units by sending action potentials along more motor neurones within the motor nerve.

8 What is a motor unit?

9 How do we control movement?

Fig. 6 Electrical stimuli and muscle contraction

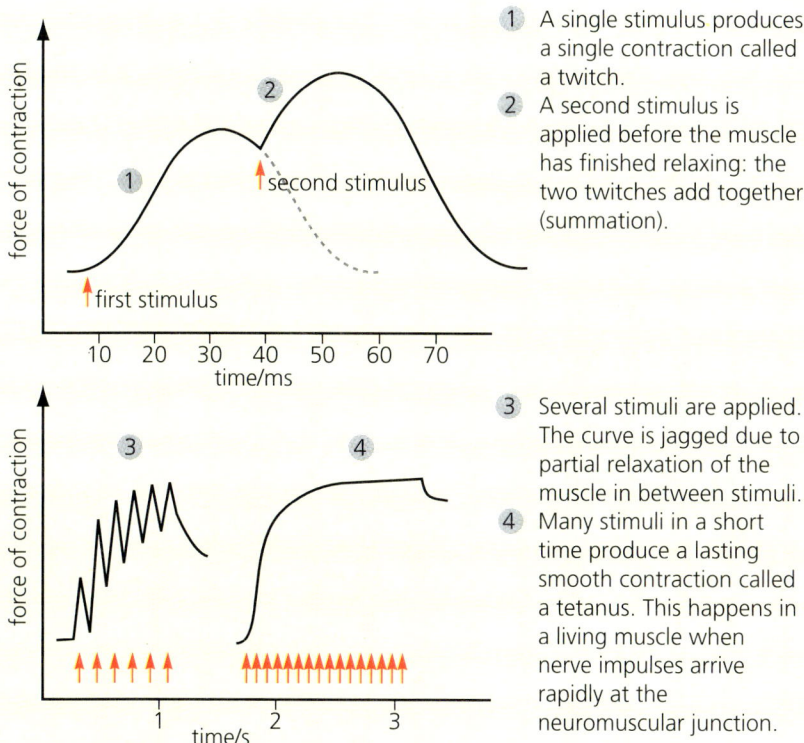

1. A single stimulus produces a single contraction called a twitch.
2. A second stimulus is applied before the muscle has finished relaxing: the two twitches add together (summation).
3. Several stimuli are applied. The curve is jagged due to partial relaxation of the muscle in between stimuli.
4. Many stimuli in a short time produce a lasting smooth contraction called a tetanus. This happens in a living muscle when nerve impulses arrive rapidly at the neuromuscular junction.

Source: Tortora and Grabowski, *Principles of Anatomy and Physiology*, Pearson, 1993

Fig. 7 A motor unit

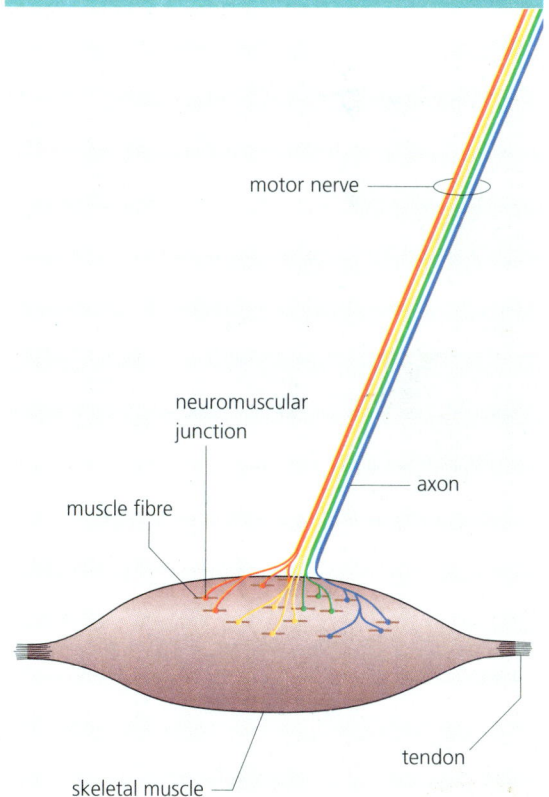

The strength of a muscle is governed by its area of cross section. Exercises such as weight lifting increase the size of muscle fibres, while inactivity leads to loss of muscle tissue. Long periods of weightlessness in zero-gravity can have the same effect as inactivity.

4.7 kg of muscle tissue. This was because some muscle fibres were destroyed to provide energy for walking and keeping warm. This only happens in very extreme conditions.

10 Study this photograph and its caption. Why are astronauts likely to suffer from loss of muscle strength?

Red and white

Muscle fibres are of two types, red and white (Table 2). Red fibres contain more mitochondria and more of the red pigment **myoglobin**, which stores oxygen.

Red or slow fibres are used for walking or jogging and white or fast fibres for sprinting and jumping or escaping from danger. Most muscles contain a mixture of the two sorts of fibre, but people differ in their red/white ratios. The proportion of red and white fibres in muscle is controlled by genes, and exercise has little effect. To this extent great athletes are born and not made. Sir Ranulph Fiennes describes his walking technique as a 'polar plod' that he could keep up for hours. His muscles are probably well supplied with red fibres.

Endurance training increases blood supply, connective tissue and the number of mitochondria in muscles. Fiennes and Stroud were able to build up their muscles before the expedition. During the walk however, Fiennes lost 3.3 kg and Stroud lost

A high-jumper needs mainly white muscle fibres for rapid action.

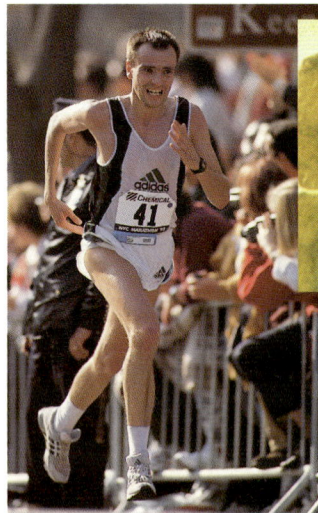

A marathon runner needs mainly red muscles fibres for endurance.

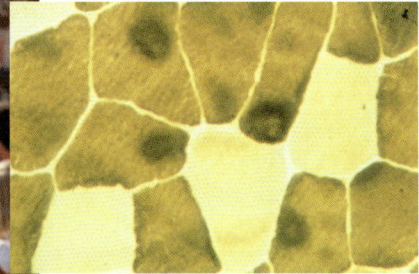

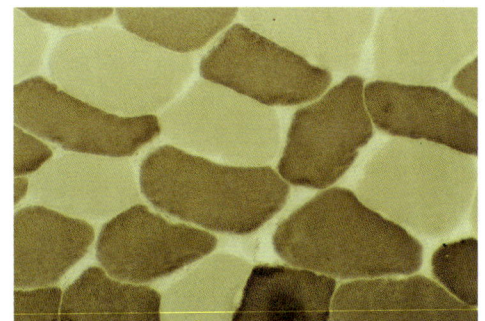

This muscle biopsy is from an averagely fit person.

Muscle type	Speed of contraction and relaxation	ATP source	Respiration	Fatigue	Motor unit	Myoglobin quantity
red	slow	carbohydrate and fat metabolism	aerobic	slow	small	large
white	fast	creatine phosphate	anaerobic	rapid	large	small

Table 2 Comparison of red and white muscle fibres

11a Sprinters run 100 m and 200 m races at high speed. Are they more likely to have mainly red or white fibres in their muscles?

b Study the photographs of activities and athlete muscle biopsies. How would you describe the muscle biopsy of an averagely fit person?

Key ideas

- Light micrographs show that muscles are striped bundles of fibres.

- Electron micrographs show that muscle fibres contain many myofibrils. Myofibrils have a banded pattern of repeating units called sarcomeres.

- Muscles contain the proteins actin and myosin, which contract using ATP energy.

- Sarcomeres contract when actin filaments slide between myosin filaments using a ratchet mechanism.

- Muscle contraction begins when Ca^{2+} ions released from the sarcoplasmic reticulum start the ratchet mechanism.

- Nerve impulses arrive via motor neurones, then travel along the sarcolemma and T-tubules and cause the release of Ca^{2+} ions around myofibrils.

- One action potential causes one twitch; a rapid sequence of action potentials causes a full contraction or tetanus of a muscle fibre.

- Motor neurones each serve a group of muscle fibres called a motor unit. The brain can select varying numbers of motor units for action.

- There are fast and slow (white and red) muscle fibres contracting at different speeds. People differ in the proportion of the two fibre types they have in their muscles.

4.4 Automatic response

Table 3 Examples of reflex actions

Stimulus	Response
treading on a pin	quick removal of limb
touching a hot or very cold surface	quick removal of limb
sudden noise	startle response ('jumping')
bright light in eye	iris constricts and pupil gets smaller
unusual chemical in nose	sneezing
tapping tendon below knee cap	knee jerk
touching side of baby's cheek	rooting reflex – turning towards the source of the stimulus

All living things respond to stimuli. A stimulus is defined as a change in an organism's environment. Information from a stimulus is detected by sensory receptors and transmitted to the central nervous system (CNS). The CNS produces a response by sending nerve impulses to the effectors (muscles).

The simplest responses to stimuli are **reflex actions**. A reflex action is a rapid, automatic, inborn response to a stimulus (Table 3).

Fig. 8 A reflex arc

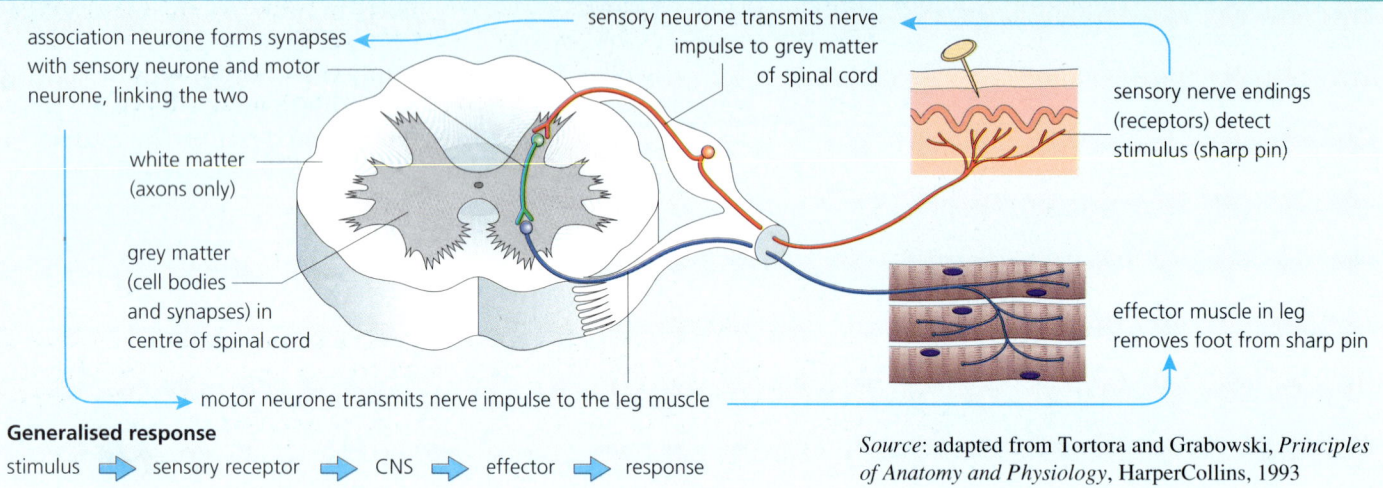

association neurone forms synapses with sensory neurone and motor neurone, linking the two

sensory neurone transmits nerve impulse to grey matter of spinal cord

sensory nerve endings (receptors) detect stimulus (sharp pin)

white matter (axons only)

grey matter (cell bodies and synapses) in centre of spinal cord

effector muscle in leg removes foot from sharp pin

motor neurone transmits nerve impulse to the leg muscle

Generalised response

stimulus ➡ sensory receptor ➡ CNS ➡ effector ➡ response

Source: adapted from Tortora and Grabowski, *Principles of Anatomy and Physiology*, HarperCollins, 1993

Reflex actions are rapid because there is a direct link between the receptor and the effector, called the **reflex arc** (Fig. 8).

A reflex arc is the shortest route linking a receptor to an effector. The nerve pathway can either go through the brain (for example, in the iris reflex or sneezing) or through the spinal cord (for example, in the knee jerk or removing a limb from a pin). Usually two synapses occur in the reflex arc, but in the knee jerk reflex there is just one. Synapses occur only in the grey matter. Reflexes do not have to be learned: the nerve pathways are built into the nervous system under genetic control. They control simple responses needed for survival or avoiding danger.

12 Why are reflex actions so fast?

Reflex actions do not need conscious thought. When we tread on a pin or touch something hot or cold, the response occurs before we are aware of the pain. The fact that we *later* become aware of pain shows that reflex arcs are connected (through synapses) to other parts of the nervous system. This allows us to keep our balance when removing the foot from a pin, and to shout for help if we are injured.

Spinal reflexes are used in walking (Fig. 9). Every muscle contains a sense organ called a **muscle spindle**, by which

movement is coordinated. When a muscle is stretched, the muscle spindle stretches and the muscle then responds by contracting.

The knee and other joints can swing backwards and forwards by spinal reflexes alone. When a baby learns to walk, several spinal reflex pathways are connected together. When all the synaptic connections have been made, the complex sequence of muscle movements used in walking can occur with little input from the brain.

13 If walking is achieved by spinal reflexes, why does a baby have to learn to walk?

Fig. 9 Muscle spindle reflex

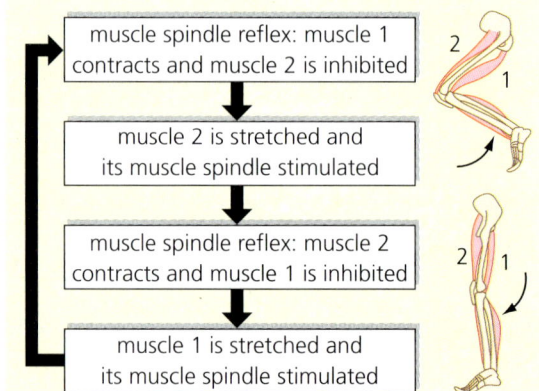

muscle spindle reflex: muscle 1 contracts and muscle 2 is inhibited

muscle 2 is stretched and its muscle spindle stimulated

muscle spindle reflex: muscle 2 contracts and muscle 1 is inhibited

muscle 1 is stretched and its muscle spindle stimulated

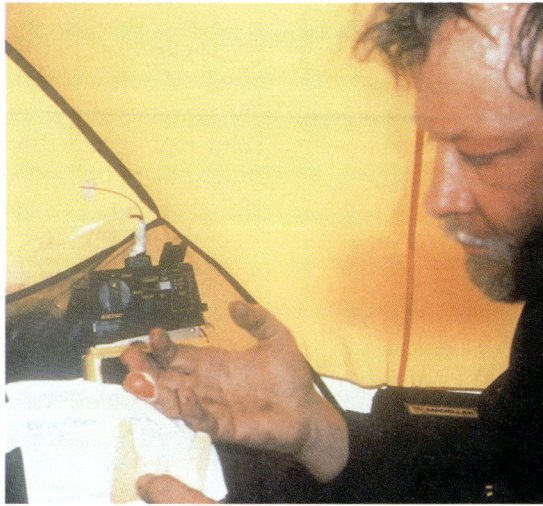

Reflexes can sometimes be prevented by the brain. For example, we can decide when to start and stop walking. The brain simply inhibits the stretch reflexes in the spinal cord. Also, we can steel ourselves not to respond to touching a hot object.

Fiennes and Stroud overcame the pain of frostbitten hands and toes by willing themselves to continue: Mike Stroud recorded in his diary towards the end of the trek, 'every day begins, continues and ends in pain'.

Key ideas

- Reflexes are fast, automatic responses that help survival.

- Reflexes are controlled by a simple loop of neurones called a reflex arc.

- Stretch reflexes are coordinated to control leg movements used in walking.

- Some reflexes can be overcome by the brain.

4.5 Brain power

The brain controls both our conscious acts and decisions and all the necessary activity that goes on within our bodies without our thinking about it, or even being aware that things are happening.

Stroud and Fiennes pushed themselves to the limits of mental as well as physical endurance. There was nothing but flat snow and ice for 85 days until they saw these mountains.

The cerebrum

Will power and decision making take place in the **cerebrum**. As Dr Stroud puts it:

With sufficient will power, the human body is able to do remarkable things. By the end of the expedition we had both shed nearly 50 kg in weight; more than double the amount we thought we would lose. Some days we were using nearly twice as much energy as we were eating. In the third month, our blood glucose concentrations sometimes fell below a quarter of their usual value, and it was surprising that we remained conscious. Somehow, our bodies adjusted to the priority of keeping going, and our muscles were able to continue pulling the sledges. After 95 days we were on the Ross Ice Shelf and had more than achieved our goal. We averaged 15.7 miles a day and raised £1.6 million for the Multiple Sclerosis Society.

Fig. 10 The human brain

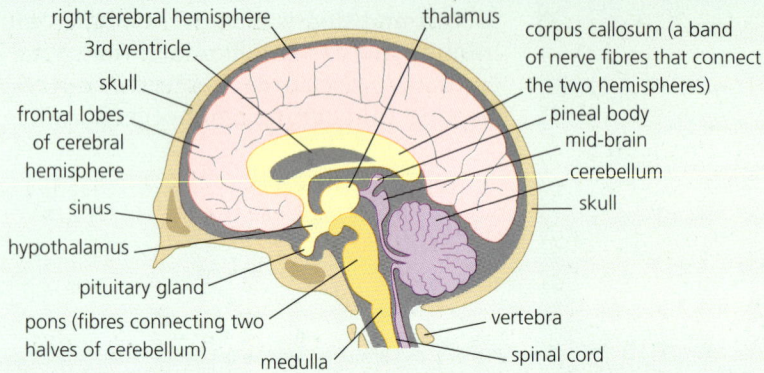

right cerebral hemisphere
thalamus
corpus callosum (a band of nerve fibres that connect the two hemispheres)
3rd ventricle
skull
frontal lobes of cerebral hemisphere
pineal body
mid-brain
cerebellum
skull
sinus
hypothalamus
pituitary gland
pons (fibres connecting two halves of cerebellum)
vertebra
spinal cord
medulla

Source: adapted from MacKean, *Introduction to Biology*, Murray, 1979

The cerebrum is divided into left and right **cerebral hemispheres**, connected by the nerve fibres of the **corpus callosum**. The cerebrum is greatly enlarged in mammals, particularly primates (monkeys and apes). The folded surface of the cerebral hemispheres is covered with a thin layer of neurone cell bodies, sometimes known as grey matter or the **cerebral cortex** (Fig. 10).

The cerebral cortex contains:

• sensory areas;
• association areas;
• motor areas (Fig. 11).

14 Where is the cerebral cortex?

Fig. 11 The cerebral cortex

Localisation of function in the left cerebral hemisphere

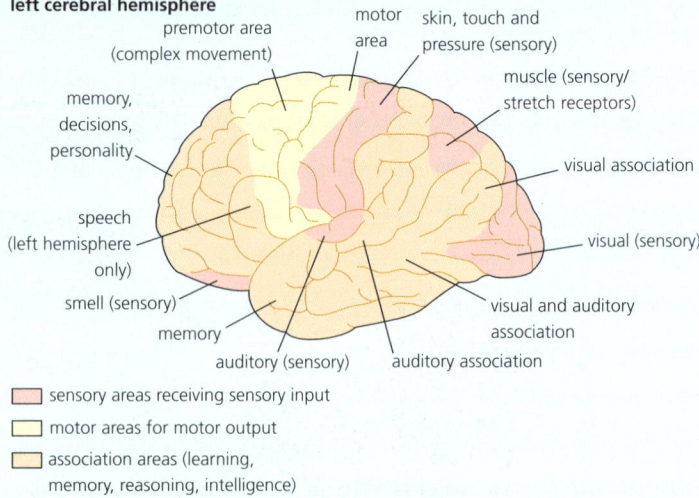

premotor area (complex movement)
motor area
skin, touch and pressure (sensory)
muscle (sensory/ stretch receptors)
memory, decisions, personality
visual association
speech (left hemisphere only)
visual (sensory)
smell (sensory)
visual and auditory association
memory
auditory (sensory)
auditory association

■ sensory areas receiving sensory input
□ motor areas for motor output
□ association areas (learning, memory, reasoning, intelligence)

Source: adapted from Rowland, University of Bath Science 16–19 , *Biology*, Nelson, 1992

What shape do you think you are?

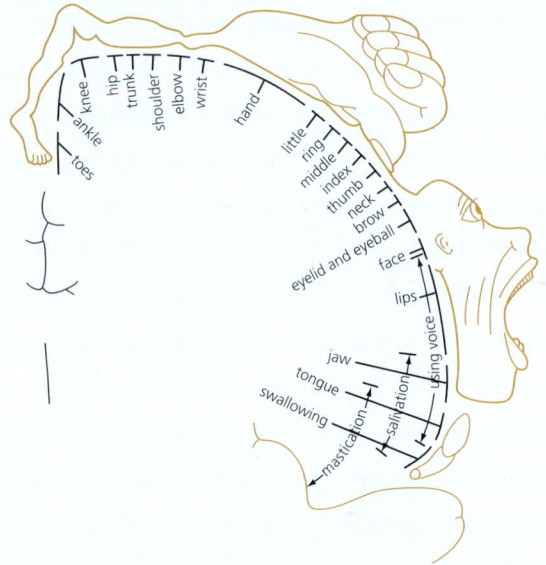

knee, hip, trunk, shoulder, elbow, wrist, ankle, toes, hand, little, ring, middle, index, thumb, neck, brow, eyelid and eyeball, face, lips, jaw, tongue, swallowing, mastication, salivation, using voice

Vertical section

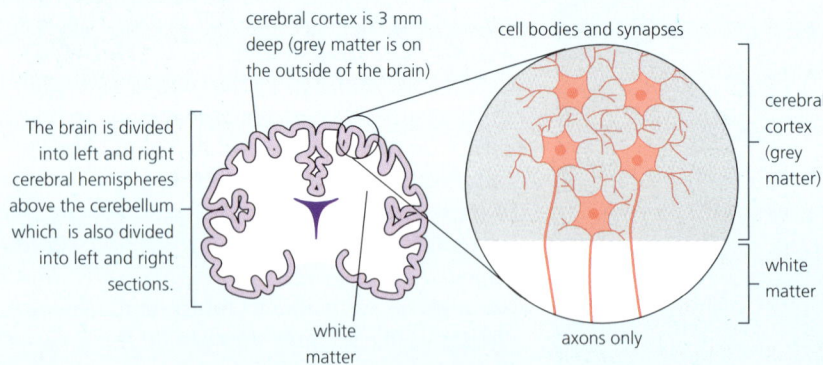

cerebral cortex is 3 mm deep (grey matter is on the outside of the brain)
cell bodies and synapses
cerebral cortex (grey matter)
The brain is divided into left and right cerebral hemispheres above the cerebellum which is also divided into left and right sections.
white matter
white matter
axons only

This diagram shows the right cerebral hemisphere sliced from top to bottom with areas for muscle control labelled. The body is represented upside down by the brain. The reason for this is not known. The drawings round the brain represent the amount of brain space given to each part of the body. More brain space is given to places where fine control of muscles is needed, such as fingers and lips. Similar drawings can be made for the sensory areas of the cerebral cortex.

Source: Penfield and Rasmussen, *The Cerebral Cortex of Man*, Macmillan, 1950

Source: adapted from Blakemore, *Mechanics of the Mind*, Cambridge University Press, 1977

The **sensory areas** receive impulses from the sensory nerves about all the stimuli reaching the body. **Localisation of function** means that each part of the skin surface is represented in a particular part of the touch sensory area.

The frontal lobes of the cerebral cortex are the main **association area** of the brain. They receive information via the brain's sensory areas, make decisions, and send out instructions through the **motor areas**. For example, when we see a familiar face, our visual association areas piece together the image of the face and compare it with faces stored in memory; recognition of the face is relayed to the frontal lobes; the frontal lobes make a decision to call a greeting and send out nerve impulses to the necessary motor areas.

Motor areas (also known as the **motor cortex**) send impulses to skeletal muscles along nerve fibres passing down the centre of the brain (the brain stem) and spinal cord. Localisation of function means that damage to a particular part of the motor cortex leads to paralysis of a part of the body. The left side of the cerebral cortex controls the right side of the body, and vice versa. Injury to the right side of the brain, perhaps from a stroke, might cause paralysis in the left side of the body.

15 What do the association, sensory and motor areas of the cerebral cortex do?

Brain imaging

Brain areas used for particular activities have been mapped by **positron emission tomography** (**PET**). Radioactive fluorine attached to glucose molecules is injected into the bloodstream. The glucose collects in active areas of the brain where the radioactive fluorine decays and emits a subatomic particle called a **positron**. (Fig. 12).

16 Which brain areas would you expect to be active in someone listening and taking notes?

Fig. 12 PET scanning

Positron annihilation

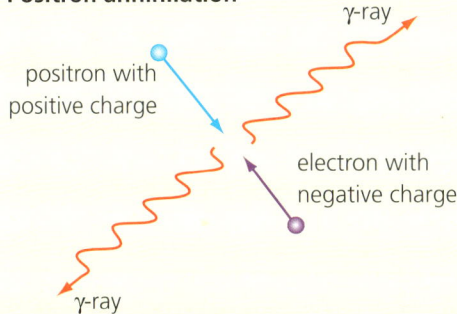

γ-ray

positron with positive charge

electron with negative charge

γ-ray

In the body, a positron travels only about 1 mm before being annihilated and producing two γ-rays that travel in opposite directions.

γ-ray detection

γ-ray emitted

positron annihilation

γ-ray detectors of the PET scanner

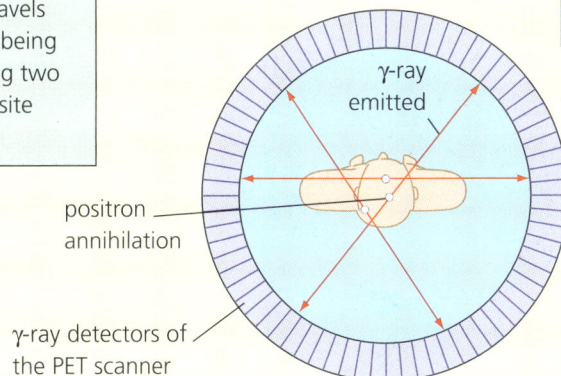

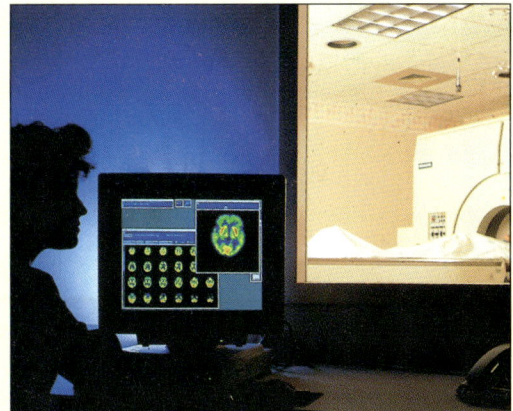

Data from the PET scanner's detectors identifies the brain area from which the γ-rays came. The brain images on the computer screen show which areas of the brain are active at the time of the scan.

Memory

Many of the decisions we make depend on memory. We have a **working memory**, which we use when we hold a telephone number in our head before writing it down, or when we keep track of the flow of a conversation. Such things are usually soon forgotten, but there is also a **long-term memory** with which we remember things about our past lives. Some of these memories can be called to the conscious mind, for example the image of the house where we live, or a familiar tune. Other memories are for motor skills, such as being able to type quickly or ride a bicycle. The cerebral cortex holds most of the memory but other regions, including the **cerebellum** and the **hippocampus**, are needed in order to store and retrieve memories from the cerebral cortex. Degeneration of the hippocampus is thought to be a major factor in Alzheimer's disease, a disease involving progressive loss of memory.

Brains that are better or quicker at reasoning and making suitable responses are generally considered more 'intelligent'. But it is important to remember that some people might have good auditory association areas and poor visual association areas, whereas others might have good visual association areas and poor auditory association areas. This means that while some people might be good at music but poor at geometry, others might have the opposite abilities. We cannot use this to say that one group or individual is more 'intelligent' than another.

17 a When you sit an examination, do you need working memory or long-term memory?

b Which do you need to answer this question?

Some functions are controlled by only one side of the brain. The left side of the brain is used for understanding and producing speech and written language, mathematics and reasoning. The right has a greater role in understanding three-dimensional shapes and music. It is sometimes claimed that people are right or left-brained, but it is a mistake to try to classify people in simple categories.

It is also difficult to make sensible comparisons between the abilities of humans and other animals. Animals often have much better sensory systems than ourselves (Chapter 5).

The cerebellum

While the cerebrum controls our conscious actions, other parts of the brain are controlling other sorts of functions. The cerebellum contains nearly half the neurones of the whole brain. It has a regular repeating structure, with many units doing a similar job.

The cerebellum receives information from the balance organs of the inner ear, from the visual system and from muscle spindles. It uses this information to control and coordinate the skilled movements needed for activities such as catching a ball, writing or singing. For instance, when we want to catch a ball, the cerebellum can compute the direction in which the body or a limb is heading and make fine adjustments. We learn to speak clearly by imitating the sound of others and correcting the sounds we make. To do this, the cerebellum must finely control the larynx muscles that are used to produce sound. Neurones in the cerebellum have axons that carry impulses to the motor cortex in order to bring about the changes in movements that are needed.

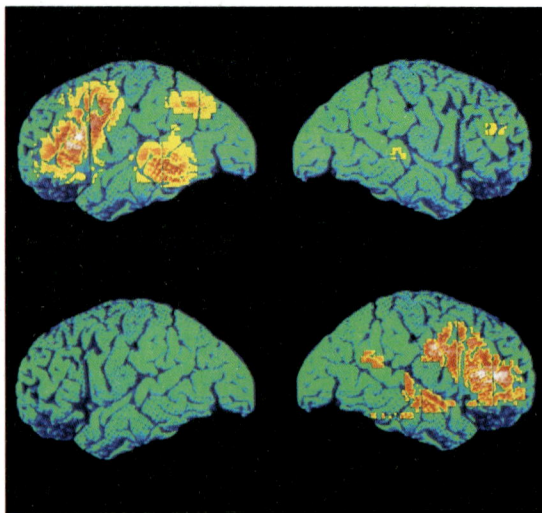

PET scans of brain activity during word recognition (red and yellow areas) superimposed on a brain image. In right-handed people, the left side of the brain is involved (top); in left-handed people, a similar pattern of activity is seen in the right side of the brain (bottom).

18 Why do people who are deaf from birth have more difficulty in speaking clearly than people who have lost their hearing later in life?

19 Mike Stroud and Ranulph Fiennes found it difficult to keep their balance across uneven ground in a 'white out' (snow storm). Why?

Table 4 Some of the actions of the autonomic nervous system

Organ or tissue	Effect of sympathetic nervous system	Effect of parasympathetic nervous system
iris of eye	widens pupil	narrows pupil
bronchi and bronchioles	widens tubes	narrows tubes
heart	increases heart rate and volume	decreases heart rate and volume
urinary system	contracts sphincter of bladder	relaxes sphincter of bladder
	relaxes wall of bladder	contracts wall of bladder
gut	prevents muscle contractions	stimulates muscle contractions
adrenal gland	releases adrenaline	

Table 5 Regulation of heart rate and breathing rate by the medulla

Information to medulla	Impulses sent via	Breathing rate	Heart rate	Arterioles
O_2 high; CO_2 low	parasympathetic system	decreases	decreases	
O_2 low; CO_2 high	sympathetic nervous system	increases	increases	
blood pressure high	parasympathetic system		decreases	dilate
blood pressure low	sympathetic nervous system		increases	constrict

Fig. 13 Motor neurones in the autonomic nervous system

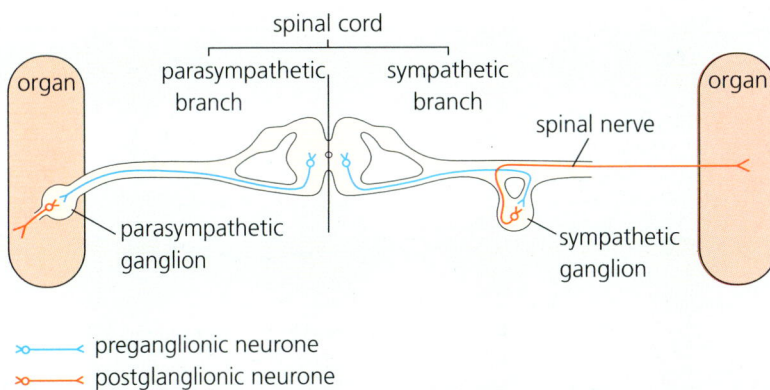

- preganglionic neurone
- postglanglionic neurone

Source: adapted from Rowland, University of Bath Science 16–19, *Biology*, Nelson, 1992

Alcohol affects the transmission of nerve impulses between cells in the cerebellum. This is why drinking a lot of alcohol affects balance and skilful movements.

The medulla

The **medulla**, situated at the top of the spinal cord, controls all the non-skeletal muscles (those which are not attached to bones). These muscles are of two types, **cardiac muscle** in the heart and **smooth muscle** surrounding the many internal tubes and ducts such as the bronchioles, the blood vessels and the gut. We are usually unaware of these internal movements, and unable to control them consciously. For example, we cannot decide to make our stomach muscles begin contracting; this happens automatically when food stretches the stomach wall.

The motor neurones controlling cardiac and smooth muscles make up the autonomic nervous system. The autonomic nervous system consists of two parts:
- the **sympathetic nervous system**;
- the **parasympathetic nervous system**.

The sympathetic system has opposite effects to the parasympathetic system (Table 4).

The medulla contains a **cardiovascular centre** and a **respiratory centre** that control the heart rate and the breathing rate according to information the medulla receives about blood pressure and the concentration of oxygen and carbon dioxide in the blood (Table 5). Adjusting the heart rate and breathing rate regulates the blood pressure and the blood concentration of oxygen and carbon dioxide.

20 a What are the two types of non-skeletal muscle?
b What is the role of the medulla?

Autonomic neurones do not go directly from the spinal cord or brain to an organ. Along the route there is a synapse in a **ganglion**. A ganglion is a place outside the central nervous system where cell bodies and synapses occur in a small region of grey matter (Fig. 13).

Sympathetic and parasympathetic postganglionic neurones have opposite effects on organs because they secrete different transmitters. Sympathetic postganglionic neurones secrete the transmitter noradrenaline and the effects of the sympathetic system are similar to the hormone adrenaline. Both systems expand the bronchioles, stimulate the heart and increase blood pressure. In this condition, the body is ready to flee from danger or attack an enemy. The sympathetic nervous system acts more quickly than adrenaline; the hormone travels in the blood stream and maintains the 'ready-for-danger' state after the sympathetic nervous system has done its job.

21 a What is a ganglion?
   b Where would you find a parasympathetic ganglion?

Parasympathetic postganglionic neurones secrete the transmitter acetylcholine. This counteracts the effects of the sympathetic system by the constriction of some muscles, and the relaxation of others.

Not the London marathon – conditions in the Sahara demand an exceptionally high intake of fluid.

It is possible to gain control of some parts of the autonomic system. For example, we learn at an early age to control the sphincter muscles of bladder and anus. Some people claim to be able to control their internal organs through meditation.

22 Which neurones secrete noradrenaline and why?

The hypothalamus

The medulla does not control all the body's physiological systems. The **hypothalamus** in the centre of the brain is responsible for temperature regulation, sleeping cycles, hormonal control and **osmoregulation**. The hypothalamus is connected to the **pituitary gland**, thus forming a vital link between the nervous system and the **hormone (endocrine) system**. Many of the functions of the hypothalamus operate through its control of hormone release from the pituitary gland. Osmoregulation, the control of solute balance (the concentration of dissolved ions) in the blood, is an example of this (Fig. 14).

Fig. 14 Osmoregulation

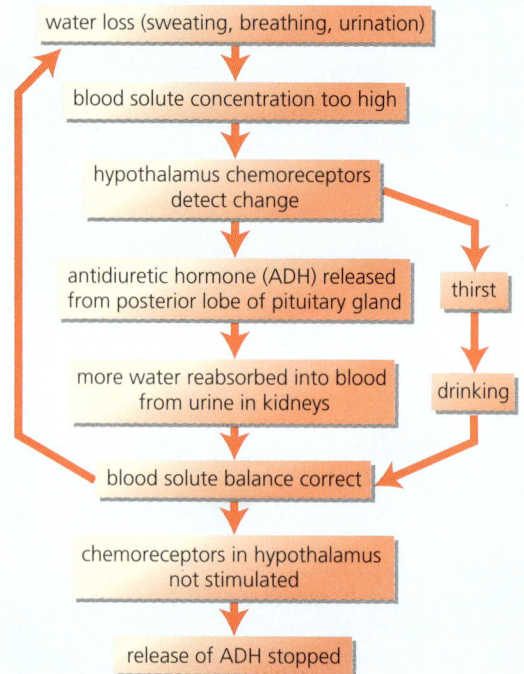

water loss (sweating, breathing, urination)

blood solute concentration too high

hypothalamus chemoreceptors detect change

antidiuretic hormone (ADH) released from posterior lobe of pituitary gland → thirst

more water reabsorbed into blood from urine in kidneys → drinking

blood solute balance correct

chemoreceptors in hypothalamus not stimulated

release of ADH stopped

Dr. Mike Stroud has researched the effects of intense physical exercise on human physiology in the coldest and hottest regions of the world. He noted that in the Antarctic he drank about 3.5 litres of fluid a day and passed about 2 litres of urine (more on *very* cold days). While competing in a week-long marathon in the Sahara, his fluid intake was 8–14 litres a day and his urine output about 2–4 litres.

23 **Study Figure 14. Why did Mike Stroud need so much more fluid in the Sahara than in the Antarctic, even though his urine output was only slightly higher?**

24 **How does the osmoregulatory system respond when enough fluid has been taken in to make up for fluid loss?**

Key ideas

- The folded outer surface of cerebral hemispheres contains grey matter and is called the cerebral cortex. Separate regions of the cerebral cortex control different sensory and motor functions.

- Within the motor areas and touch sensory areas, the body is mapped out. Damage to a small part of these brain areas leads to paralysis or numbness of the associated part of the body.

- The left side of the cerebrum largely controls the right side of the body and vice versa. Some functions are controlled by only one side of the brain. The two halves are linked by the corpus callosum.

- The cerebrum has a working memory and long-term memory.

- The cerebellum is responsible for the fine control of many skilled movements such as writing, singing, and ball catching.

- The medulla is situated at the top of the spinal cord. It controls cardiac and smooth muscles through the autonomic nervous system.

- The hypothalamus is a link between the nervous system and the endocrine system. It controls blood solute concentration, hormone balance, body temperature and sleeping patterns.

Learning to behave

Border collies are a shepherd's allies

Pippa

Joey and Ross

There are lots of ways in which dogs can be trained to help us. My border collies, save me a great deal of time and effort when I bring the sheep in for shearing, dipping, or lambing. As puppies they would run off ahead of me, lie flat on the ground and look towards me. They also liked to run round behind a group of sheep, rather than straight towards them. These movements seem to be a natural part of sheepdog behaviour. All I needed to do was teach the dogs the various commands for turning right or left, lying still, or approaching the sheep.

Sheepdog behaviour probably comes from the wolf ancestors of domestic dogs: one or two wolves in a pack run ahead to turn prey towards the other wolves. The first shepherds would have selected dogs with this behaviour. But Pippa, my Jack Russell terrier, has very different behaviour. She catches rats. Once, Pippa bred with one of my sheepdogs. The pups looked like sheepdogs with short legs. Their behaviour was intermediate too. They would round up sheep but could not be prevented from yapping at the sheep and frightening them. You can't train a dog to act against its innate behaviour patterns. Terriers don't round up sheep, and sheepdogs don't catch rats.

5.1 Learning objectives

After working through this chapter, you should be able to:

- **explain** the difference between innate and learned behaviour;

- **describe** innate behaviour such as taxes, kineses and reflex escape responses;

- **recall** that species-specific behaviour patterns occur in response to sign stimuli and innate releaser mechanisms;

- **explain** how complex behaviour patterns come from innate responses linked together;

- **describe** habituation and imprinting;

- **describe** classical and operant conditioning;

- **explain** the importance of reinforcement by reward and punishment in conditioning.

5.2 Innate and learned behaviour

Innate or inborn behaviour depends on the genetic make-up of animals whereas a trained change in behaviour during an animal's lifetime is **learned behaviour**. Innate behaviour differs from learned behaviour in several ways (Table 1).

1 Learned behaviour cannot be inherited. Why not?

Table 1 Differences between innate and learned behaviour

Innate behaviour	Learned behaviour
inherited	not inherited
not changed by environment	changed by environment
inflexible	quickly adapts to new circumstances
similar in all members of species or breed	differs between species

Migration

Birds that migrate at night, such as warblers and the thrush family, use the stars to find their way. A warbler called a blackcap has two populations breeding in Germany. The population from central Germany migrates westward to winter in Britain; the other population is from SW Germany and migrates south-westward to winter in the Mediterranean. Blackcaps from these two populations were experimentally mated and the offspring tested for their migration direction (Fig. 1).

2 Is migration in blackcaps innate or learned?

Fig. 1 Migration in blackcaps

Blackcap females have a brown head. The birds migrate at night.

Blackcaps were tested in a glass-topped container during the migration season. They could see the night sky and fluttered towards the directions in which they would normally migrate. Each bird was tested 15–20 times. The offspring of British-wintering birds migrate in the same direction as the parent birds. The offspring of crosses between the two populations migrate in a direction mid-way between those of the parent birds.

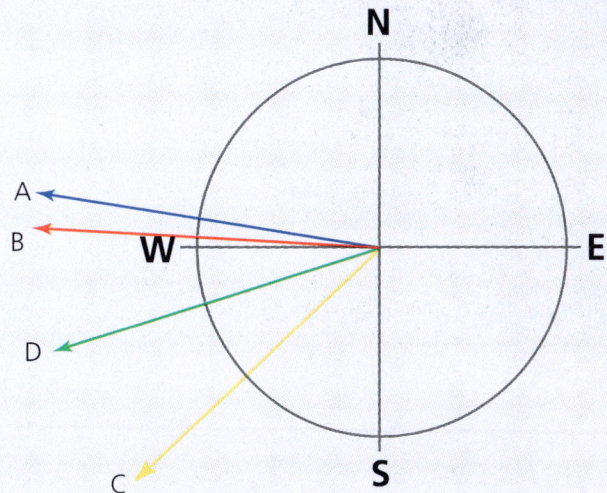

Blackcaps	Number of birds tested	Mean orientation in degrees
A adults wintering in Britain	18	279
B offspring of A	41	273
C adults wintering in the Mediterranean	49	227
D offspring of crosses between A and C	56	253

Source: adapted from Berthold et al, *Nature*, **360**: 668–9; Bird Study, **42** (1995), 89–100

Brent geese, *Branta bernicla*, wintering on an English estuary.

Fig. 2 Wintering areas of brent geese

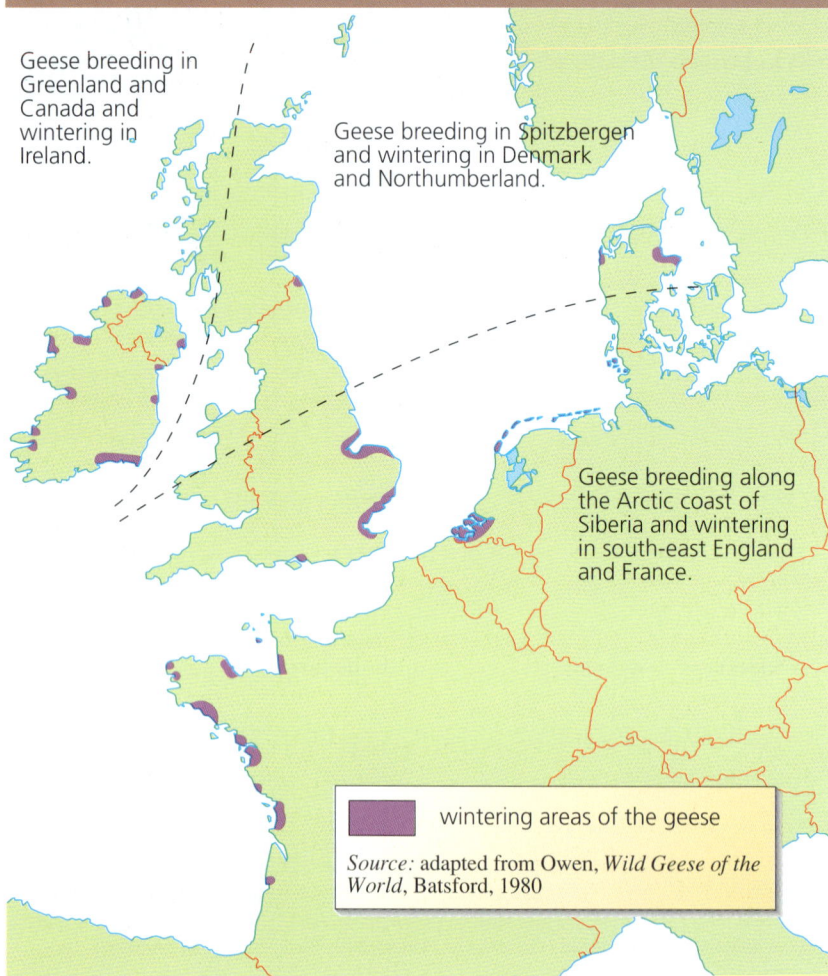

Geese breeding in Greenland and Canada and wintering in Ireland.

Geese breeding in Spitzbergen and wintering in Denmark and Northumberland.

Geese breeding along the Arctic coast of Siberia and wintering in south-east England and France.

wintering areas of the geese

Source: adapted from Owen, *Wild Geese of the World*, Batsford, 1980

In geese, family groups migrate together to the wintering grounds. The brent goose has three separate populations wintering in different parts of Britain and Europe. The young birds learn to find the traditional wintering areas by following their parents, and return to the same areas year after year (Fig. 2). Unlike blackcaps, geese live for a long time, and learned behaviour allows them to adapt to changing conditions, but it is not inherited by the next generation.

3 What is the main advantage to geese of learned migration behaviour?

Birdsong

Blackbirds can learn to imitate the sound of a telephone, and one even learned to sing the opening bars of Beethoven's violin concerto. However, its song was still recognisably a blackbird's; a blackbird can only learn within certain limits and will always sound like a blackbird. Chaffinches have local dialects, which shows that at least part of their song is learned by listening to other local birds. Chaffinch chicks reared in a sound-proof cage only sing a rough version of their song. They learn the full song by hearing other chaffinches, but they do not imitate the song of a blackbird. The genetic makeup of the chaffinch controls what it can learn: its genes put a **constraint on learning**. This is true of any species.

Mammals

In humans, only reflex actions are innate (Chapter 4). We have a nervous system with an ideal structure for learning; our brains have large cerebral hemispheres and we continue to learn throughout our lives.

The distinction between innate and learned behaviour is not always clear cut. For example, training sheepdogs is a process of modifying innate abilities through learning. Good working dogs have a blend of natural ability and readiness to learn. They are chosen for different tasks according to their innate abilities.

4 What sort of behaviour in humans is innate?

Key ideas

- Innate behaviour develops without much influence from the environment. It is controlled by the genetic makeup of the animal, and varies little within a species.

- Learned behaviour develops from an animal's experience of its environment. It allows an animal's behaviour to adapt to changing conditions but is not passed from one generation to the next in the genes.

- The structure of the nervous system determines what an animal is able to learn.

- The behaviour of an animal cannot always be easily classified as innate or learned; it is often a blend of both.

5.3 Knowing where to go

Invertebrates, unlike mammals and birds, rely on innate behaviour for most aspects of their lives, such as escaping danger, finding a suitable habitat and finding food. They do this using three types of behaviour pattern:
- a **reflex escape response**;
- a type of movement called a **taxis**;
- a type of movement called a **kinesis**.

Reflex escape response

Earthworms come to the surface of the ground on warm, damp nights. They respond to the slightest vibration by retreating rapidly down their burrows. This is a reflex escape response, and it helps them to avoid being taken by a predator such as a shrew or hedgehog (Fig. 3).

Fig. 3 Earthworm's giant axons

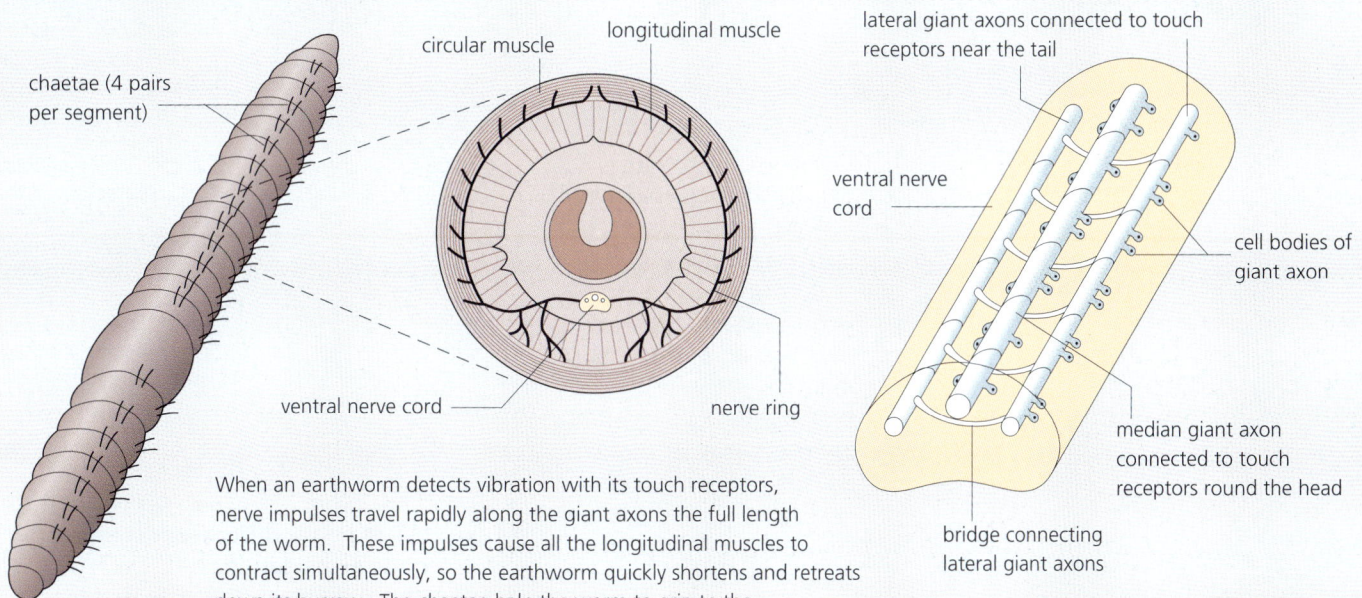

chaetae (4 pairs per segment)

circular muscle

longitudinal muscle

lateral giant axons connected to touch receptors near the tail

ventral nerve cord

cell bodies of giant axon

ventral nerve cord

nerve ring

median giant axon connected to touch receptors round the head

bridge connecting lateral giant axons

When an earthworm detects vibration with its touch receptors, nerve impulses travel rapidly along the giant axons the full length of the worm. These impulses cause all the longitudinal muscles to contract simultaneously, so the earthworm quickly shortens and retreats down its burrow. The chaetae help the worm to grip to the walls of its burrow, so it is not pulled out by a predator.

Source: adapted from Roberts, *Biology, a functional approach*, Nelson, 1982

Taxis

Fly maggots also use innate behaviour for survival. They avoid bright light which includes ultraviolet rays that kill maggots (Fig. 4). An orientation movement towards or away from a directional stimulus is called a taxis (pl. taxes). In the case of the maggot, it is **negative phototaxis** (taxis away from light). Adult flies have more protective pigments and usually move *towards* the light which warms up their bodies (**positive phototaxis**).

5 What is a taxis?

Fig. 4 Phototaxis

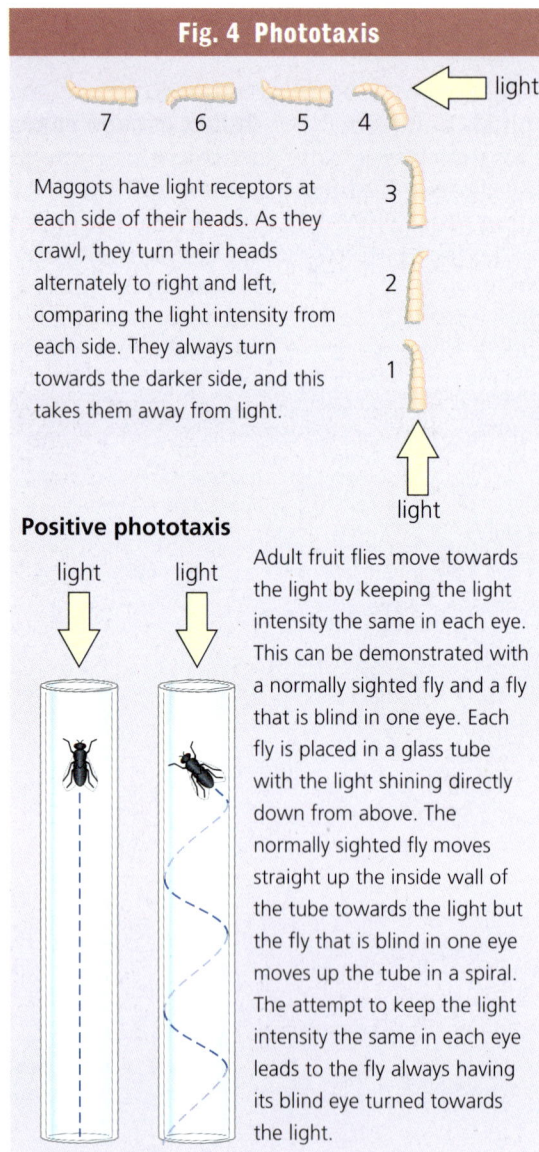

light

7   6   5   4

Maggots have light receptors at each side of their heads. As they crawl, they turn their heads alternately to right and left, comparing the light intensity from each side. They always turn towards the darker side, and this takes them away from light.

3

2

1

light

Positive phototaxis

light   light

Adult fruit flies move towards the light by keeping the light intensity the same in each eye. This can be demonstrated with a normally sighted fly and a fly that is blind in one eye. Each fly is placed in a glass tube with the light shining directly down from above. The normally sighted fly moves straight up the inside wall of the tube towards the light but the fly that is blind in one eye moves up the tube in a spiral. The attempt to keep the light intensity the same in each eye leads to the fly always having its blind eye turned towards the light.

Kinesis

Woodlice use innate behaviour to stay in a suitable environment. They live in damp places beneath logs and stones where they are not easily found by predators such as blackbirds and magpies, and are not likely to dry up. Woodlice don't move about much, but any slight drying in the environment is detected and they respond by starting to move about.

Once it has started moving, a woodlouse keeps on moving until somewhere sufficiently moist is reached, then it slows down or stops completely. In a kinesis, unlike a taxis, the animal does not go in any particular direction.

A slightly different kind of kinesis is seen in flatworms; they respond to chemical gradients. If a piece of meat is placed in a pond, there can be several flatworms feeding on it within a few minutes (Fig. 5).

Fig. 5 Kinesis in flatworms

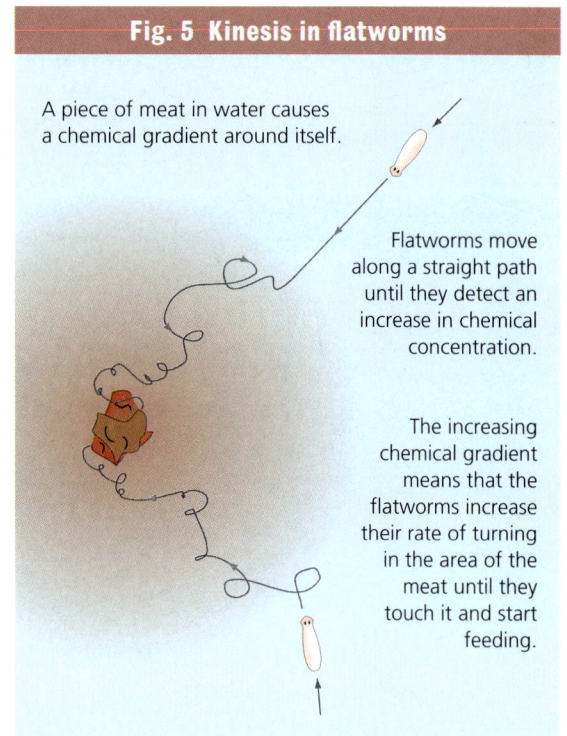

A piece of meat in water causes a chemical gradient around itself.

Flatworms move along a straight path until they detect an increase in chemical concentration.

The increasing chemical gradient means that the flatworms increase their rate of turning in the area of the meat until they touch it and start feeding.

6a What is the difference between a taxis and a kinesis?

b The direction of movement in flatworms changes in response to chemical gradients around food. Why is this kinesis rather than taxis?

Complex innate behaviour

It would be a mistake to think that all innate behaviour is simple. Honey bees use waggle dances to communicate the direction and distance of a food source to other bees. The dances are innate, and different genetic populations of honey bees have slightly different dance patterns (Fig. 6).

Fig. 6 Waggle dances

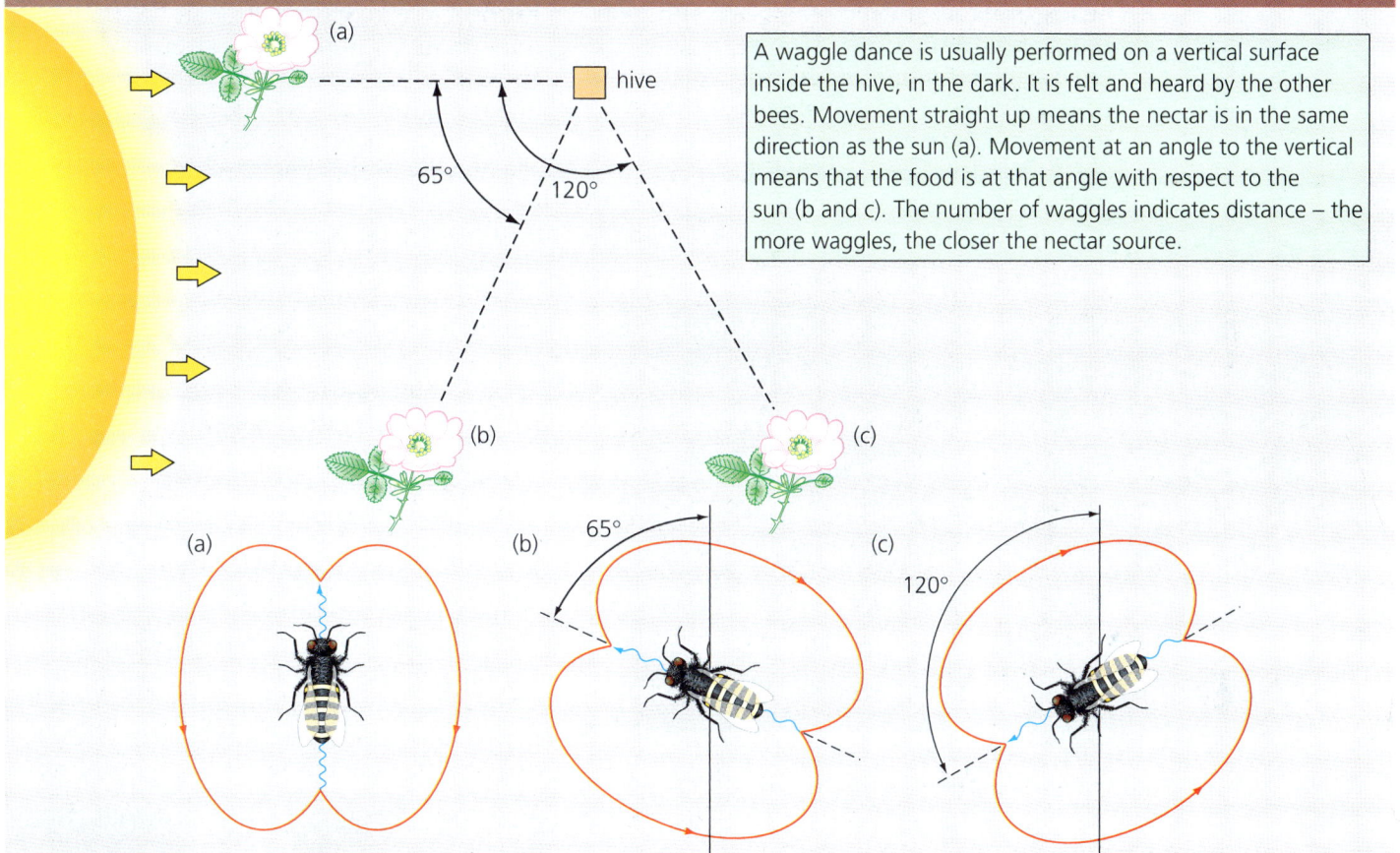

A waggle dance is usually performed on a vertical surface inside the hive, in the dark. It is felt and heard by the other bees. Movement straight up means the nectar is in the same direction as the sun (a). Movement at an angle to the vertical means that the food is at that angle with respect to the sun (b and c). The number of waggles indicates distance – the more waggles, the closer the nectar source.

Source: adapted from Von Frisch, 'Dialects in the language of the bees' in 'Readings from Scientific American' in Eisner and Wilson, *Animal Behaviour*, Freeman, 1975

Key ideas

- Invertebrates rely on innate behaviour patterns to find food and safety.

- In a reflex escape response, invertebrates move rapidly away from a stimulus that indicates immediate danger, for instance, an approaching predator.

- In a taxis, the animal detects the direction of a stimulus, such as light, and moves towards or away from it.

- In a kinesis, the animal moves in a straight line until it meets conditions resembling those it needs. It then responds by either slowing down or increasing its rate of turning until exactly the right conditions are met.

5.4 Reading the signs

If a stuffed robin or just a bunch of red feathers is placed in a male robin's territory, it attacks them vigorously. The robin does this even if it has never seen another robin before, because the red breast of robins acts as a **sign stimulus**. Sign stimuli usually have bright colours, and in this case the red colour acts as a signal or sign that makes the robin attack and drive the 'intruder' out of its territory. The sign stimulus is said to release the attack behaviour of the robin (Fig. 7). The robin's brain has an **innate releaser mechanism**.

However, robins do not always attack red feathers – the birds must be in the right **motivational state** or mood. In birds, male sex hormone increases this territorial behaviour. At migration time, the testes shrink and less sex hormone is made. Migrating robins sometimes form small groups, showing very little aggression. Even in the breeding season, a robin's tendency to attack is variable and the bird is less likely to attack an intruder near its territory boundary than one in the centre.

Q 7 Would a male robin attack a male sparrow that strayed into the robin's territory?

Behaviour triggered by sign stimuli is usually the same in all members of the species. For example, all herring gull chicks peck at the red spot on the parent's beak. This is known as **species-specific behaviour**. However, innate responses to stimuli are not always automatic and predictable. The term 'innate releaser mechanism' should be used with caution.

Fig. 7 Innate releaser mechanism

A male robin sees and prepares to attack an 'intruder' in its territory. The bright red breast feathers of the stuffed robin are a sign stimulus. The robin attacks relentlessly as it tries to drive out the 'intruder'.

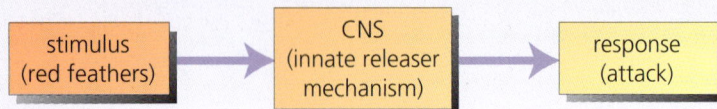

stimulus (red feathers) → CNS (innate releaser mechanism) → response (attack)

The red spot on a parent herring gull's bill is a sign stimulus for pecking. When the chick pecks, the parent gives food. Here, the red and white striped stick is a more effective stimulus than the realistic-looking model because it has better red/white contrast.

Conflict and convention

Conflicts between animals happen over territory, mates and food. When a male Siamese fighting fish sees a rival, its red and blue colours darken and it turns broadside to its rival with its fins spread out. It beats water towards its opponent with its tail. It then turns head-on and erects its gill covers. As one fish turns head-on, the other responds by turning broadside. The display continues until one gives up, goes pale, lowers its fins and slinks away. These fish only kill one another if they are in an aquarium with too little room to escape. Usually, little damage is done because the formal displays involve little contact.

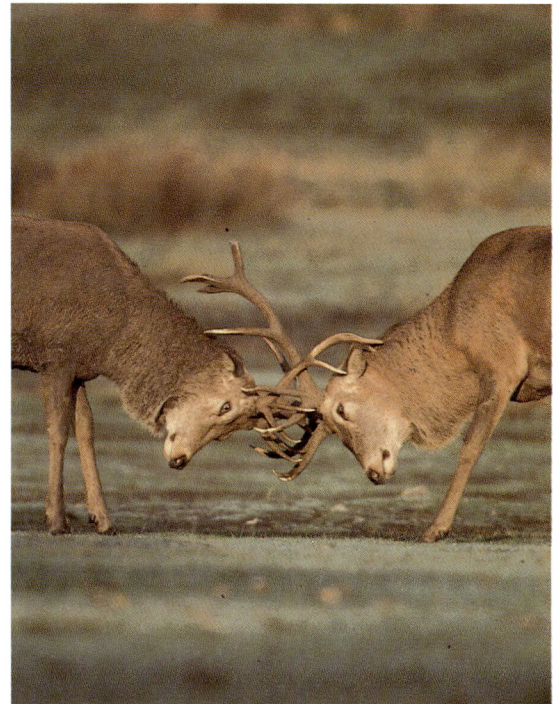

Red deer stags lock antlers when fighting over a group of hinds. Usually, the stronger stag wins without injuring his opponent.

Some animals have evolved **conventions** by which the winner is decided in advance. A male speckled wood butterfly will defend a patch of sunlight among the trees. When a female speckled wood passes by, he pursues her and attempts to mate. Other males are chased off; the male that settled in the sunlight first nearly always successfully defends his patch. This convention saves energy and little damage is done.

8 What is particularly advantageous about resolving conflict by formal display and convention?

Stickleback reproduction

Three-spined stickleback reproduction begins in the spring when male fish swim up river and defend a territory. Each develops bright breeding colours and, like the robin, attacks any intruders. Each separate part of the breeding sequence is triggered by a sign stimulus, and the whole pattern links together to form a functional sequence (Fig. 8).

Fig. 8 The three-spined stickleback courtship pattern

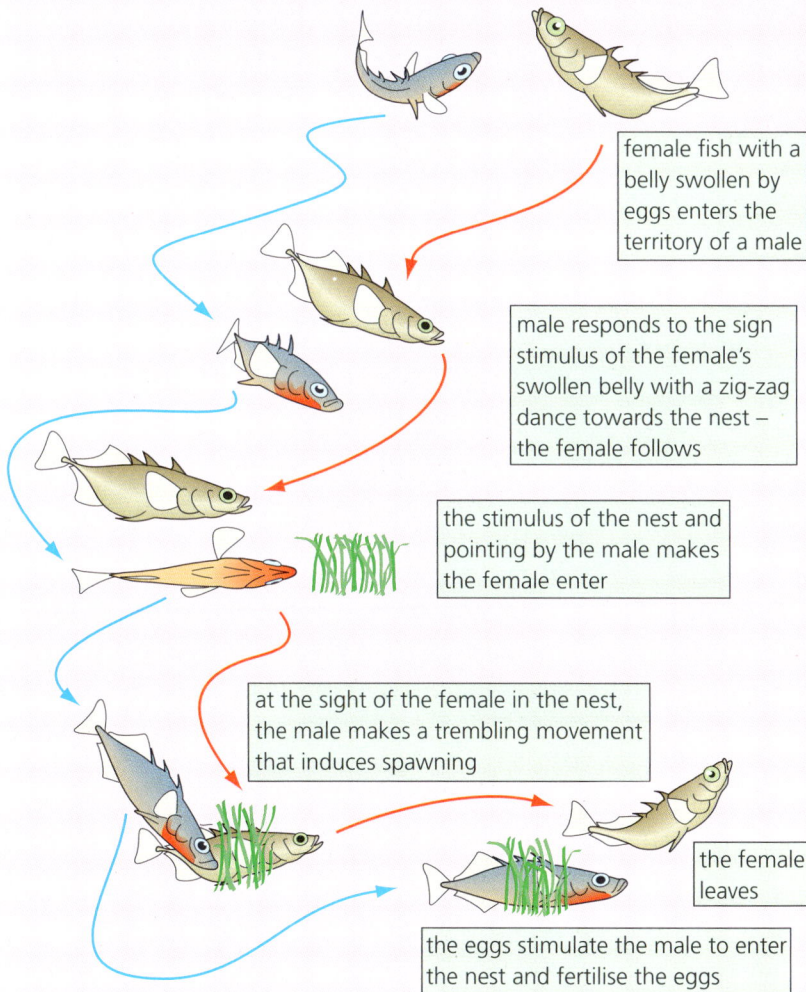

female fish with a belly swollen by eggs enters the territory of a male

male responds to the sign stimulus of the female's swollen belly with a zig-zag dance towards the nest – the female follows

the stimulus of the nest and pointing by the male makes the female enter

at the sight of the female in the nest, the male makes a trembling movement that induces spawning

the female leaves

the eggs stimulate the male to enter the nest and fertilise the eggs

Source: adapted from Tinbergen, *The Study of Instinct*, Oxford University Press, 1951

Female courtship display

Although female courtship displays are much rarer, an example is found in a British bird called a red-necked phalarope. This bird is a wader that breeds in northern Scotland. The female bird is brightly coloured in the breeding season, whereas the male is rather drab and sits on the eggs in the nest.

Exploiting responses

Because innate behaviour is inflexible, it can be exploited by other species. For example, the innate behaviour of red ants in protecting their young is exploited by the caterpillar of the large blue butterfly after its final moult, when it drops from its food plant to the ground and is rescued by ants.

Humans also exploit other animals' responses to sign stimuli. For example, fishing lures mimic the sign stimuli shown by animals that are the prey of trout, salmon or mackerel.

An understanding of sign stimuli is helpful when it becomes necessary to deter birds. Peregrine falcons are used at airports to scare gulls and lapwings from runways; small birds quickly head for cover when they see the silhouette of a hawk or falcon because the shape of the bird of prey acts as a sign stimulus, releasing the escape behaviour of their quarry.

Models of hawks or owls can be used to deter birds from flying into plate glass windows. Small birds will not come too near an owl at is daytime roost, but they will often fly around nearby **mobbing**. By altering the shape and markings of the model we can discover which features of the stimulus are important in keeping small birds away.

Q 9 a What features of an owl are likely to be sign stimuli to small birds?
   b How would you make a fair test of different owl models as bird scarers?

When found by a red ant, the caterpillar of the large blue butterfly makes a drop of sweet liquid. Ants lick this fluid, and transfer ant smell to the caterpillar. The caterpillar then folds and inflates its body, so it looks like an ant larva. The ants take the caterpillar into their burrow, as though it were a misplaced ant larva.

Once in the ants' nest, the caterpillar feeds on ant larvae. It may eat 1200 ant larvae before it pupates and finally emerges from the ant nest as an adult.

Key ideas

- Animals respond innately to sign stimuli.

- Sign stimuli are usually brightly coloured.

- Responses to sign stimuli change with motivational state; they are not automatic.

- Sign stimuli act as releasers of species-specific behaviour.

- Innate behaviour is inflexible and can be exploited by other animals.

5.5 Learning

Learning can be defined as a change in behaviour caused by experience. Examples of learning that happen fairly rapidly are **habituation** and **imprinting**.

Fig. 9 Habituation

Startle reaction in babies

when a repeating tone is sounded, the baby stops suckling and makes a 'startle reaction' to the noise

after 9 further bursts of repeating tone, the baby does not respond to the sound; it has habituated to the sound

startle reaction reappears when a different sound is made

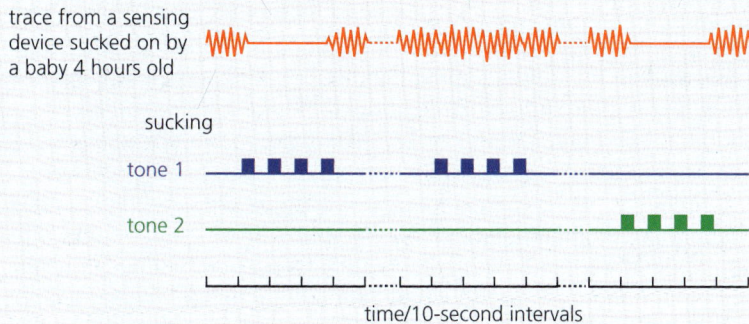

trace from a sensing device sucked on by a baby 4 hours old

sucking

tone 1

tone 2

time/10-second intervals

Source: adapted from Blakemore, *Mechanics of the Mind*, Cambridge University Press, 1977

Withdrawing reaction in hermit crabs

Hermit crabs live in adopted whelk shells. 30 crabs were tested by tapping the sides of their tanks. This was continued with each crab, on each day, until the crab failed to respond (by withdrawing into its shell). On the first day, an average of eleven taps was required before there was no response; by day 18, the majority of crabs didn't respond at all.

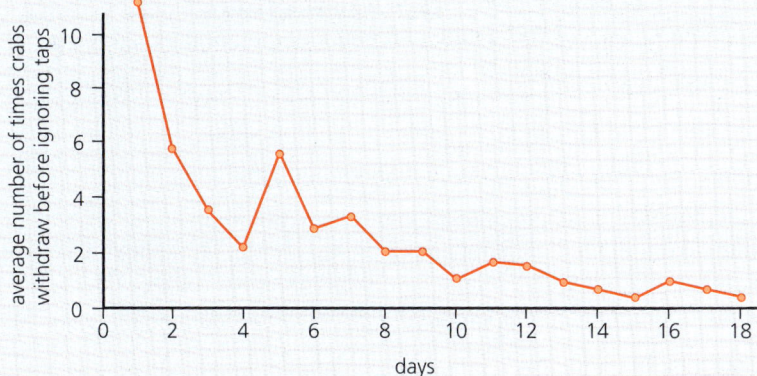

average number of times crabs withdraw before ignoring taps

days

Habituation

Bird deterrents do not work for long. Sooner rather than later, a hawk or owl model will be ignored because the birds become habituated to it. Habituation is considered to be a type of learning, because *behaviour changes*. The birds learn that the model is not dangerous and *cease to mob it*.

To qualify as learning, the change must be fairly permanent. Birds might stop mobbing an owl because they are tired; this is clearly not learning. If birds that have stopped mobbing an owl model, start to mob a cat model, it shows that they are habituated to the owl model, and are not simply tired.

Habituation can also be seen in snails. When snails emerge from their shells and stick out their tentacles, any vibration causes the tentacles to withdraw. Repeated vibrations cause less and less response. In fact, habituation occurs in most animals, including humans, and prevents sense organs and the nervous system from being saturated with useless information (Fig. 9).

10 Why is habituation considered to be a type of learning?

11 Study Figure 9. What conclusions can be drawn about stimuli causing habituation?

A scarecrow ceases to be effective when birds have become habituated to it.

Imprinting

Chicks, goslings and ducklings flee the nest a few hours after hatching. They follow the first moving thing they see, which is usually a parent bird. After that, they only follow objects that look like the first object followed. This learning process is called imprinting (Fig. 10). Imprinting ensures the young birds follow their parents, so the brood keeps together and avoids danger. There is a **sensitive period** when imprinting is extremely likely, but if it does not happen then, it might not happen at all. Imprinting in birds can happen in response to a wide range of objects varying in size from a matchbox to a human. Round and conspicuous objects are followed more readily. Adult birds often show courtship and reproductive behaviour towards the object of imprinting – this is a problem if the imprinting process has gone wrong.

Q 12 Why is imprinting considered to be a type of learning?

The magnificent Californian condor is close to extinction. In an attempt to save the species, condors are reared in captivity for later release into the wild. To ensure successful release into the wild, imprinting must be carefully controlled and, as juveniles, the birds must learn their species identity as condors. All feeding of the chicks is done using a glove puppet that looks like a condor's head, and the young condors are not able to see any view of the person feeding them.

Puppies and kittens need to be handled in their first few weeks of life if they are to become confident and friendly domestic pets. If they are not handled enough during this sensitive period, they will grow into timid or aggressive animals. Working dogs, such as sheepdogs, are not handled as much as household pets. They usually respond only to one person and may be aggressive to others.

Q 13 Why are glove puppets that look like parent birds used when hand-rearing young birds?

Fig. 10 Imprinting in domestic chicks

These takahe chicks beside a dummy mother model are about to feed from a dummy parent beak. To ensure correct imprinting, the chicks see only objects that resemble adult takahes.

Source: data below adapted from Hinde, *Animal Behaviour*, McGraw Hill, 1966

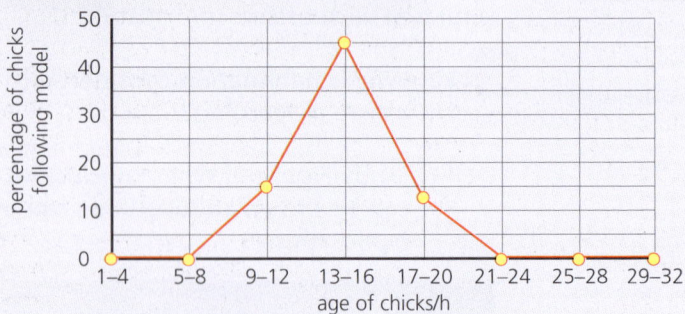

At first, the chicks' feathers are drying and they are not able to walk. If the chicks are left for over 24 hours, they will not follow a moving object at all.

Key ideas

- Learning can be defined as a change in behaviour caused by experience. The change in behaviour must be permanent, not just caused by fatigue.

- Learning can be as simple as ceasing to respond to a stimulus that caused a response before. This is called habituation.

- Some birds follow the first moving object they see after hatching. Thereafter, they follow that object and no other. This is called imprinting. As adults, some birds show courtship and reproductive behaviour towards the imprinting object.

5.6 Conditioning

The animals sharing our homes are quick to learn where food is to be found. The sound of a tin being opened brings the cat or dog into the kitchen very quickly because the animal has learned to **associate** the sound of the tin opener with food. This type of learning is called **conditioning**. There are two types of conditioning: the **conditioned reflex** and **operant conditioning**.

Conditioned reflex

Conditioning interested physiologist Ivan Pavlov, who worked in St. Petersburg in the nineteenth century. Pavlov wanted to test the effects of different types of food on the salivation reflex. A skilled surgeon, he re-routed the dogs' salivary ducts so that he could collect and measure the amount of saliva they produced when food was placed in their mouths. However, he noticed that his dogs started to salivate as soon as they heard his approaching footsteps, *before* he had given them any food. Pavlov then found that by giving the dogs a reward of food, he could train them to salivate to a stimulus such as a flashing light or a ringing bell. Pavlov called this type of learning, in which the usual stimulus (food) is replaced by a new one (light or bell), a conditioned reflex (Fig. 11). Because this was the first type of conditioning to be described, it is now also known as **classical conditioning**.

Operant conditioning

Pavlov's work encouraged others to study the way animals learn. Burrhus Skinner developed the **Skinner box**, a special apparatus for training an animal, usually a rat. By pressing a lever, the rat gains a reward of a food pellet. At first, the rat presses the lever just by accident, but soon learns to associate lever-pressing with a reward. This is called operant conditioning, because the animal is rewarded for an operation (movement) that it does naturally from time to time. The food reward makes it more likely the rat will press the lever again; in other words, the food is a positive **reinforcement** for the behaviour.

All kinds of movements can be reinforced in operant conditioning. For example, if a pigeon is rewarded every time it preens its feathers, the rate of preening very quickly increases. Reinforcers need not be food: monkeys learn for the reward of seeing another monkey; dogs respond to the reward of attention from their owner.

14 a What sort of conditioning did Pavlov's dogs show at the sound of a bell?

b What sort of conditioning is developed by reinforcement?

Fig. 11 Conditioning a reflex action

Phase 1
A hungry dog is tested for its reactions to two unrelated stimuli: food and a bell.

The dog salivates at the sight of food.

The dog does not salivate at the sound of a bell.

Phase 2
The dog undergoes a period of training in which a bell is rung whenever the dog is fed.

Phase 3
The dog now salivates at the sound of a bell alone.

Source: adapted from MacKean, *Introduction to Biology*, Murray, 1979

Punishment

Animals can also be conditioned using punishment as negative reinforcement. A stern telling-off works as a punishment and prevents a dog from jumping up to greet people, so long as the training is done consistently in the first few months.

Aversion therapy is a controversial and no longer popular practice in which punishment is administered to people trying to break an undesirable or unwanted habit.

Caterpillars of the mullein moth have a nasty taste and bright orange spots. The bad taste works as a punishment for any bird that attempts to eat the caterpillar, and the bird learns to associate the orange spots with a bad taste. Other poisonous animals use similar colours as **warning coloration.**

Learned human behaviour

Learning plays a major part in our everyday lives. For instance, association learning is used by advertisers. Some advertisements try to familiarise us with a particular sign or logo, so that we are more likely to select it from a range of similar logos. More commonly, advertisements try to make us associate their product with being sexy, successful or intelligent. We are encouraged to think that buying the product will give us these qualities.

15 Study the advertising photographs below. In each case, how is the advertiser trying to persuade us to buy the product?

Parents often use presents or pocket money to reward children for good behaviour. They also use various forms of punishment. This is rather like operant conditioning in rats, but humans can often see the motives behind other peoples' behaviour and resist being 'conditioned'. Sociologists and psychologists study the effects of reward and punishment on human behaviour. While most people agree that too much punishment can make a child withdrawn and unresponsive, some think that too many rewards might encourage selfishness. Skilled parents and teachers are good at giving the right amount of control and encouragement.

Advertisements try to persuade us to buy particular branded products.

"AT HAAGEN-DAZS WE **hold** WITH THE PRINCIPLE THAT ICE CREAM SHOULD BE THICK AND CREAMY. CONSEQUENTLY WE KEEP **tight** CONTROL ON THE AMOUNT OF AIR BEATEN INTO OUR PRODUCT."

Häagen-Dazs

Dedicated to Pleasure.

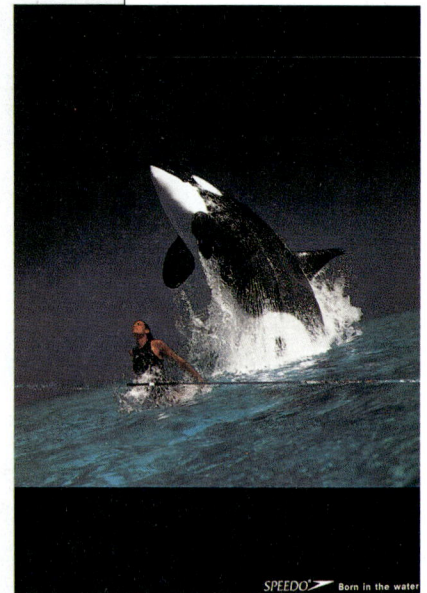

SPEEDO Born in the water

16 How might parents use conditioning in bringing up their children?

Behaviour therapy is a way of using classical conditioning to help people with a range of problems. For instance, people with phobias can be treated with a technique called systematic desensitisation. This means that they are first taught a relaxation technique, so they can relax whenever they choose. Then, a 'tolerable' sample of whatever the patient is frightened of is presented while the patient tries to remained relaxed. The severity of the fearful subject is gradually increased until the patient can face the complete object or situation without becoming distressed.

Sniffer dogs are chosen from breeds that are traditionally used to retrieve game.

Working dogs

Working dogs are a common feature of our life. For instance, besides sheepdogs, there are guide dogs for the blind, hearing dogs for the deaf, security dogs, rescue dogs, and sniffer dogs. These dogs are always trained by building on inate behaviour.

HM Customs and Excise have 73 sniffer dogs and handlers. They use rewards but no punishments to train the dogs. The dogs have 12 weeks basic training. If they reach the required standard they are then joined by a handler for a further 9 weeks.

The English springer spaniels and Labrador retrievers that HM Customs and Excise use have an innate ability to hunt and retrieve rabbits and birds. The training is started by giving the dog a soft toy. Once the dog is possessive about its toy, the toy is hidden, and the dog is rewarded by playtime with the handler when the toy is found. In the next stage, the toy is sprinkled with a drug (heroin, cocaine, amphetamine or cannabis) or with explosives. The hiding game continues and the dog learns to find the toy using the smell of whatever it has been sprinkled with. The dog shows the handler where the toy is hidden by biting, barking and scraping and is rewarded by further play.

Recently, HM Customs have started a passive dog trial. These dogs are trained to search passengers. Instead of biting or barking, the dog is trained to sit down next to any passenger carrying a suspect item.

Key ideas

- In conditioning, animals learn to associate one thing with another. Conditioning needs either a reward or a punishment to reinforce the desired behaviour. So, reinforcement can be either positive or negative.

- In a conditioned reflex, the usual stimulus is replaced by a new stimulus.

- In operant conditioning, a natural movement of the animal is reinforced so that it happens more often.

- Human behaviour is influenced by the rewards and punishments given out by society.

Investing in the future

Will human beings cause the extinction of many other species? How can we learn to share the planet's living space without reducing the diversity of life? Studies of animal behaviour should help us to see how humans and other animals can coexist and benefit each other, and we might also develop insights into our own behaviour. For example, do humans defend territories in ways similar to other animals? Is human communication similar to that of other animals? Can the social development of young primates tell us anything about our own development? For a varied population of animals and plants to survive and coexist with us in the future, we have to apply our knowledge for the benefit of all life on the planet.

This little basilisk (Jesus lizard) is so tiny and so fast that it can run on water. Like many other species, it is threatened with extinction.

The beauty of our planet seen from space hides the conflicts on its surface. There are endangered species in the air, in the seas, and on the land.

6.1 Learning objectives

After working through this chapter, you should be able to:

- **explain** the advantages and significance of territory;

- **analyse** simple courtship patterns;

- **understand** the role of hormones and pheromones in courtship behaviour;

- **describe** the behaviour patterns associated with oestrus in mammals;

- **explain** how knowledge of reproductive behaviour is useful in animal husbandry;

- **describe** parental care in birds and mammals;

- **describe** the opportunities for learning provided by extended parental care.

6.2 Defending a territory

Wherever there are open fields in Europe, skylarks are heard singing as they hover and circle 50–100 metres above the ground. Their song has inspired many poets, but scientists use it to estimate populations. The song advertises the breeding male skylark's claim to his territory; about 10 000 m² (or a hectare) of the field beneath him. Other male skylarks keep away when they hear the song, but probably have their own territory nearby.

In the 1960s, about 4 million pairs of skylarks were breeding in Britain; now there are about 2 million. We know this from the annual counts of common birds that began in 1962. The counting method is known as the Common Birds Census and it estimates the number of territories of skylarks and other birds in about 200 farmland or woodland sites around the country. Volunteer bird watchers make about 10 springtime visits to their census area and map the positions of all singing birds. From the completed census maps, changes in the numbers of each bird species are calculated and any reduction in the numbers of birds shows up very quickly.

A **territory** is defined as any defended area. The skylark's territorial defence is his song flight, which usually lasts for 2–3 minutes, though some continuous skylark songs have been measured to last over an hour. The benefits of such energetic activity differ from species to species, but include:
- defending a food source;
- defending a nest site;
- attracting a mate.

Finding food

Larger bird species usually have larger territories. A wren defends no more than 5000 m² of woodland, whereas a golden eagle defends an area of up to 90 km² (90×10^6 m²) of moorland and mountain. For these two, and many other species, the territory preserves a source of food for adults and their young. The bigger the bird, the more food it needs. Carnivorous birds tend to have bigger territories than herbivorous birds (Fig. 1). This is because less food energy is available at higher trophic levels. Usually, territories are only defended against members of the same species. Different species feed on different things (Fig. 1).

Fig. 1 Bird territories

Territory size

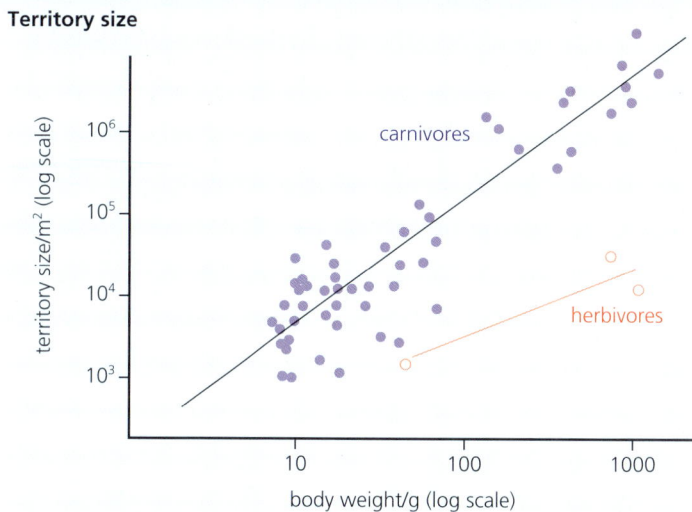

Source: adapted from an article by Schoener, *Ecology*, **49** (1968)

Territory overlap

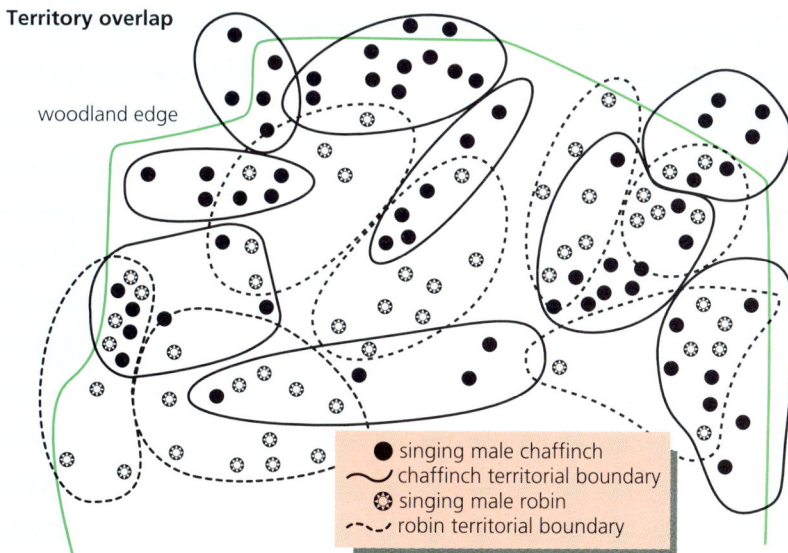

- ● singing male chaffinch
- ～ chaffinch territorial boundary
- ✺ singing male robin
- ⌁ robin territorial boundary

Symbols within a territory are the same male bird heard singing on different days. Area of map = 60 000 m².

The overlap of chaffinch and robin territories shows that chaffinches and robins do not compete with each other for territory.

Source: adapted from Common Birds Census for Short Wood, Northamptonshire, 1995

1 Why do birds of prey need larger territories than herbivorous birds?

Turtle dove down 77%

Corn bunting down 90%

It's not just the skylarks, many of our farmland birds are now declining in numbers.

Song thrush down 73%

Kittiwakes (*Rissa tridactyla*) and shags (*Phalacrocorax aristotelis*) nest on these dramatic cliffs in the Farne Islands.

Nests and mates

Skylarks often choose to breed in **set-aside farmland** (farmland that is not being used for crop-growing) where they can hide their nest among the weeds. Female skylarks are attracted to males with good nesting sites in their territories. Increasing amounts of set-aside land could help skylark populations to recover, but only if weed control is delayed until the birds have finished nesting. Once the best territories are occupied, other skylarks are forced to breed in poorer habitats where they are less successful. Territorial behaviour can, therefore, limit the size of bird populations.

Defending a nest site is even more important in cliff-nesting birds. In the Farne Islands, researchers found that shags prefer to select the broader cliff ledges that are protected from the waves but are still close to the sea. The birds defend a small territory just around the nest site, but they feed on fish out to sea. In shags, the quality of the nest site is closely related to breeding success (Table 1). The quality of the nest site was judged on a scale of 0–4 before nesting began, and breeding success was measured by the number of chicks fledged per pair.

Table 1 Nest site and breeding success		
Nest site quality	Mean number of chicks fledged per pair	n (no of pairs)
0 (poor)	0.61	191
1	0.69	402
2	1.03	704
3	1.30	566
4 (good)	1.55	116

Source: adapted from Potts et al, in *Journal of Animal Ecology*, **49** (1980)

2 What does Table 1 tell you about the relationship between nest site quality and breeding success?

3 Why do shags only defend a small area round their nest site?

Table 2 Territorial defence

Species	Territory	Method of defence
dragonfly (*Libellula depressa*)	• pond	• chasing other males
stickleback fish (*Gasterosteus aculeatus*)	• area of river bed around nest	• chasing other males displaying red throat and spines
Canadian timber wolf (*Canis lupus*)	• hunting area of the pack (10 000–12 000 km^2)	• scent marking using urine • howling
domestic cats (*Felis felis*)	• hunting and breeding area • male's territory includes territory of several females and is up to 10 times bigger	• mewing and fighting • scent marking by spraying urine • rubbing scent glands on twigs and stems

Male elephant seals (*Mirounga angustirostris*) sparring.

Territory and other animals

Territorial behaviour occurs among many animals besides birds, but is rare in invertebrates (Table 2).

Territorial alternative

Humans often defend their living area. If we are discovered in someone else's house or garden we are likely to be rebuked or even attacked. However, we may be invited in by the owner if we are a friend, or if we introduce ourselves first. Humans seem to show much more aggression when inside their cars than when walking about. Perhaps we regard our cars as moving territories.

Away from individual property, there are common areas that everybody shares, such as roads, pavements and parks, where we usually feel safe among strangers. But on a grander scale, humans go to war to defend their national area against attack from another national power. Few animals have behaviour comparable to humans in this respect, although a group of male chimpanzees in the Gombe Stream National Park in Tanzania, were seen systematically killing the males in a second group. They then took over the second group's territory and females.

A territorial dispute between male rock iguanas (*Cychlura cychlura*).

A troop of baboons (*Papio anubis*) in Tanzania. Their home range includes forest and open grassy areas.

Many social mammals have a **home range** rather than a territory. In baboon troops, the home range is the area used for feeding and covers several square kilometres. Only some parts of the home range are visited on any one day, and the baboons return to the same sleeping trees every night. Parts of the home range overlap with the ranges of other troops, but the **core area**, around the sleeping site is a not used by any other troop. Baboon troops usually avoid each other although there is no active defence of the home range or core area.

Q 4 a Do individual humans have territories or home ranges?

b Do national states regard their national areas as a territory or a home range?

Key ideas

- Some animals defend a territory. This helps them to maintain a reliable food supply, attract a mate and protect a good nest or sleeping site.

- Territories are usually defended only from members of the same species.

- Territories are defended by displays, fighting, songs, sounds and smells.

- Group-living animals remain within a home range, often overlapping with the home ranges of other groups. Sometimes, the central core area of the home range is defended from other groups.

6.3 Courtship

Courtship is behaviour used to attract a mate. Courtship enables animals to:
- approach each other closely;
- recognise their own species;
- choose a strong and healthy mate;
- synchronise breeding behaviour.

Most animals have some fear of each other and maintain a small space between themselves and their neighbours. This is known as **individual space**; it provides some safety from aggression, and reduces the risk of infection. Courtship enables the male and female animals to enter each other's individual space without triggering aggression. Many small birds and mammals find a new partner each breeding season, but swans, geese and gibbons remain paired for life. Some animals that have several mates at the same time are **polygamous**. Male animals that have several female mates at the same time are **polygynous**, examples are pheasants, peacocks, red deer and Hamadryas baboons. Female animals that have several male mates at the same time are **polyandrous**, for example, starlings and chimpanzees.

Fig. 2 Breeding control in birds

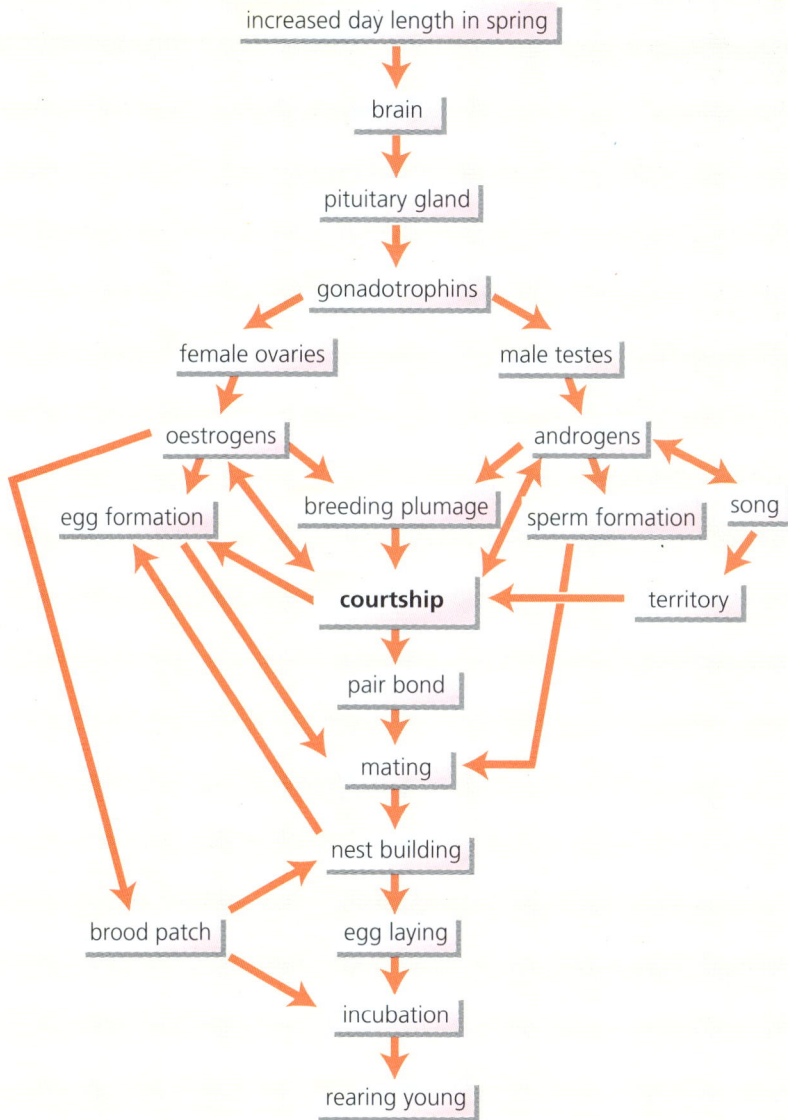

```
increased day length in spring
            │
            ▼
          brain
            │
            ▼
      pituitary gland
            │
            ▼
      gonadotrophins
         ╱        ╲
  female ovaries    male testes
        │              │
        ▼              ▼
    oestrogens      androgens
      │               │
      ▼               ▼
  egg formation   breeding plumage   sperm formation   song
            │           │                  │            │
            ▼           ▼                  ▼            ▼
                     courtship ◄────────── territory
                        │
                        ▼
                    pair bond
                        │
                        ▼
                      mating
                        │
                        ▼
                   nest building
      brood patch       │
            │           ▼
            │       egg laying
            ▼           │
                        ▼
                    incubation
                        │
                        ▼
                   rearing young
```

There are several positive feedback effects in the control of breeding. For instance, courtship behaviour stimulates the sex organs to release more sex hormone, and sex hormone increases readiness for courtship. Androgens from the male testes cause singing, and the action of singing stimulates the testes to grow more.

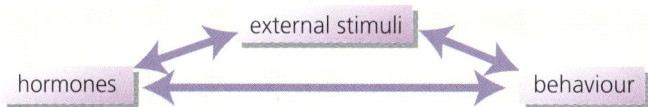

```
            external stimuli
          ◄──────      ──────►
  hormones ◄──────────────────► behaviour
```

Gonadotrophic hormones released by the pituitary in spring cause the ovaries and testes to grow. The sex organs then produce sex hormones; oestrogens in females and androgens in males. These sex hormones change the behaviour of the birds, leading to courtship, pair formation and mating.

Like birds on power lines, humans keep a small space round themselves if they can, even in crowded conditions.

Q 5 What are the advantages of maintaining individual space outside the breeding season?

Courtship and display

In birds, increasing day length at springtime triggers a chain of events controlled by **sex hormones** (Fig. 2). The female's ovaries do not begin to produce eggs without the stimulus of courtship and nest building behaviour. The male and female birds gradually develop a **pair bond** as their hormone concentrations increase and they begin breeding activities.

Females select males with the most noticeable displays, as can be demonstrated by giving swallows extra-long tail streamers; these enhanced birds are more successful at attracting mates. This is **sexual selection**, and it leads to the evolution of bigger and better courtship displays. The most dramatic male displays have evolved in polygynous species where sexual selection is strongest because only a single male animal will mate with several females. Polygynous species also show the greatest physical differences between the sexes.

Q 6 What first triggers mating and breeding behaviour in birds?

7 Why is sexual selection strongest in polygynous species?

Selecting a mate

Every species has a different **courtship display** that helps the animals to recognise a member of their own species of the opposite sex, and approach each other closely enough to mate (Fig. 3).

Fig. 3 Courtship displays

The bird of paradise shows off its tail.

The cuttlefish uses rippling colour changes.

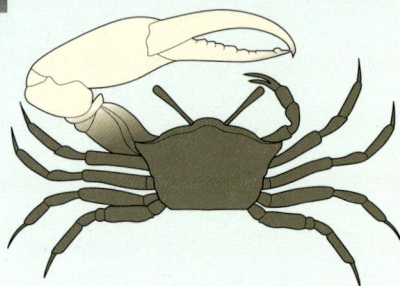

Fiddler crabs wave a claw.

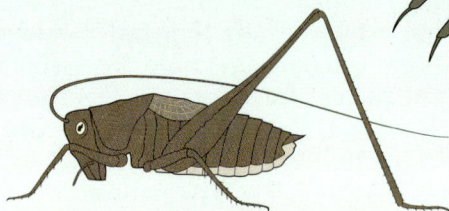

Crickets stridulate (rub wings together to produce particular sound frequencies).

Female skylarks are attracted to skylark songs but not to the songs of any other species. During courtship, the male skylark usually first sings briefly on the ground, he then hops with his head held high, bows with his tail raised, and ruffles his neck and crest feathers. All these movements help to ensure species recognition and to excite the female. Nest building, breeding and rearing a skylark brood takes about 6 weeks, so cooperation and coordination between the parent birds are essential (Table 3).

When skylarks select a mate, they are starting a major investment of time and effort. The female invests in the eggs she lays, and the male invests energy in defending the territory and feeding the young. This is **parental investment**.

The Save our Skylarks Campaign aims to raise £250 000 to fund research into skylarks' exact habitat needs, so that we can give practical, realistic advice to farmers. Skylarks need undisturbed fields with plenty of seeds and insect prey in order to rear their broods. Changes in farming practice, such as planting winter cereals and using more efficient herbicides, have probably affected skylark numbers. Set-aside land might help, provided it is not sprayed for weeds too early. Skylarks can rear up to four broods in one season, and pairs that have been successful often stay together in the same territory the next year. So, skylark numbers will increase rapidly if suitable habitat is provided.

Table 3 Cooperation in skylarks

Activity	Male/female/both birds	Time taken
nest building	female	1–2 day
egg laying	female	1 per day for 3–4 days
incubation	female	11 days
feeding young in nest	both	8–10 days
feeding young out of nest	both	15 days

Q 8 What is parental investment?

In the guppy (*Poecilia reticulata*), a small freshwater fish from Trinidad, the male courts larger females in preference to smaller ones. Larger females produce more eggs, so natural selection favours males with this size preference. Guppies breed readily in captivity, and their courtship behaviour has been studied in detail. It begins when the male approaches a female and tries to lure her away from the other fish in the shoal (Fig. 4).

Display and breeding behaviour of sticklebacks is examined in relation to sign stimuli in Chapter 5. Fish generally show less parental investment than birds; the production of eggs by the female is her major investment, but the male often has no role other than courtship and fertilisation.

9 Study Figure 4. Design a simple tick chart to record the courtship sequence in observed guppies. How would you make an ethogram from your results?

10 Explain briefly how a named invertebrate and a named bird each recognise their own species during courtship.

Fig. 4 Courtship behaviour in guppies

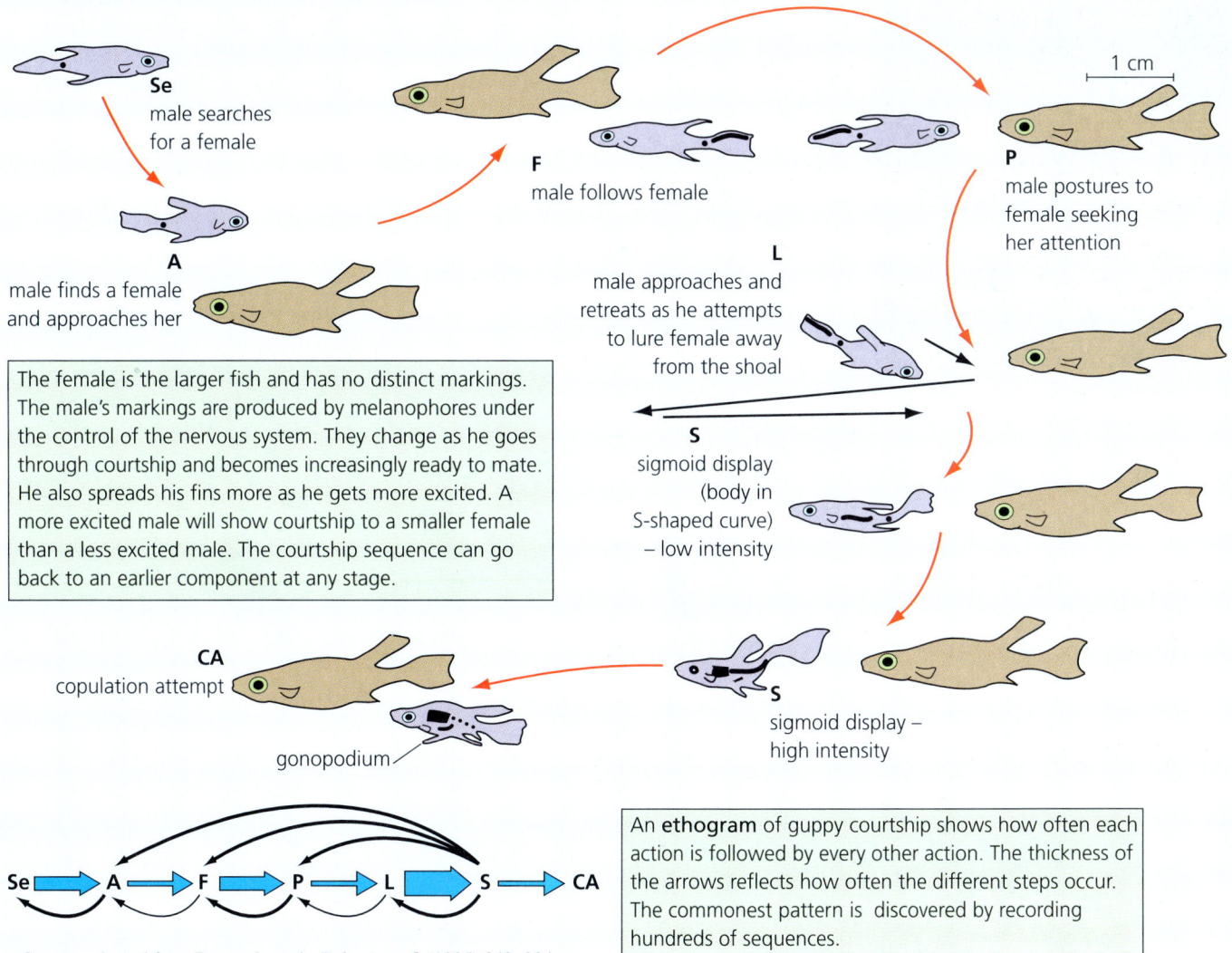

Se — male searches for a female

A — male finds a female and approaches her

F — male follows female

P — male postures to female seeking her attention

L — male approaches and retreats as he attempts to lure female away from the shoal

S — sigmoid display (body in S-shaped curve) – low intensity

S — sigmoid display – high intensity

CA — copulation attempt

gonopodium

1 cm

The female is the larger fish and has no distinct markings. The male's markings are produced by melanophores under the control of the nervous system. They change as he goes through courtship and becomes increasingly ready to mate. He also spreads his fins more as he gets more excited. A more excited male will show courtship to a smaller female than a less excited male. The courtship sequence can go back to an earlier component at any stage.

Se → A → F → P → L → S → CA

An **ethogram** of guppy courtship shows how often each action is followed by every other action. The thickness of the arrows reflects how often the different steps occur. The commonest pattern is discovered by recording hundreds of sequences.

Source: adapted from Baerends et al., *Behaviour*, **8** (1995) 249–334

Key ideas

- Courtship enables potential mates to assess each others' qualities and synchronise breeding behaviour.

- Birds often come into breeding condition in the spring. This is triggered by increased day length and controlled by sex hormones.

- There is an interaction between environmental stimuli, behaviour and sex hormones controlling reproduction.

- Courtship behaviour can be analysed by looking at the sequence of its individual components.

6.4 Chemical communication

Fig. 5 Antenna of the male silk moth

Silk moths emerging from their cocoons.

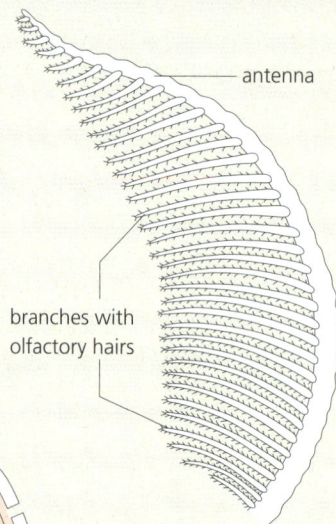

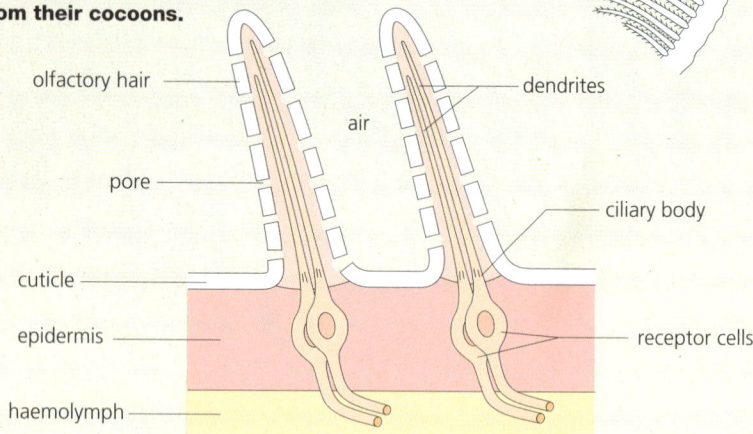

More than half the receptor cells in the moth's antennae respond only to bombycol. Molecules of bombycol diffuse through the pores in the olfactory hairs, causing a depolarisastion of the receptor cells (Chapter 1).

Source: adapted from 'Readings from *Scientific American*' in Eisner and Wilson, *Animal Behaviour*, Freeman, 1975

Communication by smell is used by many animals during courtship and territorial defence. For example, male guppies release chemicals into the water which help to stimulate female guppies. The chemicals used in this kind of communication are called **pheromones**. A pheromone is defined as a substance used in communication between members of the same species. So, pheromones differ from species to species. Compared with visual messages, pheromones have two notable advantages:

- they can move around objects;
- they continue to have an effect for a long time (they are energy efficient).

On the other hand, a visual message can often be seen from far away, is not affected by wind and shows exactly where the sender of the message is.

In insects

Unmated female silk moths (*Bombyx mori*) release the sex attractant **bombycol** from glands in the abdomen. Smell receptors in the antennae of male silk moths can detect bombycol in very low concentrations (Fig. 5). When about 200 smell receptors are stimulated at once, male silk moths flutter their wings and move towards the source of the smell.

11 a What are pheromones?
   b Name two advantages of pheromones as a form of communication.
   c What advantages does visual communication have?

Humans can use insect pheromones to control the populations of insects that damage valuable plants and crops. Mediterranean pine trees are often stripped of leaves by the caterpillars of pine processionary moths; this seriously stunts the growth of the trees. Populations of the moth can be controlled using artificial female moth pheromones inside hundreds of plastic funnels attached to the trees. The male moths respond to the pheromones and enter the funnels. Once inside the funnels, the male moths continue to move towards the pheromone source, and fall into insecticide.

12 Give a list of reasons why controlling the pine processionary moth by trapping males with pheromones is better than spraying insecticide onto the trees.

Social insects, such as ants, wasps and bees, use a very wide range of pheromones. For example, the mandibular gland of a queen honeybee makes at least 32 substances including **queen substance**. This pheromone acts as a sex attractant, attracts the workers to the queen when the bees are swarming, and prevents the workers from producing more queens in the new nest.

13 What disadvantages do pheromones have as a means of communication?

In mammals

Many female mammals advertise their readiness to mate by giving off pheromones. Bitches 'come on heat' every 6 months, when their ovaries produce eggs. When this happens, the bitch produces (in her urine) a pheromone that is very attractive to dogs. Baboons and chimpanzees display their sexual condition by colourful swellings of the genital area. The swellings reach a maximum size when the female is ovulating, and most likely to conceive. However, these signals can be difficult to see in thick vegetation, so female baboons also give off a musky scent due to pheromones when they are ready to mate.

On dairy farms, predicting the time of ovulation in cows is essential. Cows produce milk after giving birth to a calf, and will ovulate again 1–3 months after calving. It is important that the cow becomes pregnant again the first time she ovulates, otherwise there will be a longer delay before she produces milk again. If the herd is with a bull, the bull will detect the cow's condition by smell, and mate her when she is ready.

Hundreds of pine processionary moth caterpillars are emerging from this cocoon.

Male baboons compete for females who are ready to mate by chasing and fighting. The winning male follows, grooms and eventually mates with the female.

Many modern farmers use artificial insemination to save the expense of keeping a bull, but the farmer must look for changes in the cow's behaviour to tell when she is ovulating. Ovulating cows stimulate mounting behaviour in other cows and sometimes mount other cows. This tells the farmer that the cow is ready for artificial insemination. This **oestrus behaviour** only lasts 8–12 hours, so it can easily be missed (Fig. 6).

Q 14a **Why do some dairy farmers keep a bull?**
b **What is the alternative to keeping a bull?**

What about us?

In humans, the sense of smell is less acute than in many other mammals. However, the size of the perfume trade shows that smells are socially important to humans. Some commercial scents come from the pheromone glands of other mammals.

Women's vaginal secretions contain volatile fatty acids (Fig. 7). As similar substances are known to work as pheromones in monkeys, it is possible that these fatty acids are human pheromones. If so, they play a part in human sexual attraction, even if we are unaware of it.

Fig. 6 Managing milk production

Oestrous behaviour in cows

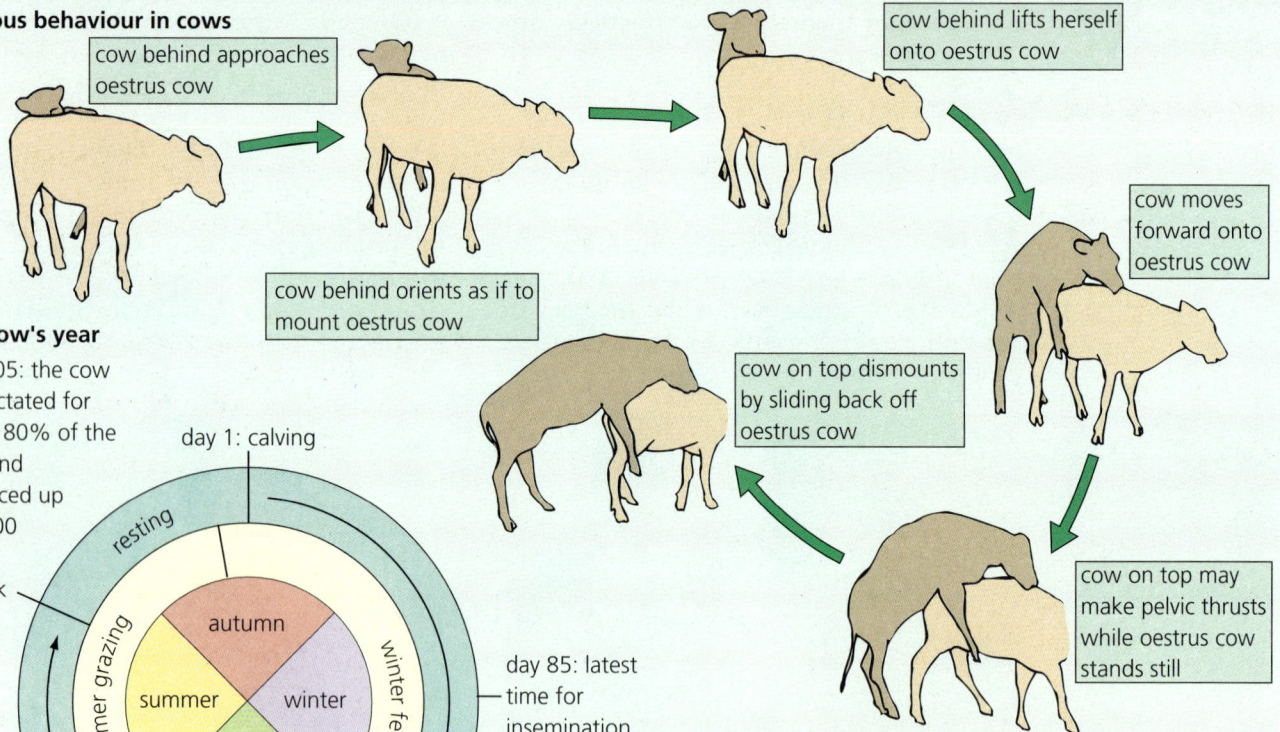

cow behind approaches oestrus cow

cow behind orients as if to mount oestrus cow

cow behind lifts herself onto oestrus cow

cow moves forward onto oestrus cow

cow on top may make pelvic thrusts while oestrus cow stands still

cow on top dismounts by sliding back off oestrus cow

There is a 90% chance that the cow that stands solidly to be mounted without walking away is in oestrus, but there is only a 30% chance that the cow that mounts another is in oestrus. Some farmers keep a cow injected with male sex hormones to encourage mounting behaviour; the farmer is then less likely to miss any cows in oestrus.

The cow's year

day 305: the cow has lactated for about 80% of the year and produced up to 5000 litres of milk

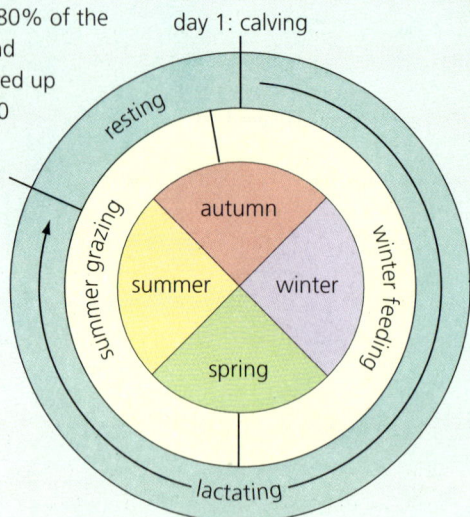

day 1: calving

day 85: latest time for insemination

resting
summer grazing
winter feeding
lactating
autumn
summer
winter
spring

Source: adapted from *The Milk Year*, National Dairy Council

Source: adapted from Esslemont et al, *Fertility Management in Dairy Cattle*, Collins, 1985

Fig. 7 Humans and pheromones

Volatile fatty acids

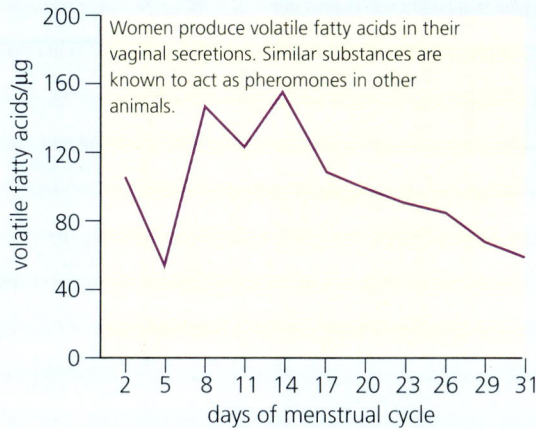

Women produce volatile fatty acids in their vaginal secretions. Similar substances are known to act as pheromones in other animals.

y-axis: volatile fatty acids/µg (0, 40, 80, 120, 160, 200)
x-axis: days of menstrual cycle (2, 5, 8, 11, 14, 17, 20, 23, 26, 29, 31)

Other possibilities

People might often be reacting to pheromones without realising it. The secretions from human sweat and sebaceous glands contain several steroids. The effect of these on our moods and behaviour are not yet understood, but there are some interesting leads.

• When a particular steroid was sprayed onto some theatre seats, the sprayed seats were occupied before the unsprayed ones.

• 'Osmone 1' is a steroid found in human sweat. The chemistry department at the University of Warwick is testing osmone 1 as a tranquilliser.

• Some steroids found in plant oils are chemically similar to osmone 1; such oils are used in perfumes and aromatherapy. Some aromatherapy oils are said to be stimulating (for example, jasmine oil) and some are thought to be relaxing (for example, lavender oil).

Source: (graph) adapted from article by Michael et al, in *Science*, **186** (1974)

Key ideas

• Pheromones are chemicals that provide a method of communication between members of the same species during courtship.

• Understanding the action of pheromones can be helpful in the control of insect pests.

• Many female mammals produce visual as well as chemical signals when they ovulate.

• Ovulating cows change their behaviour and thus alert the dairy farmer of the appropriate time to carry out artificial insemination. This enables the efficient management of milk production.

• Although we are not generally aware of it, it is likely that pheromones have a part to play in sexual attraction between humans.

6.5 Time to learn

Extended learning in birds

Young skylarks are fed by both parents until they are about 3 weeks old. By this time, they have left the nest and are fully fledged and able to fly. How young skylarks learn to sing is not known, but young male chaffinches learn the chaffinch song by hearing their male parent sing while they are still in the nest. They do not sing until the following spring, but 'remember' the song until then.

Swan and goose families stay together for the whole of their first winter. The young birds learn migration routes and the location of safe winter feeding grounds by following their parents (Chapter 5). Parent geese protect good grazing patches for their young by threatening nearby birds in the winter flock, and, because larger families usually dominate smaller ones, it helps to stay in a large family group. The adults also spend more time than their young in watching for predators, so the young can concentrate on feeding.

Canada geese have a noticeable white 'chin strap'. When they lift and shake their heads, displaying the chin strap, they are signalling to other family members that they are about to take flight to the feeding grounds.

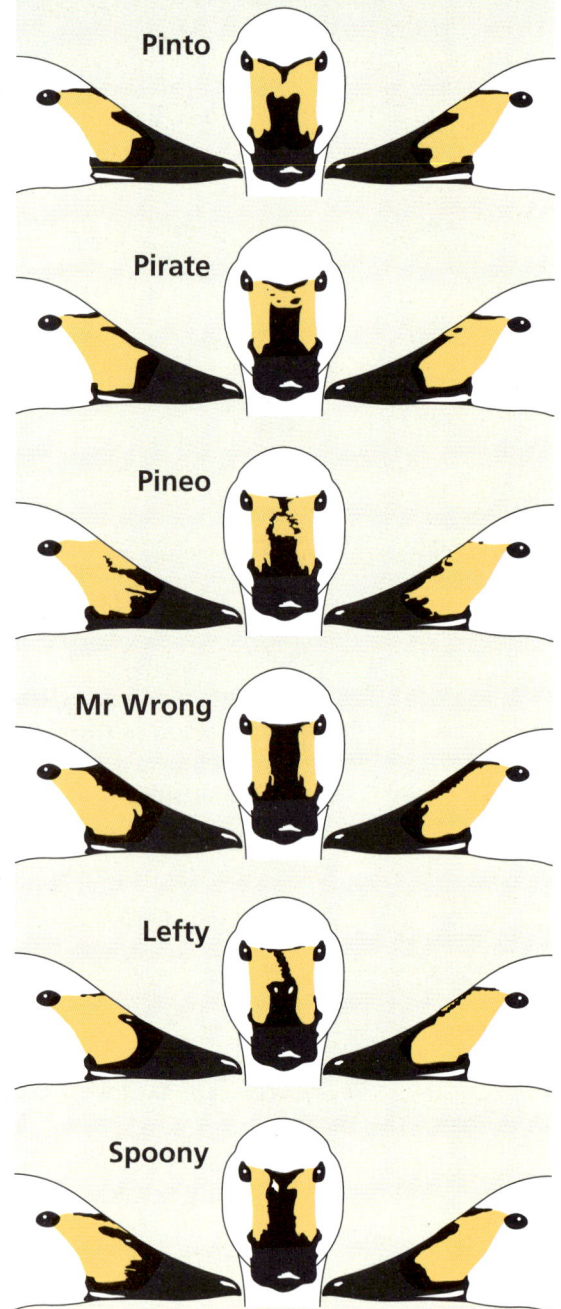

Fig. 8 Bill patterns in Bewick's swans

Pinto

Pirate

Pineo

Mr Wrong

Lefty

Spoony

'We quickly realised that the patterns of black and yellow on Bewick's swans' bills were infinitely variable. By drawing them in front and side view we could record the different patterns and give each swan a number and a name. We have now recorded the face patterns of 3,700 Bewick's swans'. PETER SCOTT

Source: adapted from Scott, *Observations of Wildlife*, Phaidon, 1980

Geese and swans can recognise members of their own families among the flock. In Bewick's swans each bird has a distinctive pattern of yellow and black on the beak (Fig. 8).

Q 15 Why do young swans and geese need to be able to recognise family members in their flock?

Extended learning in primates

The most lengthy period of immaturity occurs in **primates** (monkeys and apes). Baboons reach maturity at about 6 years of age and chimpanzees at about 12 years.

Young baboons learn from all other members of the troop, not just their own parents. A baboon infant soon learns to recognise other individuals (there may be 40 or more in a troop) and to adjust its behaviour appropriately: brothers and sisters are usually ready to protect, play or groom, but unrelated larger males might be aggressive.

The meanings of sounds and facial expressions must be learned by watching the responses of other baboons. Food sources are learned by imitation; baboons eat a wide variety of plants, insects and occasional mammals. Every habitat is different, and they all change with time.

Fig. 9 Learning to predict behaviour

To divert his pursuer, the young primate signals as if there is danger ahead, then slips away in the confusion.

There are many sources of danger, such as pythons and chimpanzees, both of which eat young baboons. Different groups have different established behaviour patterns called **cultural traditions** that are passed on by learning.

16 Why do you think primates have a longer period of immaturity than other animals?

Young primates, including humans, spend much of their time wrestling and chasing each other. This **play behaviour** helps primates to develop social responses, so that they are not too aggressive or too submissive towards others. Play might also help to develop fighting skills. Human play, including formal games, helps us to gain confidence and to learn to communicate with each other. It also helps us to learn the strengths and weaknesses of ourselves and others.

Baboons spend much time playing; play is signalled by a wide-open mouth not showing teeth, this is called the 'play face'. When humans want to play, they smile or explain in words or demonstrate, for instance by running about or throwing a ball.

Primates can learn to predict the behaviour of others in the group, and use this to gain an advantage. For example, finding food or a mate often means outwitting other members of the group. This requires the ability to predict what others will do under particular circumstances. Learning to make such predictions successfully takes time and a lot of intelligence (Fig. 9). The evolution of human intelligence may have happened partly by selection for an ability to understand the minds of others.

17 a Why is play important to young humans?
   b How do humans signal that they want to play?
   c Give two advantages of being able to correctly predict the behaviour of other members of a primate group.

6.6 Sharing the planet

Skylarks have declined in numbers but they are unlikely to become extinct. They will survive because they live in cornfields, so at least some will live alongside human populations, however many people there are. Other birds and mammals are in much more danger, particularly when they live in specialised habitats.

The bittern, a large bird of reedbeds, is now down to about 20 pairs in Britain, and attempts are being made to save it. The target is to build up the breeding population to 100 pairs by the year 2025. Knowing the territory size for a bittern enables conservationists to calculate how much reedbed is required to support this number of birds. The Royal Society for the Protection of Birds (RSPB) has recently bought 4 km^2 (400 hectares) of arable land in Norfolk, and is converting it into a reedbed for bitterns and other birds. They expect the new reedbed to support four pairs of bitterns, a small but important contribution towards the target.

18 How much reedbed will be required to support the target breeding population of 100 pairs of bitterns?

We have already lost several of our larger mammals from Britain, including boars, beavers, bears and wolves. Some people would now like to reintroduce wolves to parts of Scotland. They argue that this would help in the control of red deer, whose numbers are considered too high in some areas. Reintroducing wolves can only be done if we are sure how much space and food they will need. There was quite a debate on this subject in the pages of *Natural World* in 1995.

Wild boar (*Sus scroja*) still live in European forests.

The European beaver (*Castor fiber*) likes open streams.

The brown bear (*Ursus arctos*) climbs trees and likes to take a bath.

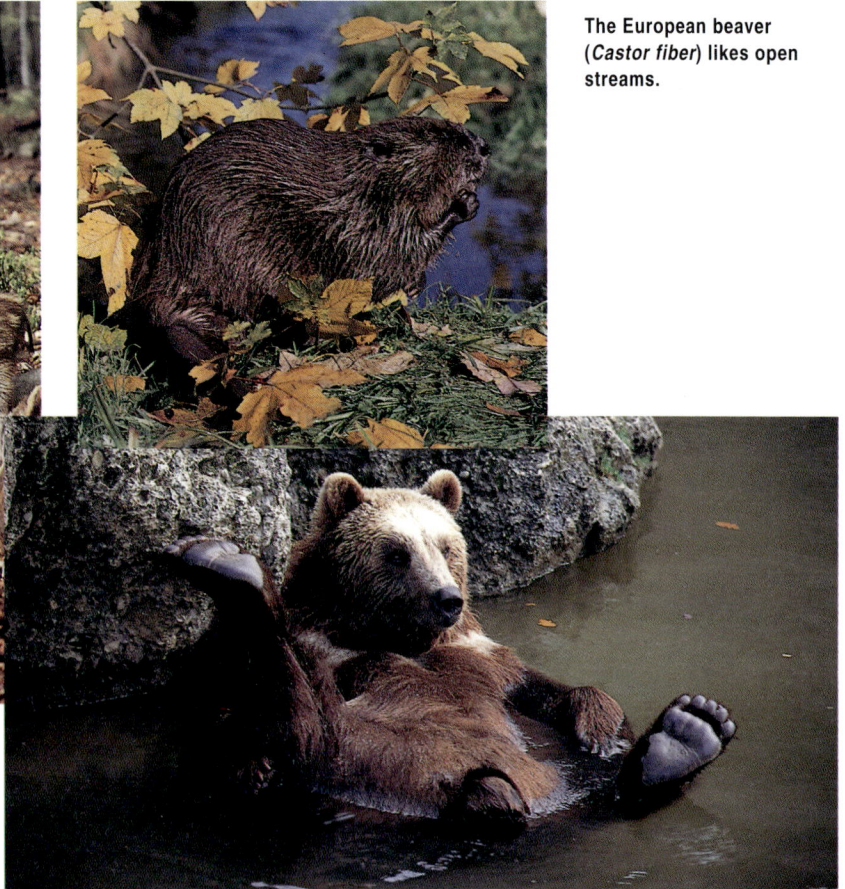

Today's public is ready for wolves. In a poll by the BBC in March 1995, 75% voted for a reintroduction. We would plan to find 500 square kilometres relatively free of sheep, and first monitor a couple of wolves as a trial to ensure all was well, then build up to three packs of about 20 wolves. In the long term, we aim for a population of about 400 throughout the Highlands. We can expect a population of 400 wolves to take about 4000 sheep a year. These figures are insignificant compared with the 1 300 000 sheep in the highlands.

Roger Panaman, Carnivore Wildlife Trust, Oxford

The last wolf was shot in Britain in 1743 on the Findhorn, Inverness-shire. Will we see them in Scotland again?

I must resist efforts for premature release. These proposals take no account of the scale of the land or of human activity in the Highlands when compared with Arizona or Texas (where wolves are soon to be reintroduced). I cannot see these arguments persuading the majority of people who live and work in Scotland. We need much more time to regenerate the land and our relationship with it.

Aubrey Manning, Chairman, Scottish Natural Heritage

I advocate that the reintroduction attempt should be made on the island of Rhum, a National Nature Reserve with no sheep but with a large herd of red deer controlled by shooting. We would learn whether wolves could manage to regulate the deer population without endangering the sheep-farming economy of the Highlands. If anything went wrong, it would be possible to recapture or kill the wolves.

D. W. Yalden, University of Manchester

Q 19 Wolf populations are threatened throughout Europe. Do you think we in Britain have a responsibility for wolf conservation?

Key ideas

- Birds and mammals usually have a long period of parental care.
- Extended parental care provides a long time for learning about the environment and about other members of the species. This gives the young a better chance of survival.
- Learning is especially important for primates. Primates living in groups must learn the identity and behaviour of other group members.
- Understanding how much space is needed to support animal populations is essential for conservation.

Primates and the global zoo

The world is a bit like a zoo – but it is far more complex. It is a dynamic system and each organism depends on others – there are no keepers to bring in fresh food or remove waste products. There are nearly 200 primate species – none are as successful as humans and some are nearly extinct. Our knowledge about them comes from studies done in the wild and from work with captive animals in zoos.

I study the bonobo in its natural habitat in Zaire. Studying bonobos in captivity also helps us to find out more about them, and might enable us to provide a future for them. What is clear is that we know very little about the bonobo and many other species, and time is not on our side in the battle against extinction.

Humans have the ability to shape the world they live in. To safeguard the future for other primates, and possibly ourselves, we need a better understanding of the biology and behaviour of our primate relatives.

The bonobo is our closest living primate relative.

7.1 Learning objectives

After working through this chapter, you should be able to:

- **recall** the main physical and reproductive characteristics of primates and relate them to living in trees;

- **explain** how primate groups can be studied;

- **describe** the social structure of gorilla groups and how this is maintained;

- **understand** how the social structure of gorilla groups offers opportunities for learning.

7.2 Primate features

Primates are mammals; they have fur and produce live young that feed on milk. They can be divided into two groups, the primitive primates called the **prosimians** and the more advanced primates called the **anthropoids**. The anthropoids are the monkeys and apes; **Old World monkeys** are those species found in Africa and Asia, while **New World monkeys** are those found in the Americas (Fig. 1).

Modern primates are a diverse group of animals and each species is adapted for life in a particular habitat. However, we know from fossil skulls and skeletons that primate

Fig. 1 The primates

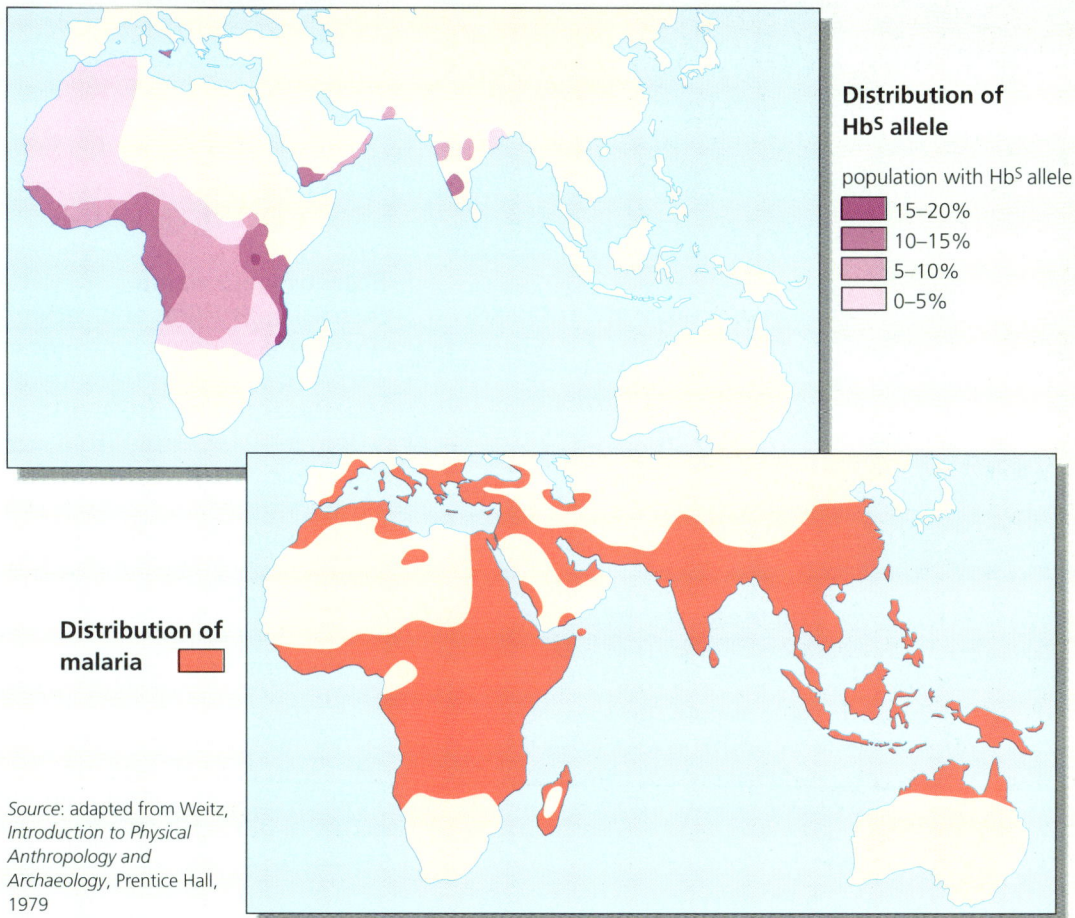

Distribution of Hb^S allele

population with Hb^S allele
- 15–20%
- 10–15%
- 5–10%
- 0–5%

Distribution of malaria

Source: adapted from Weitz, *Introduction to Physical Anthropology and Archaeology*, Prentice Hall, 1979

PPB ch9/fig7

Like all New World monkeys, spider monkeys have long prehensile tails with which they can grasp branches.

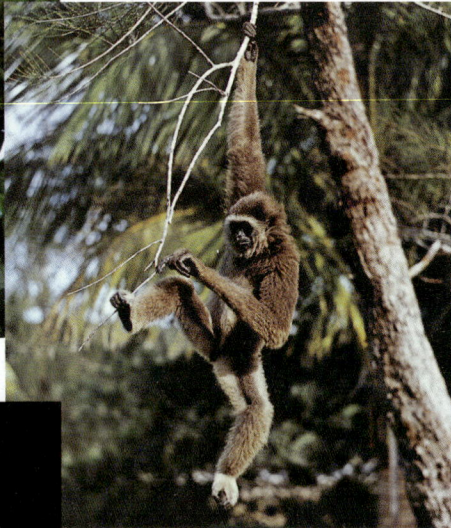

Gibbons use their powerful arms for swinging through trees and hanging from branches. They have long, curved fingers and long nails that act as hooks when they move this way.

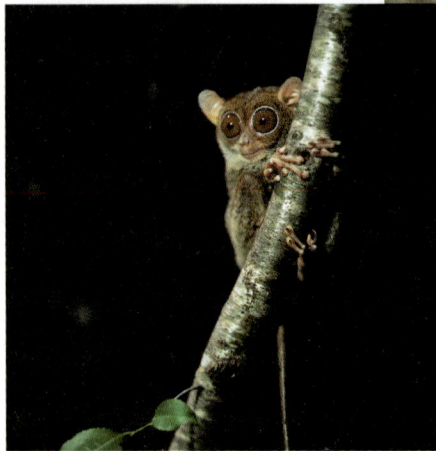

Tarsiers cling to the branches of trees with the help of pad-like swellings at the ends of their fingers. They can leap from branch to branch.

Fig. 2 Primate vision and smell

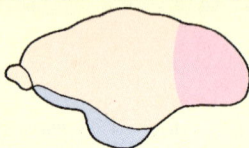

Tarsier
Tarsiers are primitive primates. They have huge eyes and, relative to the size of the brain a very large visual cortex. The **olfactory** area of the brain (to do with smell) is relatively small.

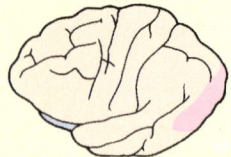

Chimpanzee
In chimpanzees, the visual cortex is relatively large but the olfactory area is substantially reduced in relation to brain size.

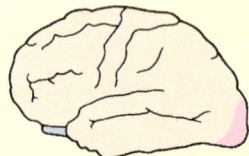

Human
Although humans do not live in trees, they still have a relatively large visual cortex, but the relative size of the olfactory region is very tiny.

■ visual cortex
■ olfactory area

Source: adapted from Young, *Introduction to the Study of Man*, Oxford University Press, 1971

ancestors were small, tree-dwelling species that lived on insects and other small animals. Several features derived from these ancestors are common to all modern primates and are adaptations for life in trees. These include:

- **stereoscopic vision** – being able to judge distance;
- **dextrous ability** – being able to manipulate things with the hands;
- some reproductive characteristics.
  Other common features include:
- **social organisation** – living in extended family groups;
- a large brain in relation to body size and, in particular, a well developed cerebral cortex.

1 List three primate features that are related to life in trees.

Stereoscopic vision

Stereoscopic vision and the ability to judge distance depend on large, forward-facing eyes and a brain with highly developed visual centres. Primitive primate ancestors would have benefited from a highly developed visual sense when moving through trees and catching insects. Although many modern primates are herbivorous, good sight is advantageous when exploring the environment and for visual communication.

Nocturnal primates have eyes adapted to life in low light. In these animals, the retina contains only rod cells, so vision is very sensitive. However, without cone cells, there is no colour vision. Species which are active in daylight have rods and cones in the retina. There is a particularly high concentration of cones, but no rods, in an area of the retina called the fovea. So, in daylight, colour vision is very good (Chapter 2). Higher primates exploit colour vision when selecting ripe fruit. In some monkeys, colourful fur is important in communication. In the evolution of apes, the development of the visual sense has been at the expense of the sense of smell (Fig. 2).

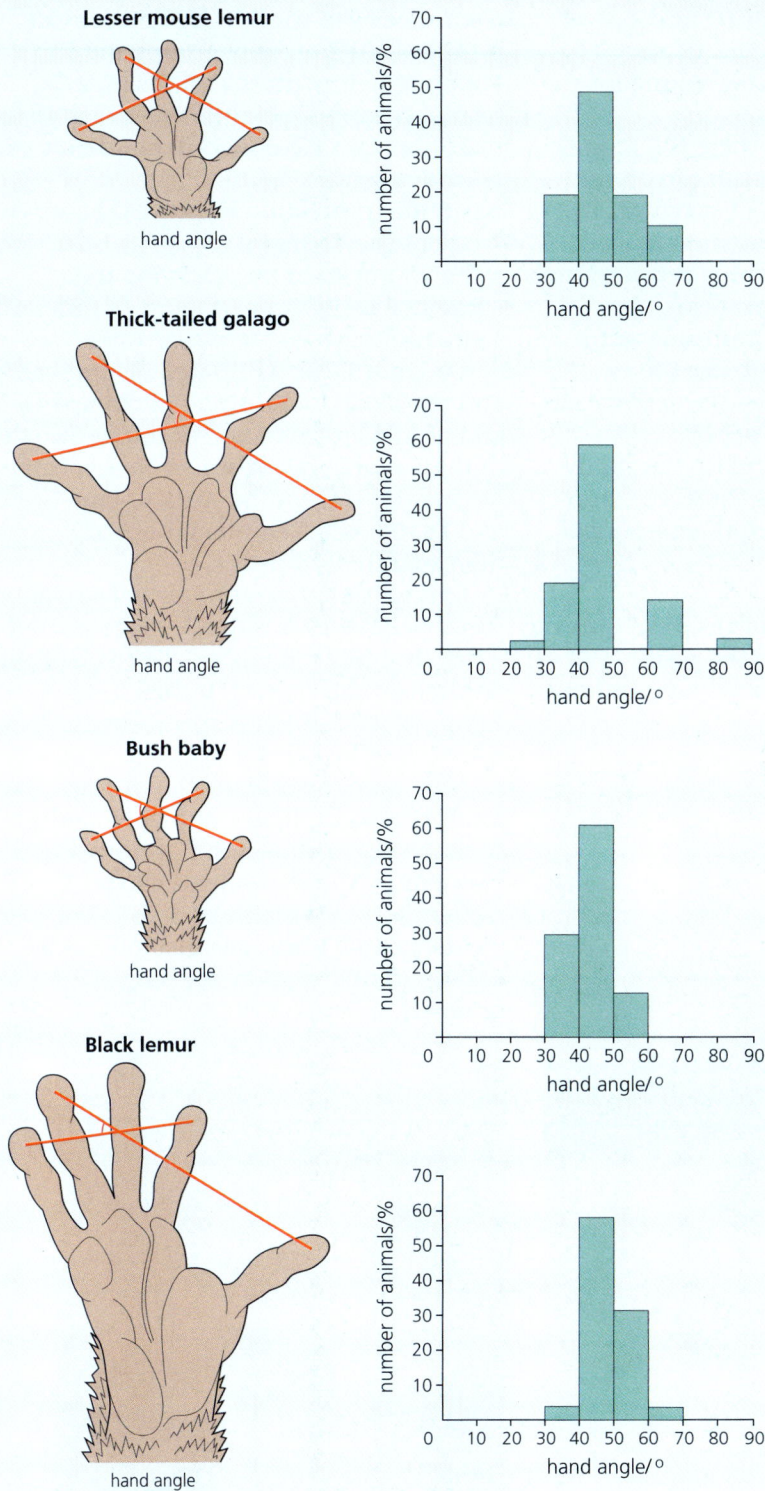

Fig. 3 Hand angles

Lesser mouse lemur

hand angle

Thick-tailed galago

hand angle

Bush baby

hand angle

Black lemur

hand angle

Source: adapted from Australian Academy of Science, *Web of Life*, 10th edn, Western Australia Education Department, 1983

Dextrous ability

Survival in the trees depends on being able to move from branch to branch safely and manipulate objects, for example, to pick and eat fruit. The dextrous ability of primates depends on the **opposable thumb**. The thumb is long and has a flexible lower joint that enables it to be folded over the palm. In most primates, the opposable thumb is found in hands and feet – they are 'four-handed'. Humans are 'two-handed' – the feet have become walking platforms with the toes all in line, and an extended heel.

Primates have two ways of holding things. They can wrap their fingers around an object to hold it firmly. This is called **prehensility**. They can also hold things between the end of the thumb and fingers. This is called **opposability**. Prehensility is useful for holding on and swinging from branches, and opposability is useful for manipulating objects.

Grip is improved by the presence of soft pads at the ends of the fingers. Primates do not have protruding claws that would hinder a good grip. The tops of the fingertips are protected by nails. Good grip is essential in young animals long before they move through the trees independently, they need to hold on to the mother's fur as she carries them through the forest.

The shape of primate hands can be compared by measuring the **hand angle** (Fig. 3).

2a **Study Figure 3. What can you conclude about hand angles in the primates shown?**
b **How is primate hand shape related to living in trees?**

Reproductive characteristics

Around three-quarters of primate species live in tropical forests, and most of these spend their life in trees. The females give birth to single offspring that have a firm grasp and cling to the mother's fur from birth. A young primate is supported in its mother's arms while taking milk from mammary glands that are sited on the mother's chest. The single pair of pectorally placed nipples can easily be reached by the young when they are supported in the mother's arms. Single offspring and ease of feeding are both advantageous in an arboreal habitat where rearing young is difficult for active primates.

In the higher primates, there is a long time between the birth of each offspring. This improves the chances of survival for the young who also gain the time to learn social skills important for the survival of the group. There is no breeding season in primates, ovulation cycles continue until pregnancy occurs. This increases the likelihood of conception. Sexual activity has become more than just an act of reproduction in primates. It also has many implications that affect the social structure of the group and strengthen the emotional bonds between animals.

Timu is an unusual gorilla. She is the first 'test-tube' baby gorilla. This reproductive technique might improve our chances of saving gorillas from extinction.

3 **Suggest how single births, firm grasp by the offspring, and the position of the mammary glands are adaptations to living in trees.**

In the wild, marmosets are constantly searching for food. They gnaw the bark off trees to get at the sticky sap underneath. A special feeding pole full of artificial sap was offered to captive-bred marmosets at Twycross zoo. However, the marmosets didn't gnaw into it. They hadn't learned the skills needed to find sap.

Social organisation

The social structure of primate groups varies from species to species. However, most primates live in family groups and the young remain with the adults for a relatively long time until they reach sexual maturity. This **extended dependency** offers two advantages:
- parental protection improves the survival rate of the young;
- the young learn by observing their parents' activities in a range of different situations.

Dr Louis Leakey initiated field research and the behavioural study of chimpanzee and gorilla groups. Most of our knowledge of the social organisation of these primates in the wild is from the work of Jane Goodall and Dian Fossey. Gorilla groups are discussed in Section 7.4.

Studying captive animals to learn about social behaviour is difficult because there is a risk that the animals are deprived of the natural stimuli on which they depend. For example, wild animals spend over 50% of their time foraging for food but zoos often provide food 'on a plate'. This means that a lot of time is unfilled – and bored animals soon develop behavioural problems. However, if these problems can be overcome, captive animal studies can also yield valuable information about social organisation as well as the biology of primates.

Fig. 4 Infant behaviour in squirrel monkeys

Dependent and independent activity

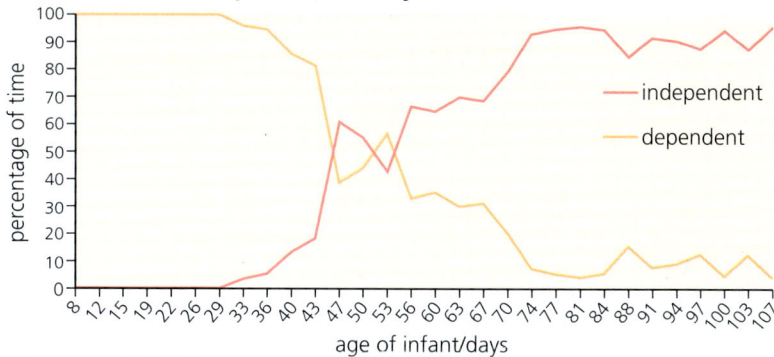

Time spent sleeping

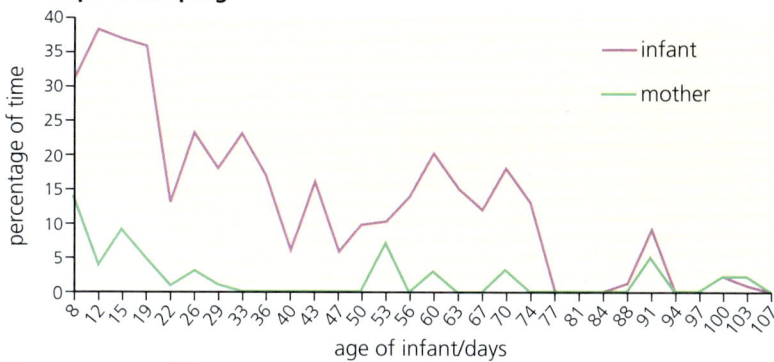

Time spent suckling

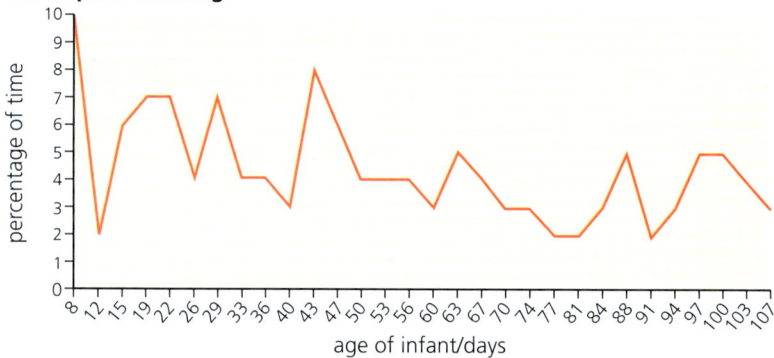

Mean activity level through the day

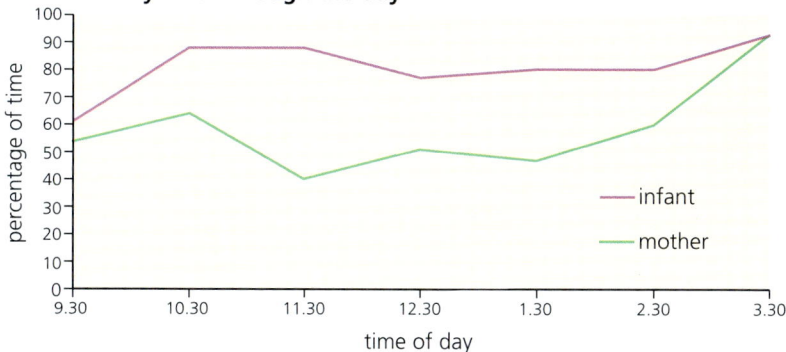

Source: Sonjha Kellock, unpublished research for BSc

Squirrel monkeys (*Saimiri sciureus*) at Twycross zoo.

Squirrel monkeys are born after 4 months gestation. They gradually become more independent and reach sexual maturity in about their third year. For the 3 years of this juvenile phase, the young monkeys are very active and playful. Adults tend to be quieter and less active. Squirrel monkeys can live for 23 years. They form loose groups of males and females with care of the young shared between the females.

4 Estimate the percentage of squirrel monkey life span taken up by (a) gestation, (b) juvenile phase (c) adulthood.

5 Assuming a 75-year life span for humans, what percentage of human life is spent in the same phases?

The Twycross squirrel monkey group consists of four captive-bred animals in one group. A student observed the female monkey and her newly born infant for 107 days and produced the graphs shown in Figure 4. Although only one female and her offspring were observed, this is the first information about young squirrel monkey development in captivity.

6 Describe the sleeping patterns of the mother and her offspring in Figure 4. Suggest a reason for the patterns.

Table 1 Behavioural firsts in squirrel monkeys		
Age in days	Behaviour first recorded	Level of dependence
8	reaching out and touching anything in reach from the mother's back	totally dependent on the mother
12	standing and walking on mother's back	totally dependent on the mother
22	pretend feeding, no food held in paws	totally dependent on the mother
29	attempted feeding, actually held some food in paws	totally dependent on the mother
33	standing beside mother	first independent activity
40	ate solid food	occasional independent activities
43	started to walk independently	independent activity increased
47	started playing, resting, grooming, touching; able to climb and run independently	independent activity increased dramatically

Young squirrel monkeys follow a pattern of behavioural firsts (Table 1).

Q 7 Give two advantages of extended dependency.

The biological features and reproductive characteristics inherited by individual primates are shaped by social interaction with others in the group and by the environment in which the animals live (Fig. 5). Knowledge of how animals live in their natural habitat is essential if we are to help endangered species in the wild or in captivity.

Bigger brains

All primates have a large brain compared with their body size. The cerebral cortex is highly developed and has specific areas that deal with visual imaging, sound interpretation and learning. Good memory and intelligent behaviour also depend on the cerebral cortex (Chapter 4).

The development of the cerebral cortex is linked to social development. Primates live in groups, interact socially, and learn from each other. Social interaction seems to be accompanied by an increase in brain size and, in particular, an increase in the cerebral cortex (Fig. 6).

Q 8 Study Figure 6. What is the relationship between relative cortex size and mean group size in primates?

Fig. 5 Individual development

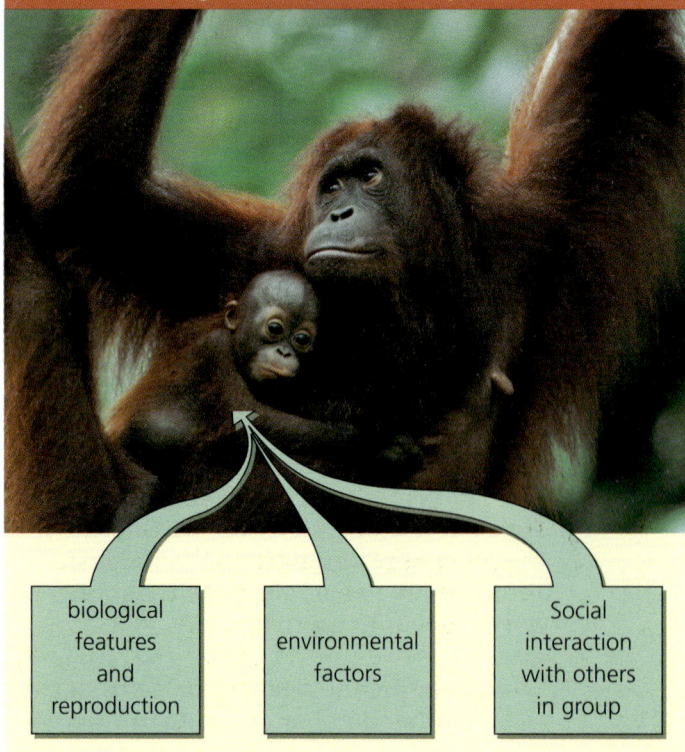

biological features and reproduction

environmental factors

Social interaction with others in group

Fig. 6 Social groups and the cerebral cortex

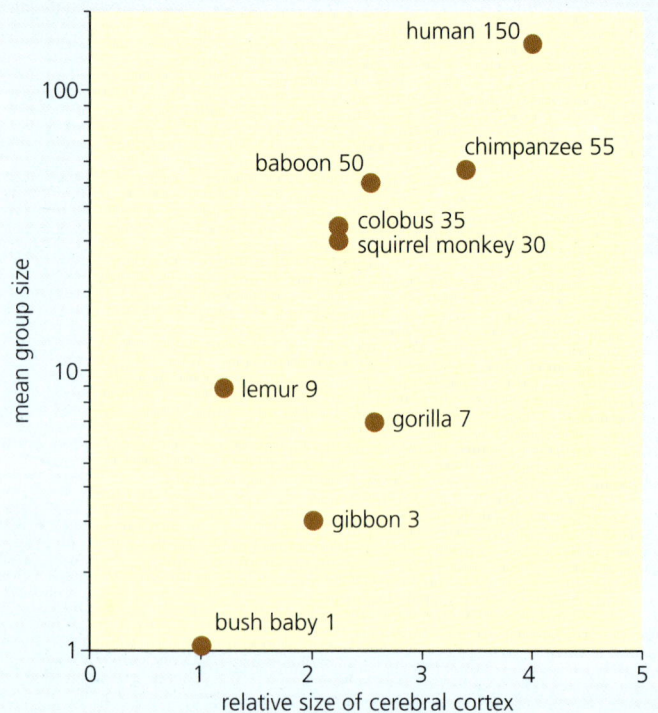

human 150
baboon 50
chimpanzee 55
colobus 35
squirrel monkey 30
lemur 9
gorilla 7
gibbon 3
bush baby 1

mean group size

relative size of cerebral cortex

Key ideas

- Primate features include: stereoscopic vision; dextrous ability; single offspring; a single pair of mammary glands on the chest; extended dependency; highly developed brains with a large cerebral cortex.

- Many of the physical features of primates are adaptations for living in trees.

- Social interaction, biological features and the environment together affect individual development.

7.3 Primitive primates

The fossil record shows that lemurs once occupied Africa, North America, and Europe, but now they are found only in Madagascar. They are prosimians, primitive social primates that mark their territory with scent produced by chest, genital and wrist glands. The scent is smeared on vegetation or, in some lemurs, impregnated in the bark of trees with the aid of sharp spurs on the wrist.

Lemurs keep in contact with each other by a range of shrill calls and use of the tail as a flag. These means of communication also announce the presence of a lemur group to other lemurs.

9 Where are lemurs found?

The ring-tailed lemur, *Lemur catta*, is adapted to life on the ground and has a mixed insectivorous and vegetable diet. It has a distinctive bushy black-and-white banded tail that can be charged with scent. Any wind then picks up and disperses the scent. Males wave their scent-laden tails to intimidate intruding lemurs. The air-borne chemical communication discourages the approach of rivals. This species also has the wet nose surrounded by bare skin that is characteristic of animals with a keen sense of smell. Ring-tailed lemur groups spend much of the day in close contact with each other and the act of scent marking is repeated by many animals in turn as the group moves along the forest floor or through the treetops.

Lemurs vary in their dependence on either sound or smell to communicate. The Indri has the loudest call of any lemur and 'sings' to proclaim its territory high in the tree tops. The Sifaka, in common with other more primitive primates, uses urine as an additional scent marker. The Indri and Sifaka are both herbivorous species that move easily through the treetops.

10 How do lemurs communicate?

Apart from the lemurs of Madagascar, the prosimians are almost all nocturnal. Examples are the tarsier in Asia, and the bushbaby in Africa. Large, forward-facing eyes give good vision in low light, but these animals rely heavily on scent and urine marking.

Key ideas

- Lemurs are primitive social primates that live in family groups and mark their territory with scent and, in some cases, urine.

- Some lemurs have large bushy tails that can be charged with scent to warn off intruders and rivals.

- Lemurs communicate by sound as well as by scent; species differ in the level to which they depend on either sound or scent.

- Other prosimians such as the tarsier and the bushbaby are nocturnal and have good night-vision. They use scent and urine marking.

7.4 Gorilla groups

In the mid-1960s Dian Fossey began almost two decades of work studying mountain gorillas. Her detailed observations offer a remarkable picture of the life and social structure of the mountain gorilla.

Mountain gorillas (*Gorilla gorilla beringei*) live in the rain forest of Zaire, Rwanda and Uganda. The climate is very wet but mild. Gorillas are the largest living primates and are herbivores; they spend nearly 70% of their time feeding on low-quality but abundant vegetation on steep, wooded slopes. The animals do not travel far, they move about half a kilometre a day in family groups. At night, they build individual nests on the ground or in the bases of trees using non-food vegetation. The younger, lighter gorillas tend to build their nests higher up in the trees. Facial expressions and body language convey a lot of information between gorillas, and gorilla sounds or **vocalisations** are used to communicate and maintain contact in dense vegetation. The males also leave scent from glands on their feet and palms as they move through the undergrowth, and a strong 'fear' scent is produced when under stress.

Observing a gorilla group

Dian Fossey was concerned that her assistants should be safe when making observations. In particular, she did not want them harmed by the gorillas. It was also extremely important that the observations provided valid data about gorilla behaviour. Consequently, she instructed her assistants to adopt certain behaviours when approaching a gorilla group:

- approach the group quietly and slowly;
- always approach with a bent back, never an upright posture;
- when close to the group, change to a crawling movement on all fours, with closed hands (knuckle-walking);
- be visible at all times;
- make throaty grunting 'contentment' noises;
- avoid eye contact particularly with the mature females or the eldest mature male;
- pretend to eat the vegetation.

Unlike the fictitious King Kong, real gorillas have a strict social code and are very passive animals.

Q 11 Suggest reasons for Dian Fossey's instructions to the observers.

Social structure

Groups of gorillas usually contain 10–20 animals and have a **hierarchy** or power structure. Each group has a dominant mature male who has grey hair on his back and flanks and is called a silverback. Only this male can mate with the mature females (a relationship known as polygyny). The silverback in any group is physically larger and heavier than the others (Table 2).

Immature gorillas remain with the group until they are sexually mature. Mature females either stay or emigrate to another group. Mature males become 'loners' or start new groups with unrelated females (Fig. 7).

Table 2 Gorilla statistics

Animals in group	Age/years	Number	Mass/kg
silverback male	15	1	170
immature males	8–13	1	110
mature females	over 8	3–4	90
young gorillas	under 8	3–6	
– young adults	6–8		75
– juveniles	3–6		55
– infants	up to 3		1–12

Q 12 a In Figure 7, what is the name of the silverback male in group 5?
b With which of his own female offspring did the silverback mate?

Fig. 7 Mountain gorillas at Visoke

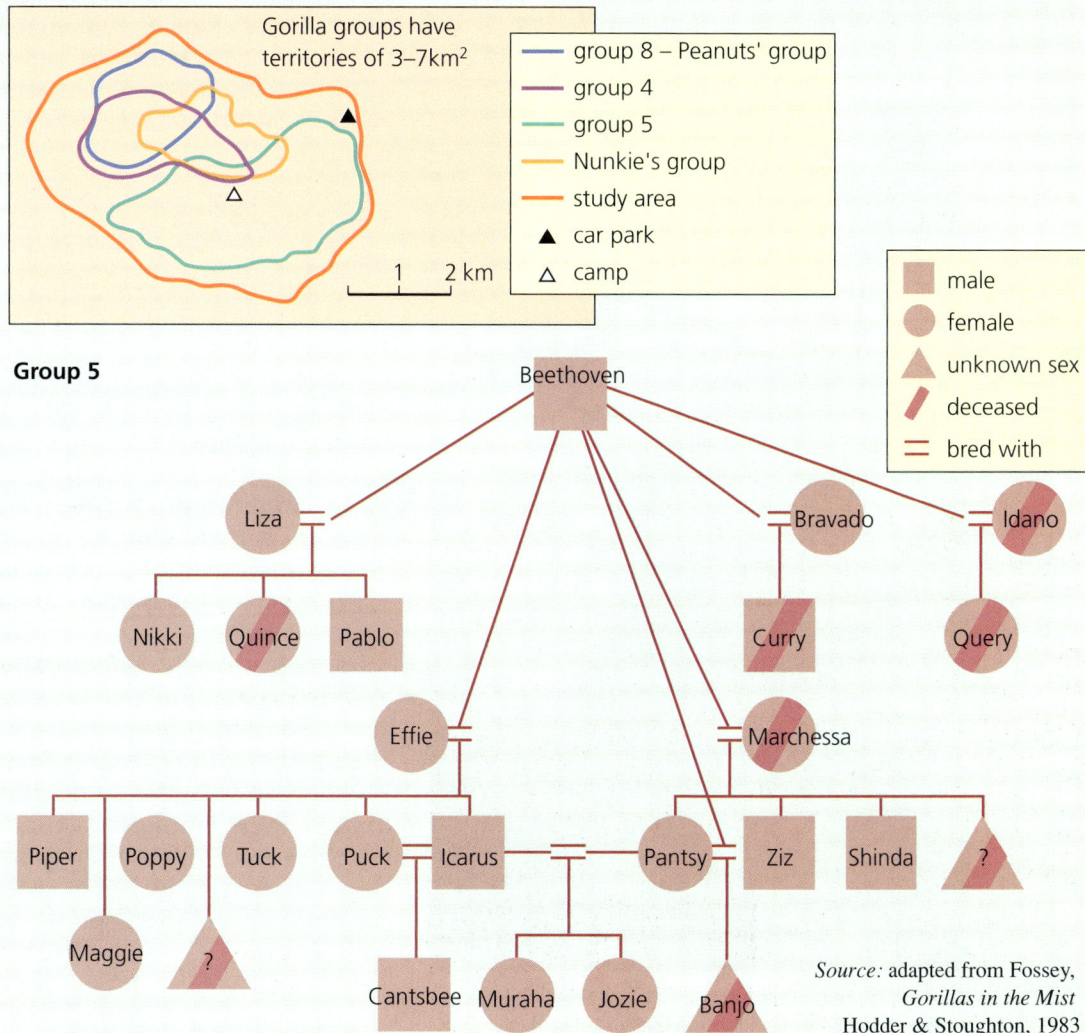

Gorilla groups have territories of 3–7km²

- group 8 – Peanuts' group
- group 4
- group 5
- Nunkie's group
- study area
- ▲ car park
- △ camp

1 2 km

Group 5

- ■ male
- ● female
- ▲ unknown sex
- / deceased
- = bred with

Source: adapted from Fossey, *Gorillas in the Mist* Hodder & Stoughton, 1983

Social activity

Grooming is an important social activity during which flakes of skin, parasites and plant debris are removed from the fur of one gorilla by another. It occurs between all gorillas in a group and is not just cleansing; it reinforces bonds between group members. Gorillas can recognise each other by sound and sight. Careful body inspection and low grunting noises seem to be very important during the long grooming sessions.

The silverback can physically exert his dominance over the others, although such activity is more often directed at intruders. The usual response of a group is to move away if disturbed, but when threatened, chest beating, roaring and charging by the silverback usually frightens off any aggressor. He is assisted by similar displays from older members of the group. This strengthens group identity by cooperation.

Male gorillas have sharp canine teeth, and bites can inflict serious damage. Signs of injury from fights with rival males are common in silverbacks.

Q 13 What does grooming achieve in gorilla groups?

Reproductive behaviour

Mountain gorilla females become sexually mature when they are 6–8 years old. They ovulate each month and are fertile for about 5 days. During this time they readily mate with the silverback and, if they conceive, will give birth 8 months later. The single offspring is cared for by the mother for about 2 years until it stops suckling. Like humans, gorillas have an extended period of dependency during which the mother helps and protects the young gorilla. This offers the young an opportunity to learn and develop social and other skills.

Mature females can give birth again after about 30 months, if they remain with the same group. Females who change groups might not conceive again for up to 60 months, a much longer **interbirth interval**. Females take time to adjust to life in a new group.

A test of observation

The interaction maps shown in Figure 8 were drawn from observations of four particular gorillas in a group: the silverback, a mature male, a mature female and a juvenile. The closeness of some other members of the group to these individuals was recorded over a period of time.

Q 14 Study Figure 8. Give a brief explanation of the presence and closeness of other gorillas around each of the gorillas studied.

Only just over 200 mountain gorillas are alive today and most of our knowledge about this endangered species came from Dian Fossey and her team of researchers. Poachers and habitat destruction are likely to drive these animals to extinction. Poachers are thought to have murdered Dian Fossey because of her enthusiastic commitment to gorilla protection and conservation.

Fig. 8 Gorilla social activity

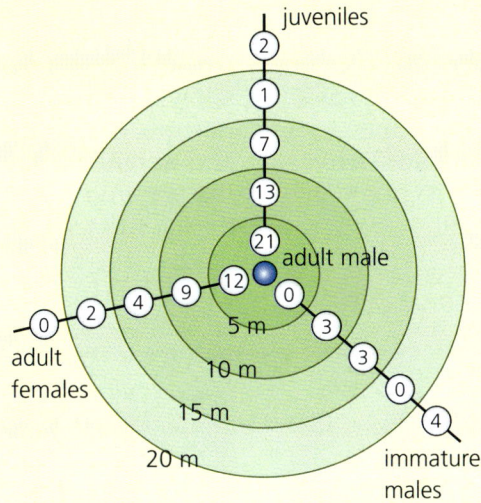

Interaction around an immature male

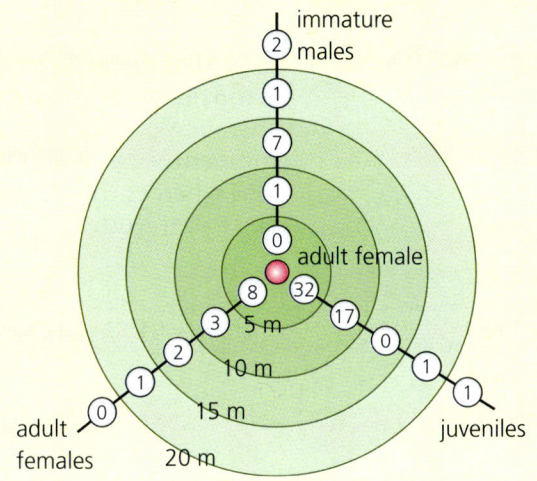

Interaction around a juvenile

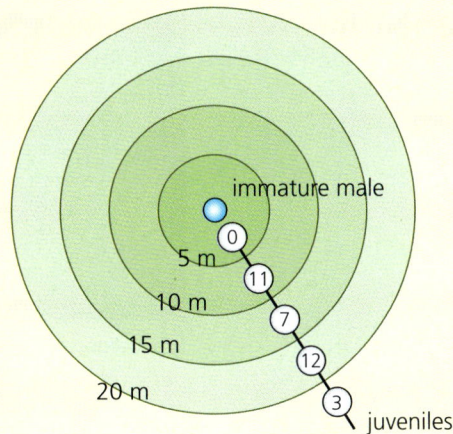

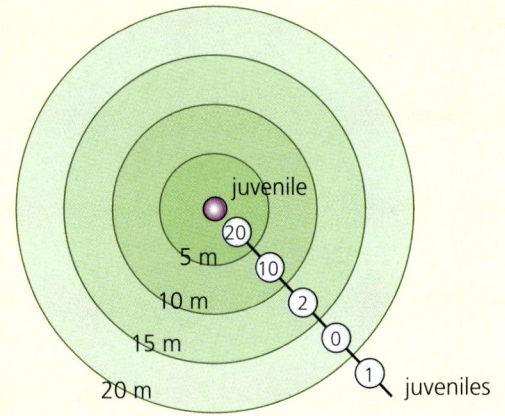

Source: adapted from Fossey, *Gorillas in the Mist*, Hodder & Stoughton, 1983

Key ideas

- Gorillas are herbivores and live in social groups.

- Gorillas groups have a hierarchical social structure with a single large dominant mature male (the silverback) and several smaller mature females. This is called polygyny.

- Gorilla groups have large ranges with plenty of low-quality food.

- Grooming, eye contact, and vocalising are important in the social structure of gorillas.

- Gorillas have single births and an extended period of dependency.

7.5 Bonobo groups

The bonobos studied in the wild by Jo Thompson are close relatives of the chimpanzee and were only recognised as a separate species in 1929. In Zaire, the chimpanzee occupies the more northerly areas whereas the bonobo has its home in the central western area (Fig. 9).

Bonobos are slender and weigh 40–45 kg when fully grown. Their limbs are long, and allow a sort of crouched upright walking posture. Males and females are a similar size and weight and live in groups containing mature animals of both sexes; they do not have a strict social hierarchy like the gorilla.

Fig. 9 Bonobo range

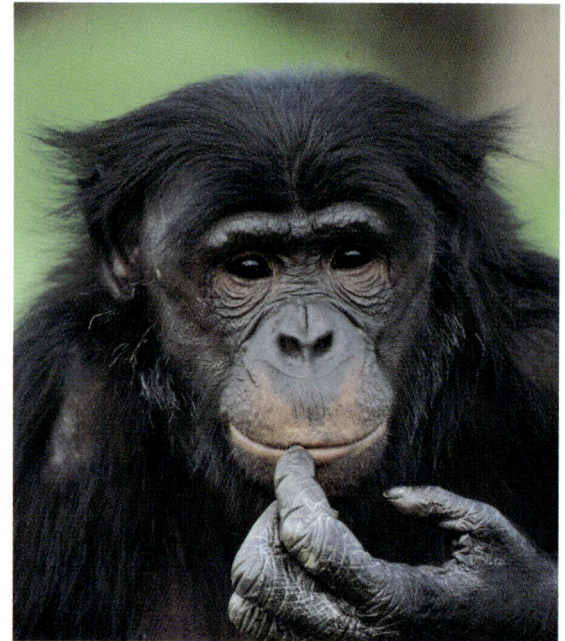

N

0   km   300

CENTRAL AFRICAN REPUBLIC

SUDAN

CONGO

Zaire

Lake Tumba

Lake Mai-Ndombe

Lake Albert

Kisangani

Lake Edward

UGANDA

Lake Kivu

RWANDA

BURUNDI

Kinshasa

Kibombo

TANZANIA

Lake Tanganyika

ANGOLA

Lake Mweru

ZAMBIA

Atlantic Ocean

AFRICA

ZAIRE

///// bonobo range
(disontinuous wild
population in forest
fragments scattered
throughout)

In bonobos, hair tufts frame a face with attentive gazing eyes.

Bonobos live for about 45 years and reach sexual maturity at about 8–10 years old. Females continue to be able to conceive until they are about 25. However, bonobos engage in sexual activity before sexual maturity, and this seems to be important in cementing relationships within the group. In females, very conspicuous, large pink sexual swellings are signals that they are receptive to mating. These are present for 23 days of the 35-day cycle, although conception can only occur for about 4 days. Females readily mate with all mature males in the group – this is called multimale polygyny.

Q 15 Give three differences between the social structure of bonobos and gorillas.

The future is very uncertain for the bonobo. There are several thousand of these animals in the wild, and just over 100 in captivity. Bonobos live in humid forests, but small and rapidly expanding human villages are not far away, and the main threat to bonobos is human activity, particularly in destroying the forest habitat. In Zaire, logging, land use, hunting, population growth, and civil unrest all add to the problem. However, any strategy to save the bonobo by protecting its environment must also take account of the needs of the villagers.

16 Why do expanding human populations have a big effect on other primate species?

The gorilla and the bonobo are among our closest animal relatives, yet we seem to be in danger of wiping them out. In zoos around the world these animals are cared for and studied, but these captive individuals might soon be all that remain of the species.

17 Do you think animals like the bonobo and the gorilla should be protected in the wild or kept in controlled environments like zoos? Why?

Key ideas

- Bonobos are close relatives of the chimpanzee.
- Male and female bonobos are similar in size and weight and do not have a strict social hierarchy. They live in mixed groups with a number of mature males.
- Expanding human populations are both a direct threat to bonobos (for example, by killing them) and an indirect threat (for example, by destroying habitat).

The pace of change

I'm researching human evolution for my social biology module. Humans are very different from their primate relatives. For instance, we use speech and language to communicate.

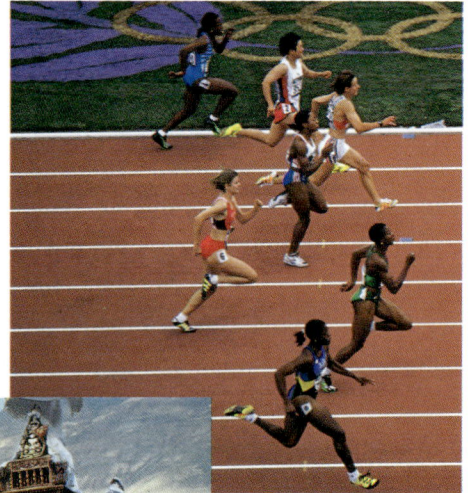

We are adapted for movement on two legs.

We have large brains and are intelligent. Our social structure is very complex and involves ceremony.

Human cultures and lifestyles vary enormously around the world. This is a festival in Peru.

8.1 Learning objectives

After working through this chapter, you should be able to:

- **recognise** the characteristics of a hunter–gatherer way of life;

- **explain** how fossils provide evidence about diet, brain size and locomotion;

- **describe** the principles of dating fossils;

- **recognise** that sparse fossil remains can be interpreted in different ways;

- **describe** the features of simple stone tools;

- **explain** how stone tools could be made;

- **suggest** uses for stone tools by looking at their features;

- **explain** the importance of the control of fire to early humans;

- **distinguish** between speech and language;

- **explain** the importance of language in complex societies;

- **distinguish** between fact and conjecture in reconstructing our cultural past.

8.2 Life as a hunter–gatherer

There are still a few small communities in the world who have a hunter–gatherer way of life. This means they live in extended family groups who cooperate in hunting meat and gathering vegetable food to ensure they survive the hostile conditions. The !Kung bush people of south-western Africa are such a community (Fig. 1).

Populations of the !Kung remain constant due to the balance of birth and death rates. Although life expectancy is relatively short in this harsh environment, control of fertility is more important. Extended suckling inhibits ovulation in women breast-feeding their babies. On average, a !Kung mother does not have another child for 4 years after the birth of her previous baby. Mothers leave the group to give birth alone. Babies with obvious disabilities could not be supported by the group and so are not brought back with the mother.

The !Kung are nomadic and move camp regularly. They must carry all their belongings, about 12 kg each. They use fire, and make a variety of effective hunting weapons, such as spears or bows and arrows and have a rich oral culture. Although they build shelters and make implements for preparing food, few of these are taken with them when they move camp.

Q 1 What is meant by the term 'hunter–gatherer'?

Food is collected every day. It is a social focus and is shared, there is no trading or food storage. There is a clear division of labour in the group (Table 1). The men hunt every day and travel up to 16 km from the camp, either in pairs or on their own. The women forage for plants for distances of up to 5 km from the camp. They usually carry their babies in the leather cape or kaross and are accompanied by children who can walk. All child-care activities fall to the women in the group.

Q 2 What factors limit the growth of the !Kung populations?

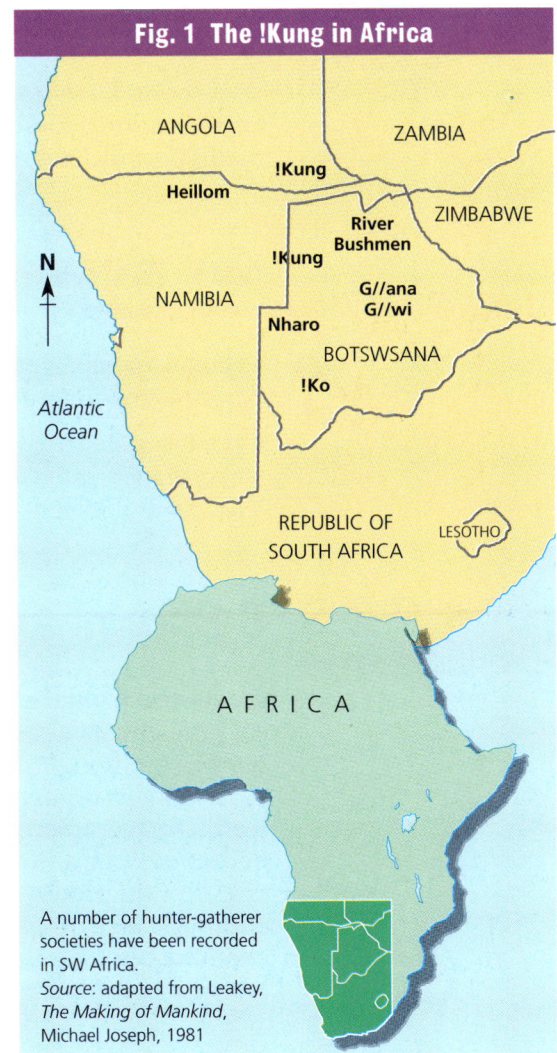

Fig. 1 The !Kung in Africa

A number of hunter-gatherer societies have been recorded in SW Africa.
Source: adapted from Leakey, *The Making of Mankind*, Michael Joseph, 1981

Table 1 Division of labour among the !Kung		
	Men	Women
Obtaining food	• hunting (about 21 hours per week)	• gathering (about 12 hours per week)
Food-related activities	• making weapons • exploration (up to 19 hours per week)	• collecting water and fuel • preparing food (up to 19 hours per week)
Total time spent in food-related activity	• about 40 hours per week	• over 40 hours per week

Source: adapted from Leakey, *The Making of Mankind*, Michael Joseph, 1981 (research by Lee)

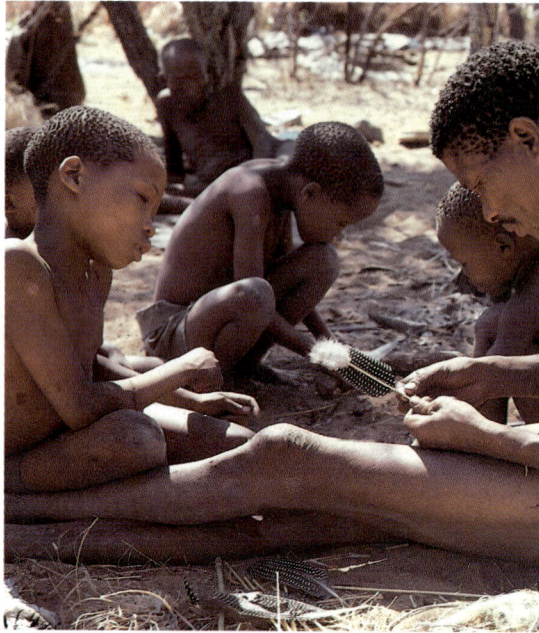

The !Kung speak an explosive 'click-language'. Clicks are shown as exclamation marks and accents.

A sound knowledge of the environment and of seasonal food sources is necessary to maintain a diet substantial enough to provide sufficient nutrients and energy for an active lifestyle. The !Kung are short and weigh less than 50 kg, yet they use over 12 000 kJ per day. This is about the same energy requirement as a northern European living in a much colder climate, and weighing 25% more. !Kung knowledge includes the habitats and food value of over 200 plants, and the herd movements of grazing animals. The mongongo nut is a particularly important food and contributes over half of the daily dietary needs.

Many fossil human ancestors have been discovered in Africa, and it is tempting to draw parallels between the hunter–gatherer life of the !Kung people and early African humans. This would not be justified because there have been too many changes. For instance, the climate is not the same as it was in the time of early humans and food plants are likely to be very different. The !Kung have bigger brains than early humans, and they have speech and language skills.

Key ideas

- Hunter–gatherers are nomadic and have few possessions. They do not store food.
- Hunter–gatherer groups depend on cooperation and division of labour to forage for plants and hunt for animals.
- Precise knowledge of the environment and a great diversity of food sources are required to ensure an adequate supply of food throughout the year.

8.3 Common ancestry

A common primate ancestor

All primates alive today, including humans, have a common primate ancestor in the past. Modern primates have many physical features in common, but these similarities do not indicate a timescale for the process of evolution. However, cellular chemicals and nuclear DNA do help in estimating evolutionary timescales. This is called **molecular evidence**.

The more closely related two species are, the more cellular chemicals they have in common and the fewer the differences between their nuclear DNA. Differences in DNA structure are caused by mutation. In the 1980s, researchers realised that with an estimate of the *rate* of mutation, it would be possible to use molecular evidence to estimate when two species separated from a common ancestor (Fig. 2).

Fig. 2 Primate ancestry

premolecular evidence   postmolecular evidence

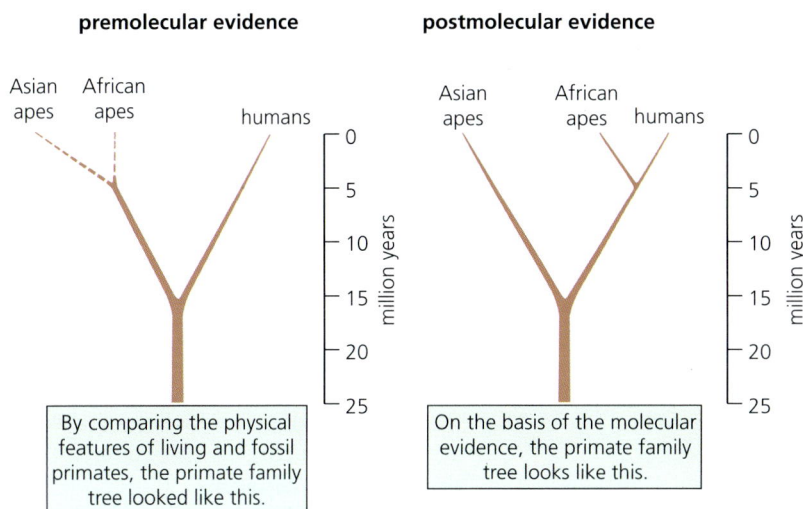

By comparing the physical features of living and fossil primates, the primate family tree looked like this.

On the basis of the molecular evidence, the primate family tree looks like this.

Source: adapted from Leakey, *The Making of Mankind*, Michael Joseph, 1981

Fig. 3 Inheritance of mitochondrial DNA

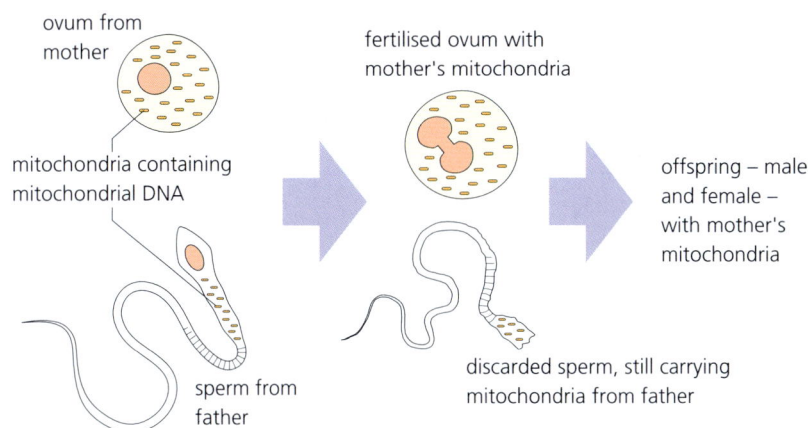

ovum from mother

mitochondria containing mitochondrial DNA

fertilised ovum with mother's mitochondria

sperm from father

discarded sperm, still carrying mitochondria from father

offspring – male and female – with mother's mitochondria

Source: adapted from Lewin, *Human Evolution, an Illustrated Introduction*, 2nd edn, Blackwell Scientific, 1989

3 Study Figure 2. How has the evolutionary relationship of humans and apes changed in the light of molecular evidence?

A common human ancestor

In general, sexual reproduction combines genetic material from two parents. However, not all of our genes are inherited as a mix from both parents. A fragment of DNA that codes for mitochondria is inherited only from the mitochondria in the egg cell (Fig. 3).

The mutations in mitochondrial DNA around the world enable us to calculate when and where this DNA first appeared in *Homo sapiens*. We can actually give a date and a place for a woman known as 'mitochondrial Eve' from whom this DNA was inherited. Our 'ancestral mother' was somewhere in Africa fewer than 200 000 years ago. This is a very short time for modern humans to have spread all over the world and scientists are still arguing about these ideas and the research methods used.

4 Some recent research seems to indicate that mitochondria from the sperm *do* enter the egg at fertilisation. Give two ways this will affect the 'mitochondrial Eve' theory, if it is true.

Key ideas

- Molecular evidence can be used to compare the relatedness of primates alive in the world today.

- Human mitochondria contain DNA that is inherited only from the mother and has provided evidence that all living humans are all very closely related; they share a common ancestor who lived as recently as 200 000 years ago.

8.4 What can fossils tell us?

Information about locomotion

The way an animal moves, its diet and its brain size are all reflected in its skeleton. Chimpanzees have a range of movements. They can swing through the trees, using their arms. They can grasp branches with their hands and equally well with their feet. On the ground, they have a stooped walk, taking their weight on flattened feet and the knuckles of their closed hands – this is called knuckle walking. They can also walk in a waddling fashion on their hind limbs. In contrast, adult humans always walk upright on their hind legs. This is called bipedal movement or **bipedalism**.

A skull can provide information about locomotion. The **foramen magnum** is the hole in the skull through which the spinal cord passes. Its position indicates the position of the skull relative to the spine. A skull which has a central foramen magnum must have been centrally balanced on the spinal column, a position that indicates bipedal movement. The skull and skeletal structures of humans and chimpanzees reflect differences in locomotion (Fig. 4).

Q5 Which parts of a fossil skeleton would you need in order to decide whether or not it was bipedal?

Fig. 4 Skeletons and movement

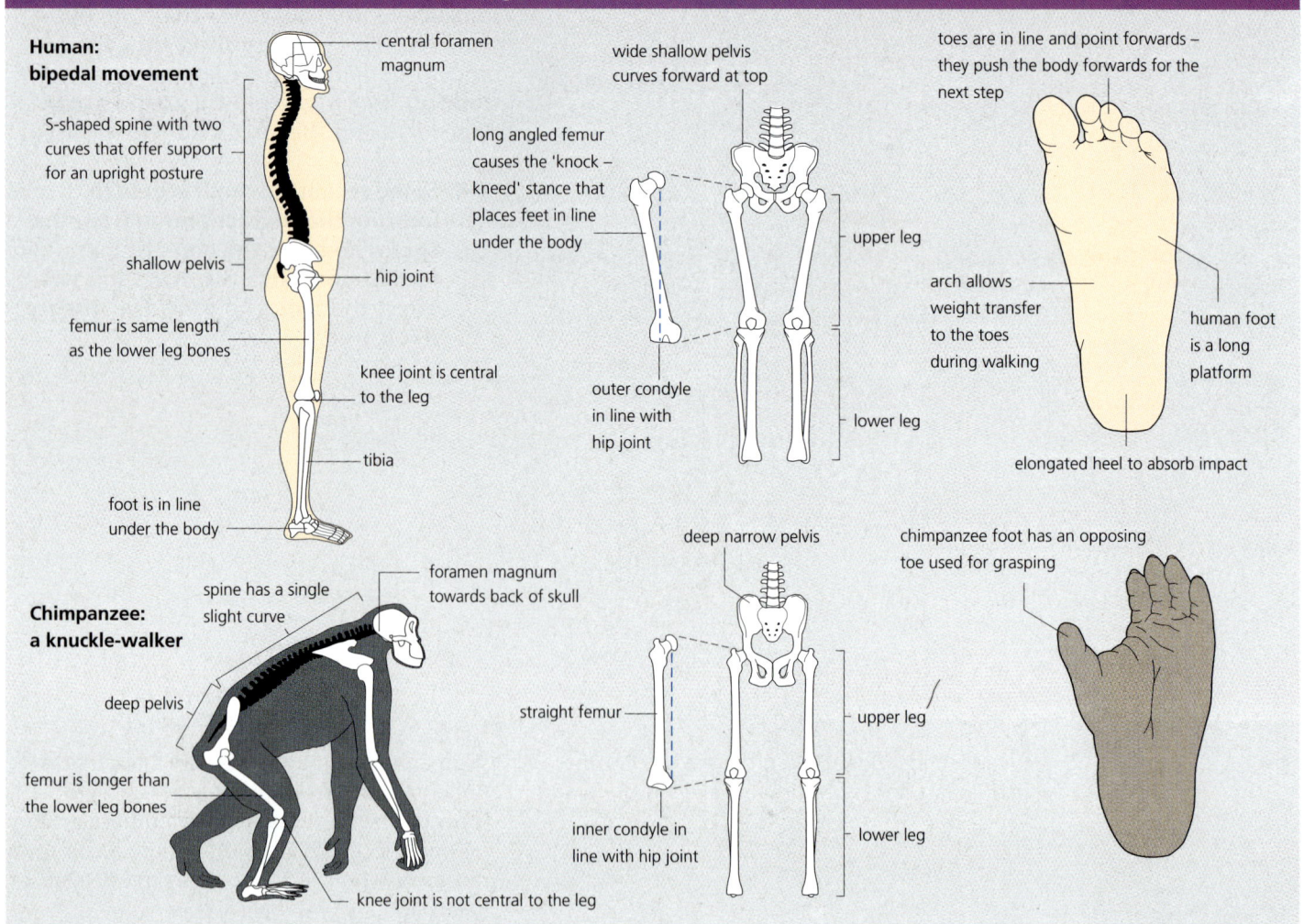

Human: bipedal movement

- central foramen magnum
- S-shaped spine with two curves that offer support for an upright posture
- shallow pelvis
- hip joint
- femur is same length as the lower leg bones
- knee joint is central to the leg
- tibia
- foot is in line under the body

- wide shallow pelvis curves forward at top
- long angled femur causes the 'knock-kneed' stance that places feet in line under the body
- upper leg
- outer condyle in line with hip joint
- lower leg

- toes are in line and point forwards – they push the body forwards for the next step
- arch allows weight transfer to the toes during walking
- human foot is a long platform
- elongated heel to absorb impact

Chimpanzee: a knuckle-walker

- spine has a single slight curve
- foramen magnum towards back of skull
- deep pelvis
- femur is longer than the lower leg bones
- knee joint is not central to the leg

- deep narrow pelvis
- straight femur
- upper leg
- inner condyle in line with hip joint
- lower leg

- chimpanzee foot has an opposing toe used for grasping

Information about brain size

The brain is surrounded by bone. Measuring volume of the 'brain box' or cranium gives an indication of brain size. Brain size is also related to other features of the skull. For instance, a chimpanzee has quite a small brain compared to the size of its body; it also has a low forehead and prominent brow ridges. Humans have relatively large brains, and to accommodate them have a large domed cranium, a vertical forehead and no brow ridges.

6 Suggest how you could get an estimate of brain volume from a complete fossil skull using dry sand.

Information about diet

The size, shape and spacing of teeth are good indicators of diet in an animal (Fig. 5). For instance, large equally sized incisors which protrude outwards and the presence of a **diastema**, a gap between the canine and incisors, is very common in animals that live on soft vegetable matter such as fruit. Large jaws with big molars covered by thick enamel are found in animals that only eat tough fibrous vegetation. Very large canines and jagged molars are often indicators of a meat-eating or **carnivorous** animal. However, in some species, large canines are for defence or dominance. Modern human **dentition** (tooth type and arrangement) is typical of an **omnivorous** animal, that is an animal that eats both plants and animals.

7 Study Figure 5. Briefly describe the differences between the jaws and teeth of chimpanzees and humans.

8 What sorts of information can fossil bones and skulls provide?

Fig. 5 Teeth and jaws

Human

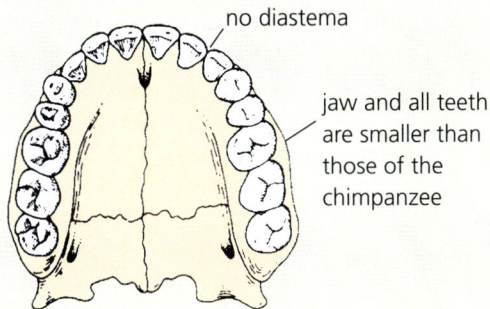

no diastema

jaw and all teeth are smaller than those of the chimpanzee

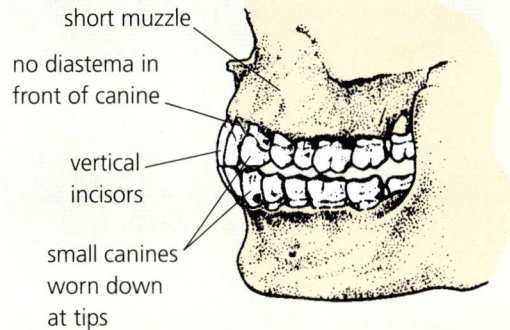

short muzzle

no diastema in front of canine

vertical incisors

small canines worn down at tips

Chimpanzee

diastema

canine to hold or tear food

incisors to bite off food

premolars

to grind and chew food

molars

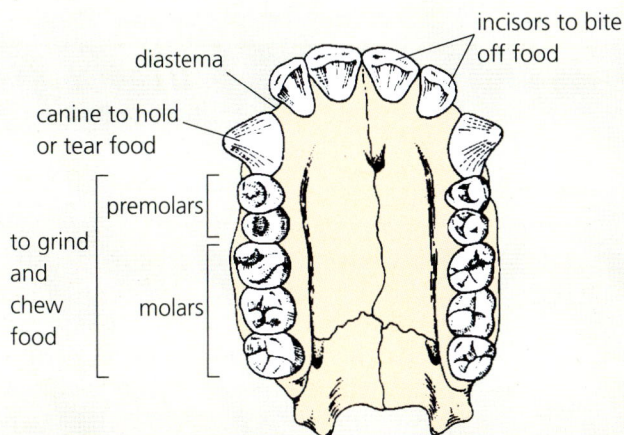

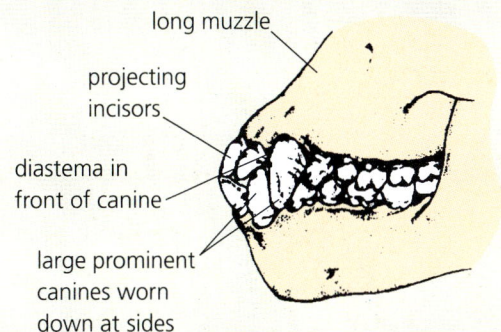

long muzzle

projecting incisors

diastema in front of canine

large prominent canines worn down at sides

Source: adapted from Leakey and Lewin, *Origins Reconsidered*, Little, Brown & Co, 1992

Key ideas

- Fossilised spines, femurs and pelvic bones tell us about the animal's locomotion, in particular whether or not the animal was bipedal.

- A skull that has a central foramen magnum indicates bipedal movement.

- Fossil skulls reflect differences in brain size (shown by the size of the cranium and angle of the forehead).

- Fossilised jaws and teeth provide information about diet.

8.5 Dating the past

It is important to know how old a fossil is. There are two main groups of methods: **relative dating** and **absolute dating**.

Relative dating

Stratigraphy is the study of rock layers and is based on the principle that older rocks lie below younger ones. Although the picture is confused by folding and faults, rock layers can be listed in age order. So, the relative date of a fossil can be judged according to the rock layer in which it was found.

Absolute dating

Finding the definite age or **absolute date** of a rock or fossil is more difficult, but some minerals have built in 'clocks'. Absolute dating can be carried out by:
- carbon dating;
- potassium dating;
- fission tracks.

Once a fossil is buried, ^{14}C (carbon 14) slowly changes to ^{14}N (nitrogen 14). However much ^{14}C is present, it takes 5730 years for half of it to change to ^{14}N. So,

5730 years is called the half-life. A specimen that contains carbon and is between 10 000 and 50 000 years old can be dated by measuring the amount of ^{14}C present.

Similarly, potassium dating measures the decay of ^{40}K (potassium 40) to ^{40}Ar (argon 40). The half-life is 1.3 billion years. Age can be determined to an accuracy of ± 200 000 years but specimens must be more than 500 000 years old for this method to be reliable.

Fission tracks are minute etched traces left by decaying radioactive atoms such as ^{238}U (uranium 238). They are found in volcanic rocks. The age of the rock is determined by linking the number of tracks to the decay rate of the radioactive elements. Specimens must be over 10 000 years old and must contain volcanic glass or the mineral zirconium to record the tracks.

9 Which technique you would use to date (a) a specimen that is about 4 million years old and (b) a fossil that is less than 40 000 years old?

Key ideas

- There are two types of dating, relative and absolute.

- Relative dating depends on comparing geological strata.

- There are three absolute dating techniques, all depend on the slow decay of one element to another. The different methods are suitable for specimens of different ages.

114

8.6 Human ancestors

There are many sites around the world where fossils and other remains have been discovered, but Africa is home to the oldest finds. This suggests that the story of human evolution began there.

The Great Rift Valley runs for 3000 miles along the length of Africa. It is here that many important discoveries have been made (Fig. 6). Bones and teeth can be preserved in sediments. This is because rather than rotting like the rest of the body, or becoming a meal for a scavenger, bones might settle in mud at the bottom of rivers or lakes and, over millions of years, become covered with other deposits. Most are lost forever, but some fossils are eventually exposed, and some of those are discovered before they are washed away by the torrential rain of the wet season.

Apes and humans are thought to have evolved from a common ancestor. Early hominid fossils are more similar to apes than to modern humans: they have a snout, a sloping forehead and a small brain. Later finds are more like modern humans. Two genera of hominids have been described, *Australopithecus* and *Homo*.

Fig. 6 Hominid fossil sites in Eastern Africa

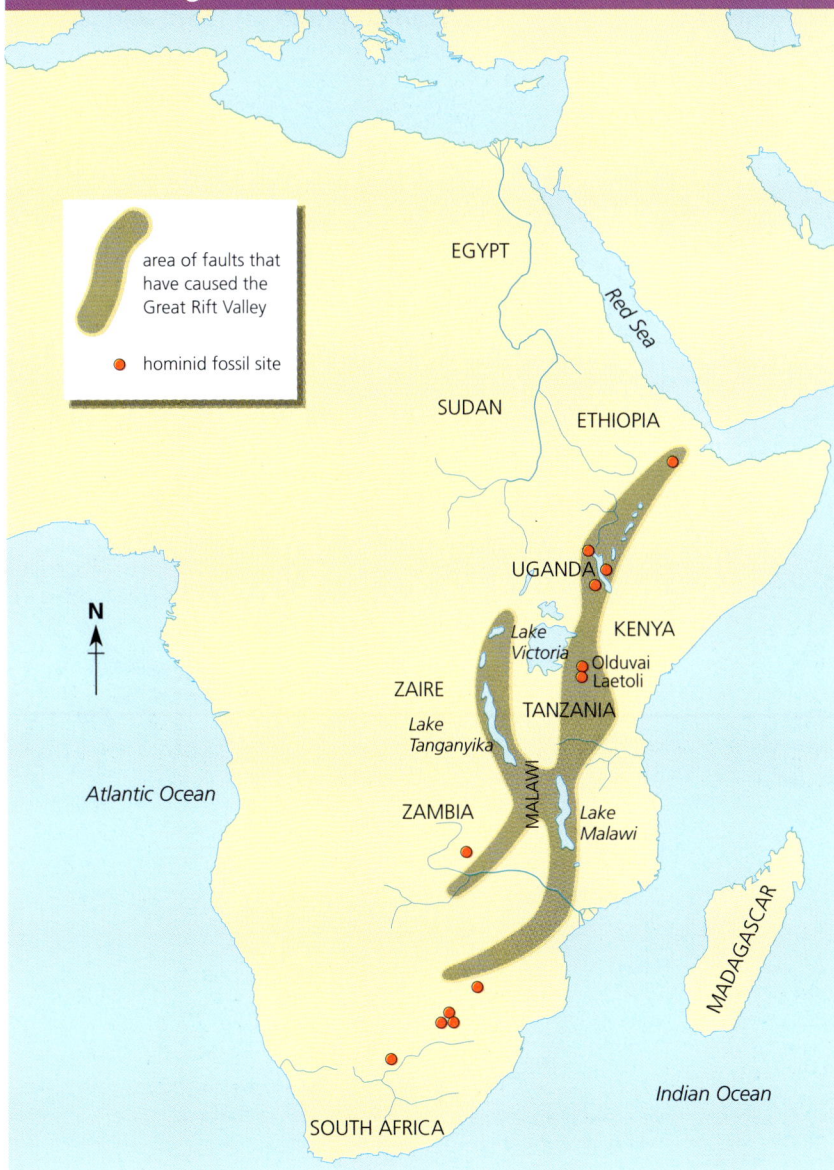

area of faults that have caused the Great Rift Valley

hominid fossil site

EGYPT

Red Sea

SUDAN

ETHIOPIA

UGANDA

KENYA

Lake Victoria

Olduvai
Laetoli

ZAIRE

TANZANIA

Lake Tanganyika

N

MALAWI

ZAMBIA

Lake Malawi

Atlantic Ocean

MADAGASCAR

Indian Ocean

SOUTH AFRICA

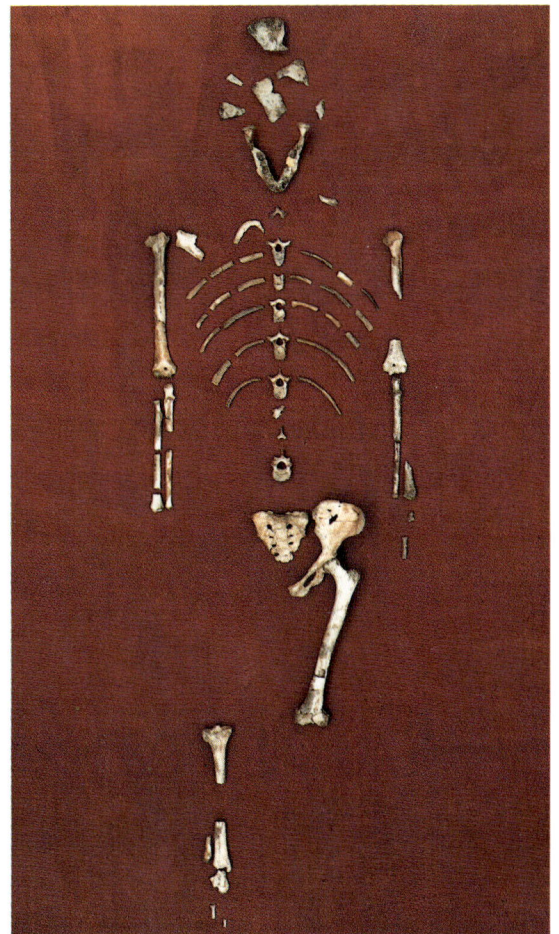

'Lucy' is a 40% complete skeleton of a female *Australopithecus afarensis*; she is almost 4 million years old.

Genus *Australopithecus*

We think that species of *Australopithecus* lived between 5 and 1.5 million years ago. The different species all possessed features between those of apes and humans. In particular, these individuals were bipedal and had a small brain (Fig. 7).

10 Study Figure 7. Compare the skulls in terms of: (a) profile of the face, (b) shape and size of the brain box (cranium), and (c) teeth and jaw shape.

Australopithecus afarensis and *A. africanus* are 'light bodied' and are described as **gracile**. *A. robustus* is 'heavy bodied' and described as **robust**. *A. robustus* has large molars with thick enamel to eat a diet of tough vegetation.

Genus *Homo*

There are three species of the genus *Homo* in the fossil record. *H. habilis*, *H. erectus* and *H. sapiens* (in that order) span some 2.2 million years to the present day and are believed to form an evolutionary sequence because they show an increasing brain size compared with their body size, a flatter face with a reduction in the size of the teeth and a smaller, compact, lower jaw (Fig. 8).

11 Study Figure 8. Describe the trends from *H. habilis* through *H. erectus* to *H. sapiens* in (a) profile of the face, (b) shape and size of the brain box (cranium), and (c) teeth and jaw shape.

Fig. 7 Skulls of genus *Australopithecus*

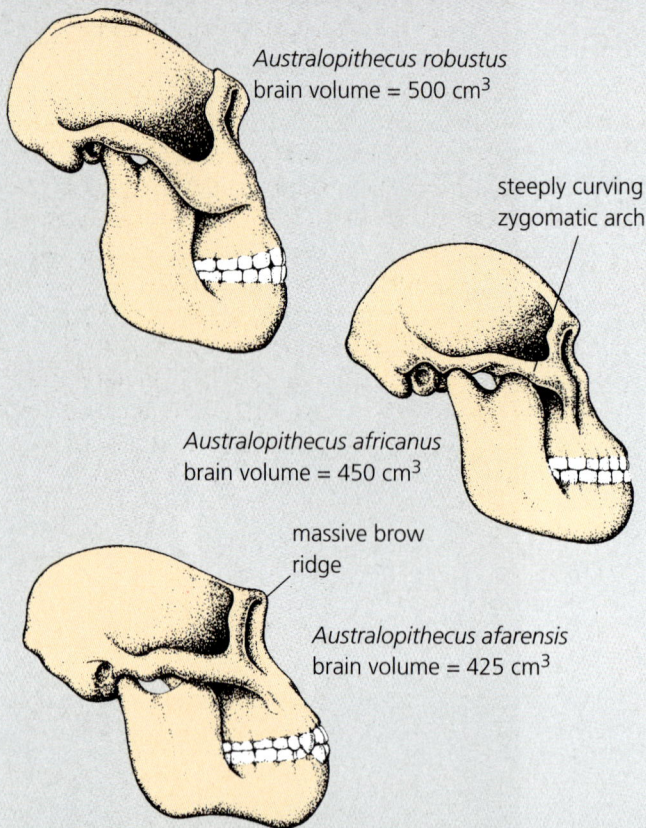

Australopithecus robustus
brain volume = 500 cm^3

steeply curving zygomatic arch

Australopithecus africanus
brain volume = 450 cm^3

massive brow ridge

Australopithecus afarensis
brain volume = 425 cm^3

Source: adapted from Tomkins, *The Origins of Mankind*, Cambridge Social Biology Topics, Cambridge University Press, 1984

Fig. 8 Skulls of genus *Homo*

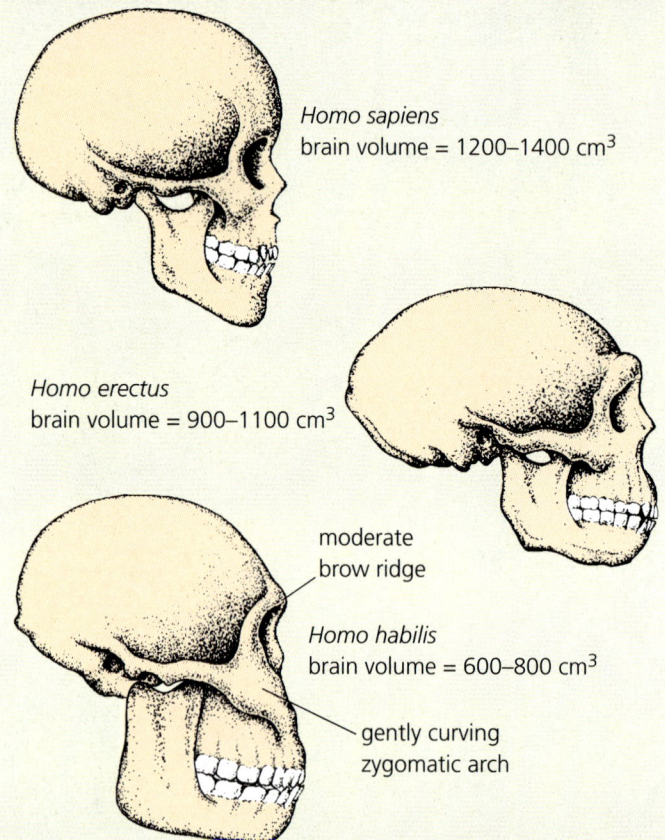

Homo sapiens
brain volume = 1200–1400 cm^3

Homo erectus
brain volume = 900–1100 cm^3

moderate brow ridge

Homo habilis
brain volume = 600–800 cm^3

gently curving zygomatic arch

Source: adapted from Tomkins, *The Origins of Mankind*, Cambridge Social Biology Topics, Cambridge University Press, 1984

Fig. 9 Two views of the fossil record

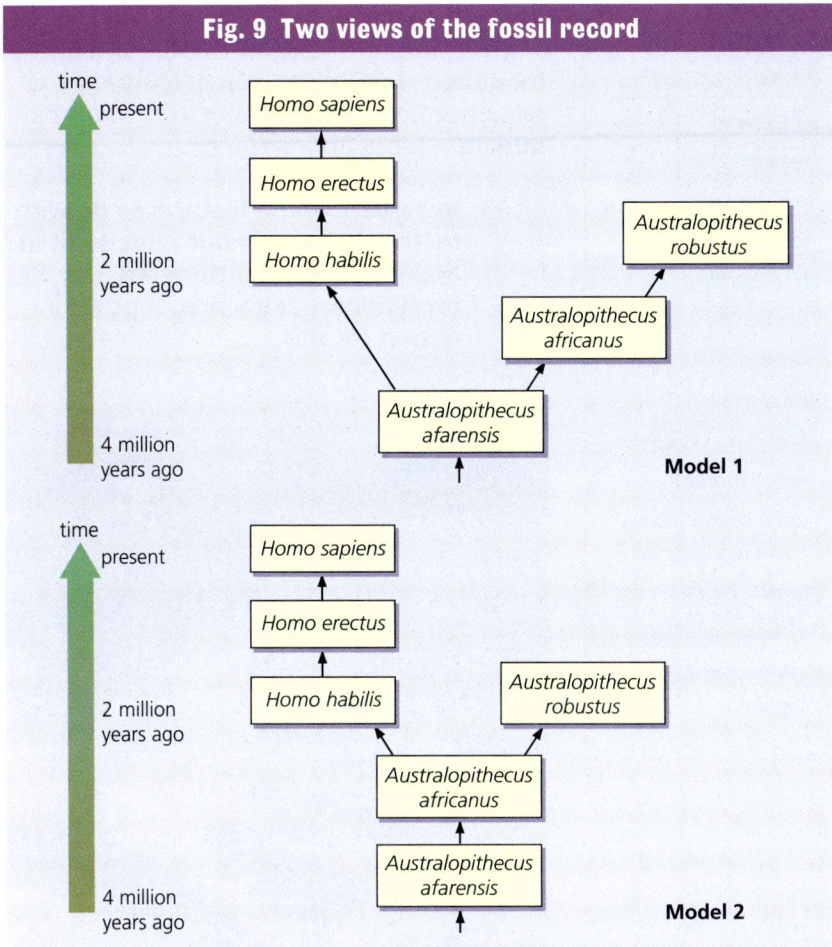

Model 1

Model 2

Fig. 10 A modern view of evolutionary relationships

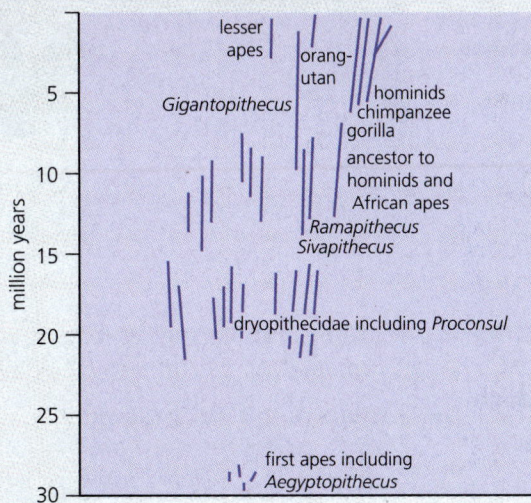

Source: adapted from Lewin, *Human Evolution, an Illustrated Introduction*, 2nd edn, Blackwell Scientific, 1989

H. habilis never left Africa, but it is with *H. habilis* remains that we find the first evidence of tool making. *H. erectus* seems to have been the first hominid to leave Africa for the colder climates of Europe and Asia. *H. erectus* has a fossil history of 1 million years. *H. sapiens*, the modern human, only appears in the fossil record from around 200 000 years ago.

It is important to remember that fossil finds can be interpreted in different ways and are always open to reinterpretation. We can begin to build a picture of human evolution over the last 4 million years by piecing together information gleaned from fossil finds. But fossils represent only a few individuals; many elements of the evolutionary picture are missing (Fig. 9).

12 Look again at Figures 7 and 8. Which features of the skulls could suggest the ancestor of *Homo habilis* was
(a) *Australopithecus afarensis*
(b) *Australopithecus africanus*?

Evolutionary trees and bushes

Until recently, anthropologists regarded primate evolution as a direct route to modern humans, and usually depicted it as a tree with relatively few branches, some of which were evolutionary 'dead ends'. As more evidence has accumulated, the path of evolution has become *less* clear, and the usual depiction now resembles a bush rather than a tree. The links between many of the branches are debatable and there are probably still many undiscovered links between the apes and *Homo sapiens* (Fig. 10).

Table 2 Fossil evidence for an evolutionary sequence			
Feature	Ape-like ancestors	*Australopithecus*	*Homo*
Teeth	thin enamel	thick enamel	thin enamel
	large incisors		reduced incisors
	incisors stick out	incisors more vertical	vertical incisors
	prominent canine	less prominent canine	reduced canine
	large diastema	small diastema	no diastema
	small molars	large molars	small molars
Jaws	fairly robust	very robust	not robust
	prominent snout	reduced snout	flatter faced
	front of jaw square	front of jaw V-shaped	front of jaw U-shaped
Skull shape	sloping forehead	less sloping forehead	vertical forehead
	small brain volume relative to body size	larger brain volume relative to body size	very large brain volume relative to body size
	brow ridges	reduced brow ridges	no brow ridges
Other	long arms, short legs	longer arms	short arm, long legs
	short femur	longer femur	very long femur
	knuckle walkers	more bipedal	bipedal

For the moment, *Australopithecus afarensis* and *Australopithecus africanus* are thought to form part of an evolutionary sequence from ape-like ancestors to human humans (Table 2).

Q 13 In Table 2, some features to do with teeth and jaws do not seem to fit in as part of a straightforward evolutionary sequence. Suggest a reason for this.

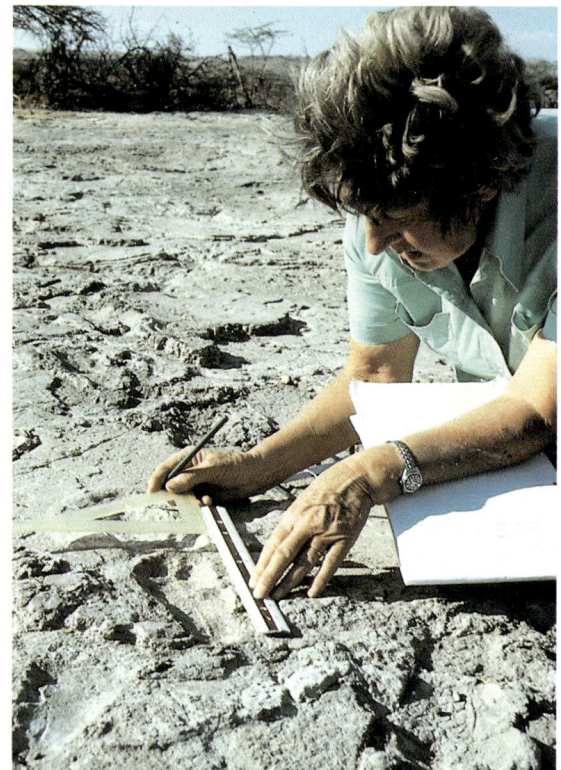

Mary Leakey found these 3.7-million-year-old footprints at Laetoli in 1976. An adult about 1.3 m tall walked along with a youngster alongside and a smaller individual walked behind in the footprints of the adult. All were bipedal. Their footprints were left in damp ash from a volcano. After the rain, the footprints set and were covered by dust.

Key ideas

- Human evolution probably began in Africa. Very few of the millions of individuals who lived are fossilised and fewer still are found.

- Fossil finds can be interpreted in different ways.

- Two hominid genera are recognised in human evolution, *Australopithecus* and *Homo*.

- There are three species in genus *Homo*. *H. habilis*, *H. erectus* and *H. sapiens*.

- In the past it was believed that evolution was a straightforward process leading to the appearance of *Homo sapiens*. With the increasing amount of fossil evidence, the picture has become less clear, although some evolutionary sequences seem to be clearly defined.

8.7 Tool technology

Tools are used by many animals to do particular tasks. This chimpanzee is using a stick to fish for termites.

Humans use tools to make new tools – this is taking tool technology a step further than other animals.

Oldowan tools are simple choppers and scrapers.

The physical changes of human evolution are accompanied by the development of new skills and by cultural change. Physical changes can be assessed from fossil evidence, and missing links conjectured or estimated according to what we already know. However, the skills and abilities which our human ancestors acquired can only be inferred from tools and other artefacts found with human remains.

It seems that during the last 2 million years of human evolution, a new skill developed: the ability to make tools. Thousands of stone tools have been found at archaeological sites and we can work out from the tools themselves how they were made and used. The finds are grouped into **tool cultures**, which represent distinct phases in the development of tool-making skills. So, we can build a picture of how tool manufacture developed.

Each tool culture lasts for thousands of years, so finds are spread over large areas. Tool cultures do not necessarily happen at the same time all round the world. However, there does seem to have been communication between hominid groups, and a spread of ideas between them. As each tool culture gives way to the next, the types of tool and the skills needed to make them show increasing complexity.

The most significant tool cultures are:
• Oldowan;
• Acheulian;
• Mousterian;
• the upper Palaeolithic cultures.

Oldowan tools

These crude chopping or scraping tools are typical of the first tool user, *Homo habilis*. A hard pebble can be shaped at one end by a series of blows against another stone. This produces a jagged cutting edge with a round base that sits comfortably in the hand.

Acheulian tools

Development of the Oldowan technique led to the production of Acheulian hand axes that are pointed, symmetrical and have two cutting edges. These tools are also sharper because material is removed from both sides of the cutting edges leaving thinner profiles.

Mousterian tools

Mousterian tool culture produced fine work due to the development of a 'retouching' technique. This involved making very delicate blows to flake off small pieces of stone along the cutting edges to produce even sharper blades. Using good quality flint, a variety of arrow heads and spear points were made along with hand-held scrapers for butchering meat.

Their ability to make these effective tools and weapons gave *Homo erectus* and early *H. sapiens* the chance to colonise the globe as hunter–gatherers.

Upper Palaeolithic tools

There were several tool cultures during the upper Palaeolithic period. They all have a two-stage production process (Fig. 11). This process makes extremely efficient use of flint, a rare commodity in some areas. So, a community using this technique would have a distinct advantage over one using more wasteful methods.

14a Study Figure 12. Which tool culture do the drawings (i), (ii) and (iii) represent?

b Match the tool drawings (iv), (v), (vi) and (vii) to the descriptions.

Acheulian hand axes (right) have two double-sided cutting edges.

This flint spear point is from an upper Palaeolithic tool culture. Upper Palaeolithic tools also include fine intricate arrow heads and knives. The development of tools was very rapid in this period, and increasing use was made of bone and ivory for the more delicate implements.

Using a stone tool such as a Mousterian scraper (left) to slice meat from a bone leaves telltale cut marks that clearly distinguish the remains of hominid meals from bones accumulated by hyenas or other carnivores. On the right is a Mousterian hand axe.

Fig. 11 Upper Palaeolithic tool manufacture

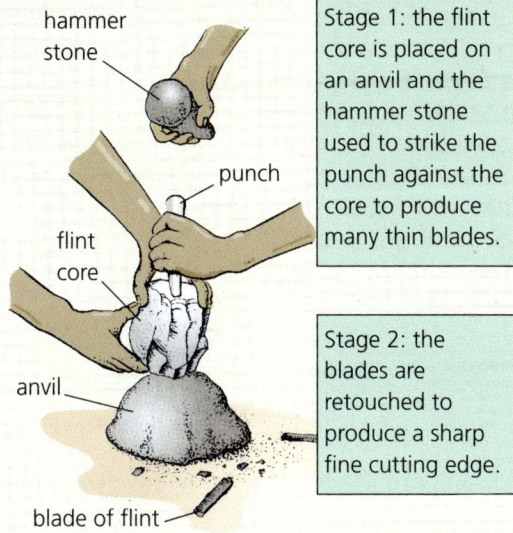

hammer stone

punch

flint core

anvil

blade of flint

Stage 1: the flint core is placed on an anvil and the hammer stone used to strike the punch against the core to produce many thin blades.

Stage 2: the blades are retouched to produce a sharp fine cutting edge.

Source: adapted from Tomkins, *The Origins of Mankind*, Cambridge Social Biology Topics, CUP, 1984

Social implications

Tool making requires cooperation to find suitable flint and, by upper Palaeolithic times, to work the flint to make tools (look again at Fig. 11). So, these activities offer a social focus as well as being essential for survival. But there may be other social implications. Tools are generally functional, but some blades have been found that show artistic decoration, others show no signs of use, and some are so thin and delicate that they are translucent. We can speculate that these might have been made for ceremonial purposes or to perhaps to demonstrate tool-making skills.

Burning questions

Bush fires, lightning and volcanoes could all provide a natural source of fire. Evidence from China and from Africa suggests that *Homo erectus* was the first hominid to use fire.

Fig. 12 Tool cultures

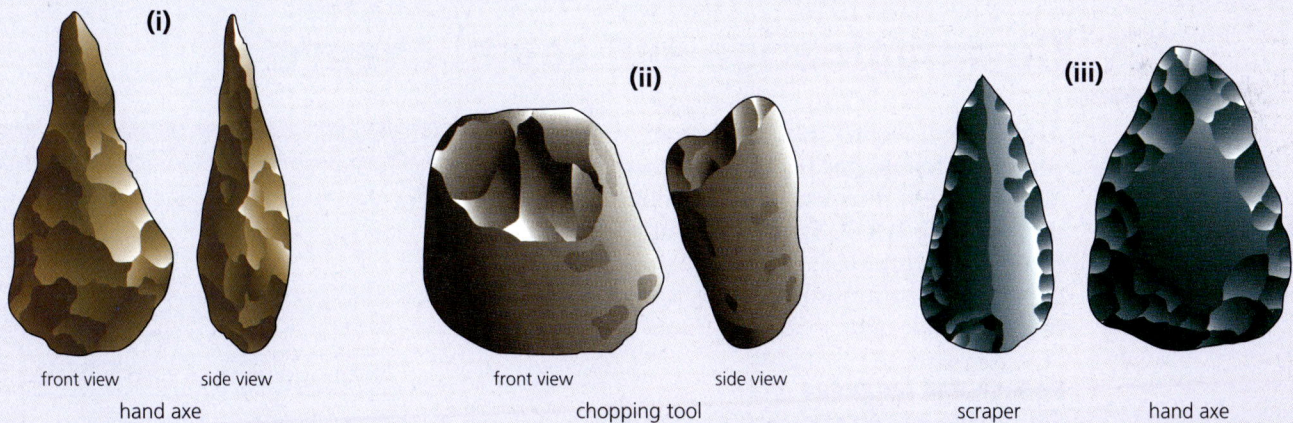

(i)

front view   side view

hand axe

(ii)

front view   side view

chopping tool

(iii)

scraper   hand axe

Source: adapted from Forma and Linstead (eds), *Teachers Pack*, Science Teachers Association of Western Australia (STAWA), 1984

Upper Palaeolithic tools

uses
barbed blade
arrowhead
scraper
pointed drill

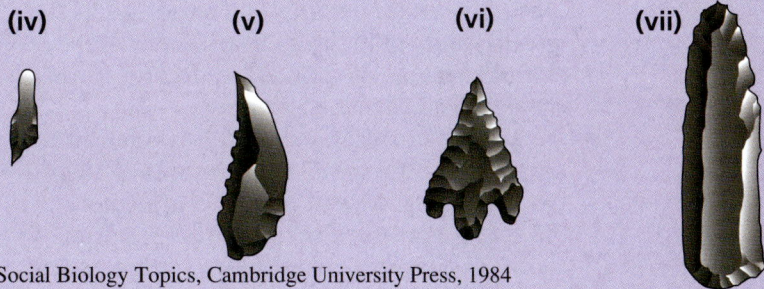

(iv)   (v)   (vi)   (vii)

Source: Tomkins, *The Origins of Mankind*, Cambridge Social Biology Topics, Cambridge University Press, 1984

Fire provides warmth in cold periods and many cave dwellings contain deep layers of ash. Heating the ends of sharpened sticks hardens the wood, so such sticks can be used as spears and digging sticks. Burnt bones and charred plant material suggest that fire was also used to cook food. Eating, and perhaps tool-making, around a fire would encourage interaction and social development. Fire also provides light and can frighten away predators.

However, fire presents several problems for a hunter–gatherer group. Collecting and storing enough fuel to maintain even a modest fire takes time, and it is difficult to move fire to a new home site.

15 List five uses of fire for a group of hunter–gatherers.

Key ideas

- Stone tools reflect the development of new skills in each tool culture. The tools are most sophisticated in the later upper Palaeolithic tool cultures.

- It is possible to explain how tools were made and their uses by studying their shape.

- There is evidence to suggest that the first use of fire was by *Homo erectus*.

- Fire was probably used as a source of heat and light, as well as for cooking. Fire is believed to have promoted social change.

8.8 Communication and social change

As the fossil record shows, the pace of physical change has been fairly slow – anatomically, humans have not changed much in the last 30 000 years. But the human development of mental skills, particularly communication by speech, far outstrips that of any other primate (Fig. 13).

Speech and language

Chimpanzees can vocalise (make noises with the voice box) and have their own language for communication, but they do not have a suitable physical structure in the mouth and throat areas to produce the complex range of sounds needed for human speech (Fig. 14).

Kanzi is a chimpanzee who has learnt to understand the basics of human language. She can use a symbol board to interpret messages conveyed through the pictorial symbols that represent objects, subjects and abstract ideas. She can also convey her own messages to humans by pointing to

Fig. 13 Speaking, hearing and understanding

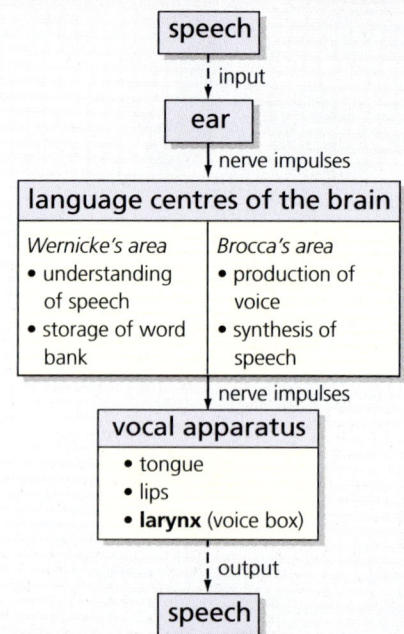

speech
↓ input
ear
↓ nerve impulses

language centres of the brain

Wernicke's area	*Brocca's area*
• understanding of speech	• production of voice
• storage of word bank	• synthesis of speech

↓ nerve impulses

vocal apparatus
- tongue
- lips
- **larynx** (voice box)

↓ output
speech

Fig. 14 Vocal apparatus

The larynx, tongue and mouth are all involved in producing sound. The brain sends nerve messages to muscles in these regions to alter the pitch, loudness and duration of sound.

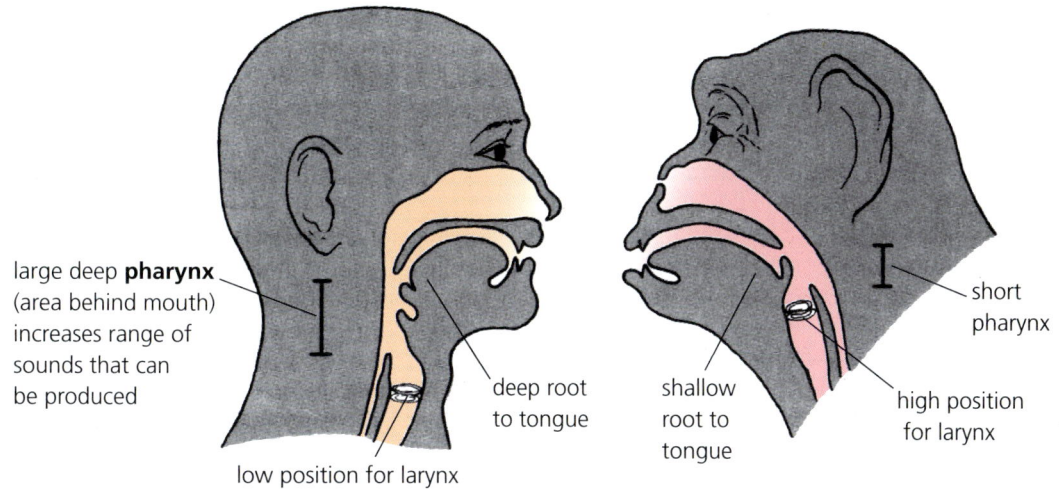

large deep **pharynx** (area behind mouth) increases range of sounds that can be produced

deep root to tongue

low position for larynx

shallow root to tongue

short pharynx

high position for larynx

Source: adapted from Leakey and Lewin, *Origins Reconsidered*, Little, Brown & Co, 1992

Kanzi and her communication board.

symbols. So, Kanzi not only understands basic human language, she can communicate with humans too. In fact, her human language skills are roughly as good as a 5-year-old human child.

Q 16 The latest version of the symbol board has electronic pads that produce voice sounds when touched. Does this mean Kanzi uses speech to communicate with humans?

Human languages are very complex; children take many years to develop the abilities needed to communicate using language. Human communication is very sophisticated. It allows individuals to exchange information and describe their own experiences. So, a community can share ideas, divide tasks and pass on information or skills. This is of enormous importance in allowing the development of complex societies since otherwise every individual would have to learn everything from scratch, and only limited cooperation would be possible. In short, a structured and complex society cannot develop without speech and language.

The communication revolution

We have evidence from the past of tool making, use of fire, and activities such as cave painting. But we do not have any evidence about how our ancestors lived, about their ability to communicate, or about how their societies were organised. These things have to be conjectured from the artefacts that remain.

Humans have now moved into an age of electronic communication. This has revolutionised the sharing of ideas and the storage of information. The speed of transmitting information and the number of people who can gain access to ideas has increased enormously. Is this a catalyst for even further and faster change?

17 Do you think electronic information technology will alter the social evolution of humans? Give evidence to support your ideas.

Key ideas

- Human mental development, especially the use of speech and language, is far faster than physical change.

- Language depends on a recognised sequence of signals to convey meaning. Only humans communicate by speech.

- Speech is complex vocalisation and requires suitable vocal apparatus.

- Speech and language mean humans can work together efficiently, and can transfer skills and ideas.

- A sophisticated and structured society depends on communication.

- We can only conjecture about the level of communication and social organisation of our human ancestors.

Planning how we look

This young woman from Yemen has her hands and forearms decorated for a wedding.

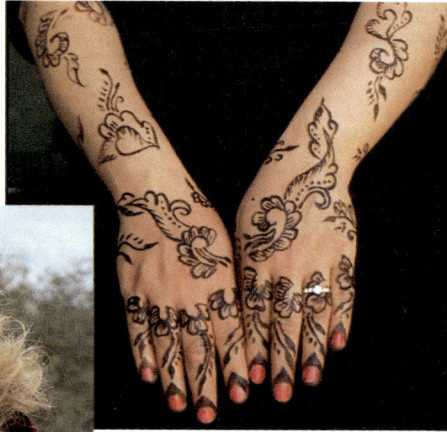

An Aboriginal man of the Walpiri Tribe from the Tanami desert.

How we look says a lot about us. Humans have a high level of self-awareness and all human cultures have beauty treatments. In complex societies, visual signals convey messages in everyday situations and at special ceremonies. Ceremonial adornment is often particularly intricate and formal. Beautification usually focuses on the head and face, but any part of the body can receive such attention. Changing appearance is usually temporary, although tattoos or cosmetic surgery can have permanent effects. Is it wise to change the way you look just to follow fashion – and will you ever be able to catch up with the latest trend?

Elaborate hairstyling and make-up were features of Punk culture among young Britons.

This Toposa man from south Sudan has traditional facial decoration.

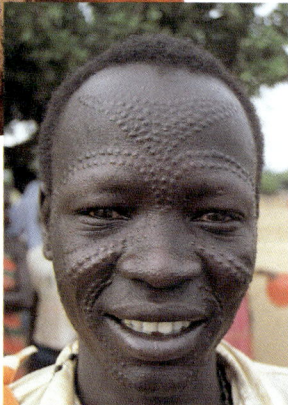

9.1 Learning objectives

After working through this chapter, you should be able to:

- **describe** the two main types of variation;

- **recall** that variation is caused by genes and the environment;

- **interpret** evidence from genetic studies, especially twin studies;

- **describe** the relationships between the global distribution of skin pigmentation, sunlight intensity, rickets, and skin cancer;

- **explain** how skin pigmentation is linked to vitamin D production;

- **explain** how the sickle cell gene may give protection against malaria;

- **explain** how developments in medicine, gene therapy and genetic counselling might affect the frequency of genetic disease.

9.2 Varying features

Inheriting genetic material as a mix from both parents leads to variation in the features of the offspring because the genetic material can be combined in different ways. We all vary in our facial features, but there are two types of variation. Features that can easily be put into distinct categories show **discontinuous variation**, whereas features that fall somewhere in a range of possibilities show **continuous variation** (Table 1)

Human ears might or might not have a 'Darwin's point' (a small lump protruding from the inside top curve of the ear). This feature shows discontinuous variation and has two distinct physical forms or **phenotypes**: Darwin's point is present or absent. There is no 'half-way house' and all ears can be put into one class or the other.

However, ear length (measured in millimetres) can vary in adults from 50 mm to 80 mm. So, ear length is a continuously variable feature; there are no distinct phenotypes. To handle data of this type, length classes must be created. Each measurement of ear length is put into the appropriate length class and the total number of examples in each class is can be shown as a histogram. A distribution curve can be drawn from the histogram (Fig. 1).

Q 1 a **What is the difference between continuous and discontinuous variation?**
   b **Give an example of each type of variation.**

Mr Spock's ears are a feature showing continuous variation because all Vulcans have ears that are pointed to some extent. All ears (human or Vulcan) show continuous variation in the wide range of ear sizes.

Fig. 1 Normal distribution curve

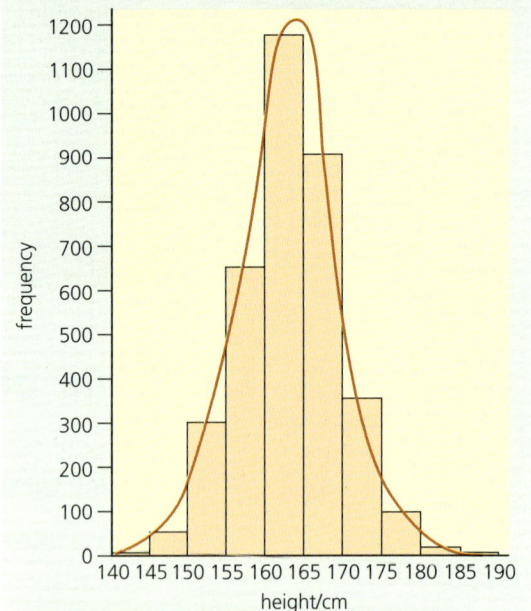

This histogram and curve show the frequency of particular heights in a sample human population. The pattern is called a **normal distribution curve** and is typical of any characteristic that shows continuous variation.

Table 1 Features and variation	
Features showing discontinuous variation	Features showing continuous variation
presence of a gap between front teeth	skin coloration
freckles	hair colour
cleft chin	face width/length
tilted or straight nose	nose width/length
cheek dimples	height

As a beauty therapist, I can make temporary changes to the way hair looks but I can't do anything about hair loss. There seems to be a gene for baldness in men, but the genetics is complicated. Shock can also cause hair loss – this is an environmental effect.

Genes and environment

Variation in hair colour is continuous, but how hair looks is affected by styling, tints, treatments and age. As we age, our hair changes colour, sometimes goes grey or even falls out completely. The hair we have is a product of the environment and the genes we inherited from our parents. Beautification, whether temporary or permanent, is an environmental change and cannot be passed on to our offspring.

Q2 Why can't even permanent changes such as cosmetic surgery be passed on to our offspring?

Table 2 Relatives and shared alleles

Relatives	Minimum proportion of alleles in common/%
monozygotic twins	100
dizygotic twins; brother–sister; brother–brother; sister–sister; father–child; mother–child	50
grandparent–grandchild; aunt/uncle–niece/nephew	25
first cousins	12.5

Source: adapted from Tomkins, *Heredity and Human Diversity*, Cambridge Educational, 1992

Researchers have studied twins to try and compare the effects of genetic and environmental factors on continuous variation. Identical twins begin life as a single fertilised ovum called a **zygote**. This divides to form two genetically identical individuals known as **monozygotic twins**. Non-identical or **dizygotic twins** are conceived from two separate ova fertilised by different sperm cells.

Non-twin brothers and sisters are known as siblings. Each sibling is conceived from an individual ovum fertilised by an individual sperm. Siblings are genetically different and are also different ages.

We inherit our genes in equal parts from both parents, so we can work out the proportion of alleles that certain relatives have in common (Table 2).

Q3 a Why do dizygotic twins look different from each other?
b Explain why identical twins are always the same sex.

One way of looking at the differences between related children is to compare their scores on a series of simple intelligence tests. Intelligence is only one of the many possible comparisons, but it illustrates the difficulties involved in this type of study. The assumption behind twin studies is the idea that differences between dizygotic twins living in the same environment are due to genetic factors since the environment is as near identical for each twin as possible. It is argued that differences between monozygotic twins (who are genetically identical) must be due to environment. Thus, it is expected that monozygotic twins reared apart differ from each other more than monozygotic twins reared together. Differences between non-twin siblings – for whom both genes and environment differ – are due to the interaction of genes and environment.

At first sight, results appear to support these ideas. However, there is controversy about the interpretation of the results because it can be argued that the *interaction* of genes and environment might cause any of the observed differences. It is not merely a matter of considering genes *or* environment. These studies also provoke a lot of controversy because of the small sample sizes and the arguments about the definition and measurement of intelligence (Table 3).

A **correlation** between two factors implies an underlying connection. So, the correlation between smoking cigarettes and the likelihood of dying from lung cancer implies a link between the two. A correlation coefficient is a mathematical measure of the likelihood of a link existing. Table 3 shows the average correlation coefficient between certain pairs of individuals and the results of the intelligence tests. The nearer the correlation coefficient is to 1.0, the more likely it is that a link exists between the relationship of the individuals and the similarity of their test results.

Table 3 Correlation between related pairs and the similarity of their intelligence test results

Relationship	Correlation coefficient
monozygotic twins reared together	0.86
monozygotic twins reared apart	0.72
dizygotic twins reared together	0.60
siblings reared together	0.47
siblings reared apart	0.24
cousins	0.15

Source: adapted from Gross, *Psychology, the Science of Mind and Behaviour*, 3rd edn, Hodder & Stoughton, 1996 (research by Bouchard and McGue, 1981)

4 a What does Table 3 suggest about the similarity of test results and how closely related two people are?
  b In Table 3, what does the effect of environment seem to be?

Fig. 2 Dermal ridge counts and relatedness

Fingerprints

simple arch

tented arch

loop

whorl (symmetrical)

whorl (spiral)

whorl (double loop)

Scattergram

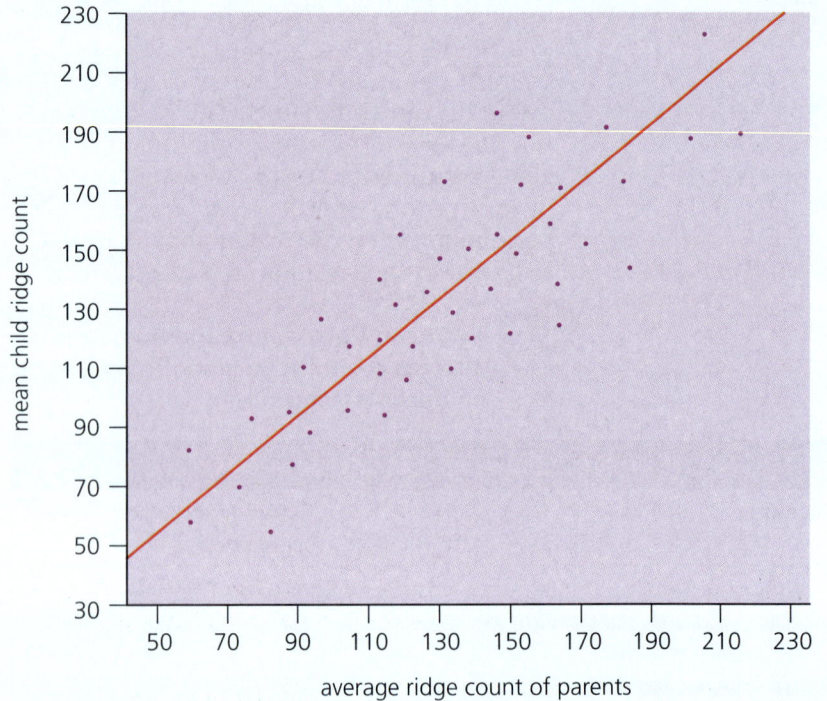

mean child ridge count / average ridge count of parents

Source: adapted from Forma and Linstead (eds), *Human Biology Laboratory Manual*, Science Teachers Association of Western Australia (STAWA), 1984

The development of individuals as intelligent beings involves a complex sequence of changes. Environmental influences are very difficult to assess in isolation from genetic factors and almost impossible to measure.

The contribution of genes and environment to a variety of physical features such as fingerprints has also been investigated (Fig. 2 and Table 4). However, the fact that an observed correlation closely matches the theoretical correlation does not necessarily mean that the theory is correct. The correlation could be due to some other factor. Here, the observed correlation is marginally less than the theoretical correlation (Table 4) which suggests that in this case the environment has little effect – but it does have *some* effect as the scattergram shows (Fig. 2).

There are many obvious ways in which environmental factors do affect the phenotype of individuals (Table 5).

5 a How do Figure 2 and Table 4 support the hypothesis that 'fingerprint patterns are controlled by genetic factors'?

b Is there any evidence in this data to suggest that environmental factors could be involved in ridge counts of individuals?

c Using Table 5, describe how one disease and one other environmental factor can affect the phenotype of an individual.

Table 5 Some effects of the environment on phenotype

Environmental factor	Effect on phenotype
exposure to sun	darkening of the skin
continued exposure to sun	thickening and ageing of the skin
excess food (energy) intake	obesity – increased fatty tissue
lack of food	impaired growth
calcium deficiency	deformed bones – a calcium deficiency disease called rickets
rubella virus	can cause deafness and blindness in a child infected as a fetus

Table 4 Correlations between ridge counts for related pairs

Relationship	Observed correlation	Theoretical correlation
mother–child	0.48	0.5
father–child	0.49	0.5
sibling–sibling	0.50	0.5
parent–parent	0.05	0.0
monozygotic twins	0.95	1.0
dizygotic twins	0.49	0.5

Source: Levitan, *Textbook of Human Genetics*, Oxford University Press, 1988

Key ideas

- Variation can be categorised as continuous or discontinuous.

- Continuous variation is the type of variation in which a feature varies within a range of values.

- Discontinuous variation is the type of variation in which a feature falls into a distinct class.

- Twin studies have enabled some comparison of the parts played by genes and environment in determining variation, but it is almost impossible to study the effect of one without the other.

- Variation is a product of genes and environment acting together.

- Phenotype can be affected by environmental factors.

9.3 Sun and skin

The sun produces electromagnetic radiation of many wavelengths. Wavelength is measured in nanometres (nm). The part of the spectrum called **ultraviolet radiation** (**UVR**) is divided into three sections:

- UVA (long wavelength ultraviolet) 315–400 nm;
- UVB (mid-wavelength ultraviolet) 295–314 nm;
- UVC (short wavelength ultraviolet) 200–294 nm.

The ozone in the upper atmosphere absorbs cosmic rays, X-rays, and the UVC, but UVA and UVB easily reach sea level. UVR has two important effects on skin:

- it stimulates skin to produce vitamin D;
- it damages pale skin, but darker skin is protected (Fig. 3).

Vitamin D

Skin produces vitamin D when exposed to UVB. Vitamin D is essential for bone growth and bone strength; lack of Vitamin D causes rickets, a disease in which the bones are soft. People with rickets often have legs that are curved rather than straight because the leg bones bend under the weight of the body. Rickets makes it very difficult to walk.

Pale skins, which are found in the more temperate regions, are able to make more vitamin D than dark skins (Table 6).

Table 6 Vitamin D production in skin

Skin type	Vitamin D production after exposure to a measured amount of UVB / arbitrary units
white	100
black	16

Q 6 Why is vitamin D important to health?

Fig. 3 Wavelength of UVR and sunburn

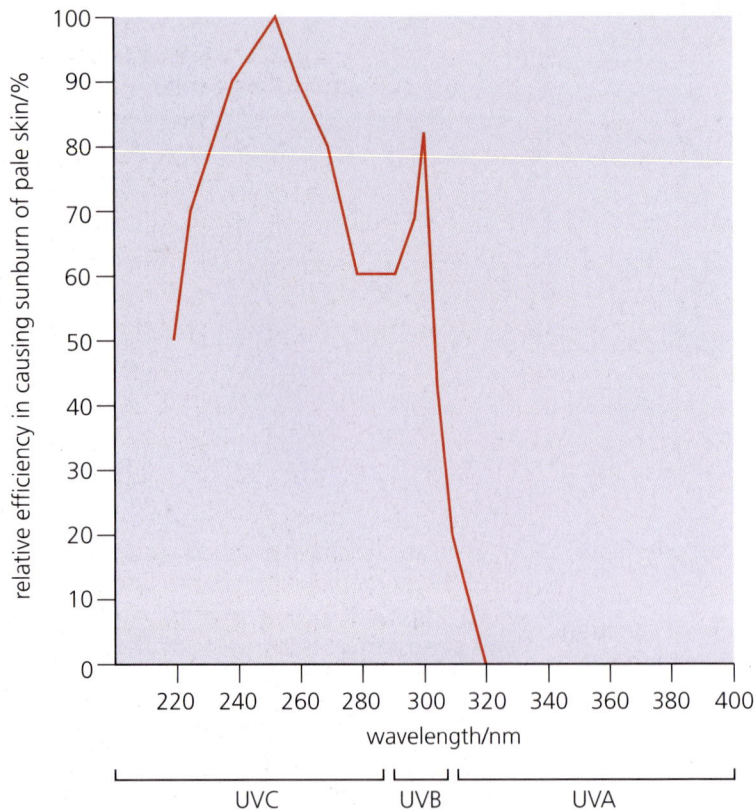

Graph: y-axis "relative efficiency in causing sunburn of pale skin/%" (0–100), x-axis "wavelength/nm" (220–400). Regions labelled UVC, UVB, UVA.

Prolonged exposure to strong sunlight darkens and ages pale skin. People with black, brown or olive skins are affected less. This Tibetan woman has spent most of her life outdoors.

Harmful effects of the sun

Skin comes in a range of different colours – skin colour is a feature showing continuous variation. Pale skins can be damaged by sunburn. The efficiency of each wavelength of UVR in causing sunburn can be assessed using artificially generated radiation (Fig. 3).

7 From Figure 3, what is the relationship between UVR wavelength and sunburn?

The colour of skin depends on the amount of a pigment called **melanin**; the more of it there is, the darker the skin colour.

Skin contains cells called **melanocytes** that produce structures called **melanosomes**. The melanosomes contain the melanin. Dark skin contains no more melanocytes than pale skin, but the melanosomes are larger and more evenly distributed. Melanin can absorb ultraviolet radiation that would otherwise harm DNA in the nuclei of the skin cells. Each melanocyte can protect the nuclei of several surrounding cells (Fig. 4).

When skin is exposed to the sun, it produces more melanin and so darkens. This is most obvious in very pale skins and is called tanning, but all except the darkest of skins show a significant colour change in response to sunlight. Melanin production is stimulated particularly by UVB, which also causes sunburn. UVA is less efficient at stimulating melanin production, but also less likely to cause sunburn. When exposure to sunlight is reduced, the extra melanin is eventually broken down and the skin lightens again.

The **epidermis** or upper layer of the skin is a thin, dead protective surface. It can form a further barrier to ultraviolet radiation by doubling in thickness after exposure to the sun. However, both melanin production and skin thickening take time. Over-exposure to the sun in the meantime damages the delicate tissues below the epidermis. The result is reddening and blistering – sunburn.

8 a What function do melanocytes have?
b What section of UVR stimulates melanin production?

Fig. 4 Melanin and protection

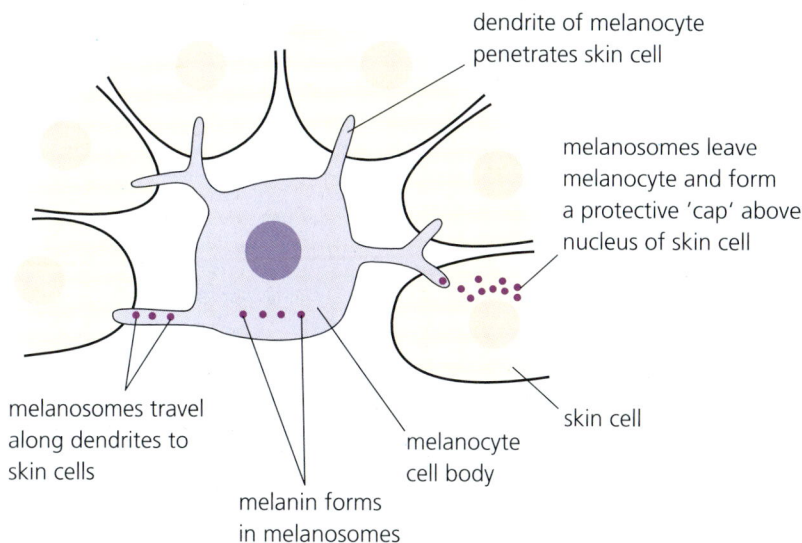

dendrite of melanocyte penetrates skin cell

melanosomes leave melanocyte and form a protective 'cap' above nucleus of skin cell

melanosomes travel along dendrites to skin cells

melanin forms in melanosomes

melanocyte cell body

skin cell

Source: adapted from Burton, *Essentials of Dermatology,* Churchill Livingstone, 1990

Movement of people round the globe has led to a mix of pale and dark skins in most parts of the world.

Worldwide pigmentation

The fossil evidence suggests that human ancestors spread out from Africa to other parts of the world. Since then, there have been many migrations of peoples to start new lives in other countries, and all degrees of skin pigmentation are found all round the world. However, it is possible to generalise about the underlying worldwide distribution of skin pigmentation and sunlight: the nearer the equator, the 'stronger' the sunlight and the more intense the local skin pigmentation (Fig. 5).

Q 9 Study Figure 5. Describe the relationship between sunlight intensity and the worldwide distribution of skin pigmentation.

People living in sunnier regions would benefit from a skin rich in melanosomes that protected them from ultraviolet radiation and sunburn. In fact, a dark skin would be essential for hunter–gatherer life in equatorial areas. In more northerly temperate areas, where light levels are lower and more clothing is required, the ability of pale skin to produce vitamin D more efficiently would be an advantage. The evolution of paler skin as human ancestors moved out of Africa into more temperate climates is an example of **natural selection**.

The total number of alleles in a population is called the **gene pool** of that population. Over many generations, natural selection by environmental factors makes the features that give individuals an

Fig. 5 Sun, skin pigmentation and UV light intensity

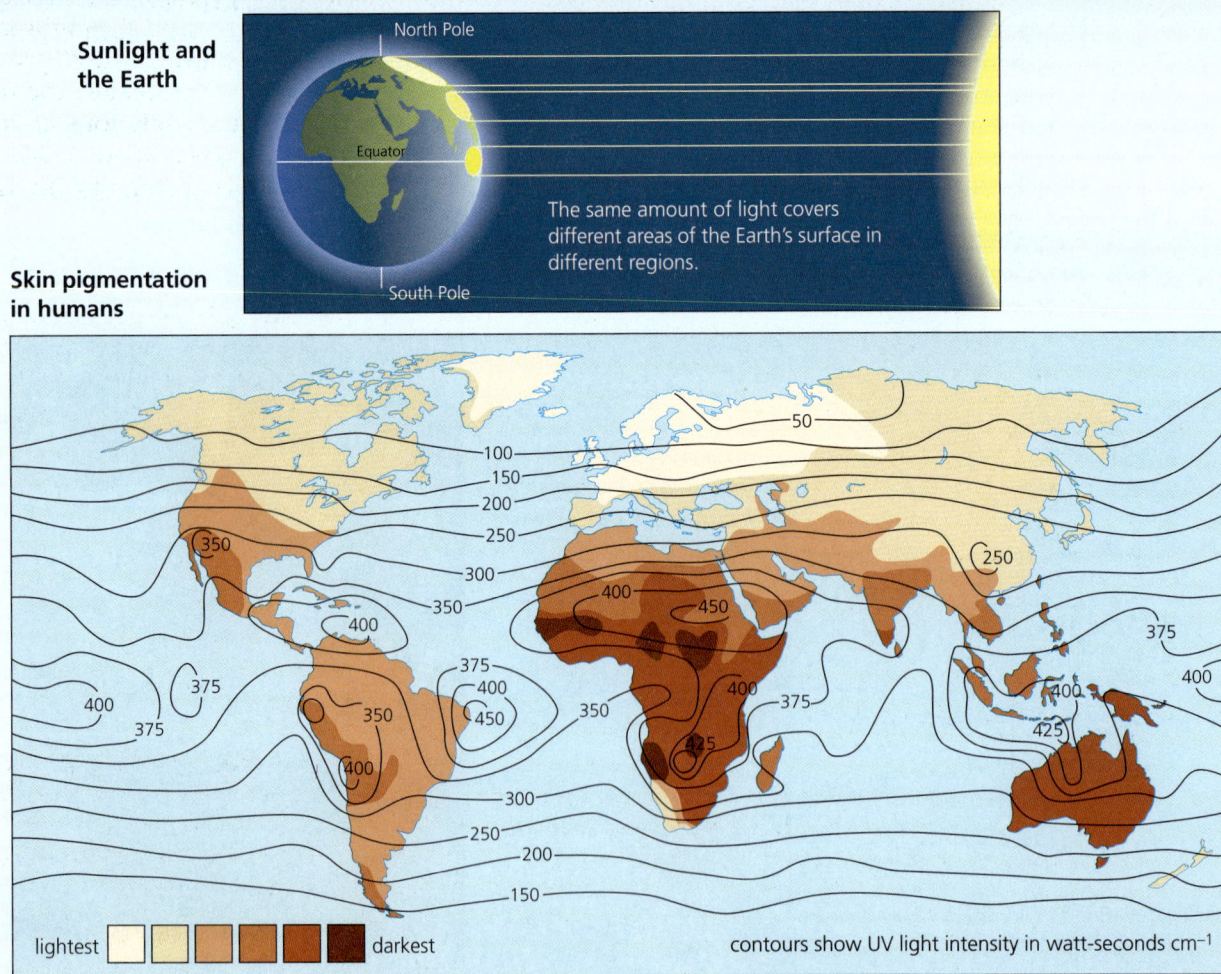

Sunlight and the Earth

North Pole

Equator

South Pole

The same amount of light covers different areas of the Earth's surface in different regions.

Skin pigmentation in humans

lightest ☐☐☐☐☐☐ darkest

contours show UV light intensity in watt-seconds cm⁻¹

Source: adapted from Jolly and Plug, *Physical Anthropology and Archaeology*, 2nd edn, Knopf, 1979 and Goldsby, *Race and Races*, Macmillan, 1971

In Australia there is a strong campaign to protect yourself from the sun. You are urged to slip on a shirt, slop on some sun lotion, and slap on a hat.

Table 7 Skin cancer and place of residence in white-skinned people

Location	Number of cases per 100 000 people per year
United Kingdom	28
Africa (Cape)	133
Texas	168
Queensland (Australia)	265

advantage in particular situations more frequent. So, the alleles that code for these features become more common in the gene pool.

10 Explain how human ancestors migrating north into temperate climates could have evolved paler skins.

11 What is meant by the term 'gene pool'?

New life, new problems

In our modern world many pale-skinned people live in very sunny climates or travel to sunny holiday locations. Although tanning protects the skin, repeated exposure to sunlight ages the skin and increases the risk of skin cancer, especially in older people. There are many forms of skin cancer, from **benign** slow growing lumps, to rapidly spreading **malignant** forms that invade other tissues and rapidly cause death. There is a relationship between where on the globe people with pale skins live and the number suffering from skin cancer (Table 7).

12 Study Table 7. What is the relationship between where people live and the number of skin cancer cases in white-skinned people?

Key ideas

- Pale skins produce more vitamin D on exposure to measured amounts of UVB than dark skins.

- Skin pigmentation is due to the size and distribution of melanin-containing cells called melanocytes.

- Melanin protects skin cell nuclei from ultraviolet light.

- Worldwide distribution of skin pigmentation is a product of natural selection. As dark-skinned human ancestors moved to temperate climates, selection favoured the evolution of paler skins.

- Pale skins are more prone to skin cancer than dark skins.

9.4 Selection in action

Haemoglobin is the body's oxygen carrier (Fig. 6). Normal haemoglobin is called haemoglobin A. A sickle cell mutation in the gene that codes for the β chains makes a slight but very important change in β-chain structure; this mutant haemoglobin is called haemoglobin S. The type of haemoglobin is determined by a pair of alleles, so there are three **genotypes** (Table 8).

Fig. 6 Haemoglobin structure

Four polypeptide chains make up the haemoglobin molecule. Each chain contains 574 amino acids.

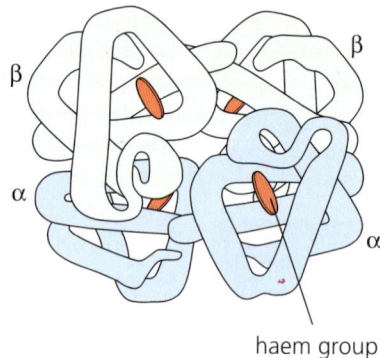

haem group

Each chain is attached to a haem group that can combine with oxygen.

Table 8 Haemoglobin genotypes

Genotype	Phenotype
Hb^AHb^A	normal
Hb^AHb^S	sickle cell trait
Hb^SHb^S	sickle cell anaemia

Table 9 Genotypes of children in a malarial area

	Total number	Hb^AHb^A genotype	Hb^AHb^S genotype
children with malaria	100	96	4
children without malaria	100	82	18

Source: Levitan, *Textbook of Human Genetics*, Oxford University Press 1988 (research by Giles, 1967)

Villagers in Yemen where the Hb^S allele is widespread but, unlike central Africa and southern Asia, there is little malaria.

People with genotype Hb^AHb^A have normal red blood cells. Individuals with the Hb^SHb^S genotype have red blood cells that become sickle shaped when there is a shortage of oxygen, for instance during exercise. This is because the structure of their β chains changes in these conditions. The fragile cells break up easily so there are not enough cells to carry oxygen to the tissues. Without proper medical care, people who have the genotype Hb^SHb^S often die as young children. People with the Hb^AHb^S genotype have mainly normal red blood cells, but about a third of their red cells are sickle cells. These people survive because there are enough red blood cells to supply the tissues with oxygen.

There is a positive side to having sickle cell trait in a malarial area. The malarial parasites, *Plasmodium* spp., spend part of their life cycle in normal human red blood cells, but cannot grow well in sickle cells. So, having some sickle cells in the bloodstream offers some protection against malaria (Table 9 and Fig. 7).

Fig. 7 Distribution of the Hb^S allele and malaria

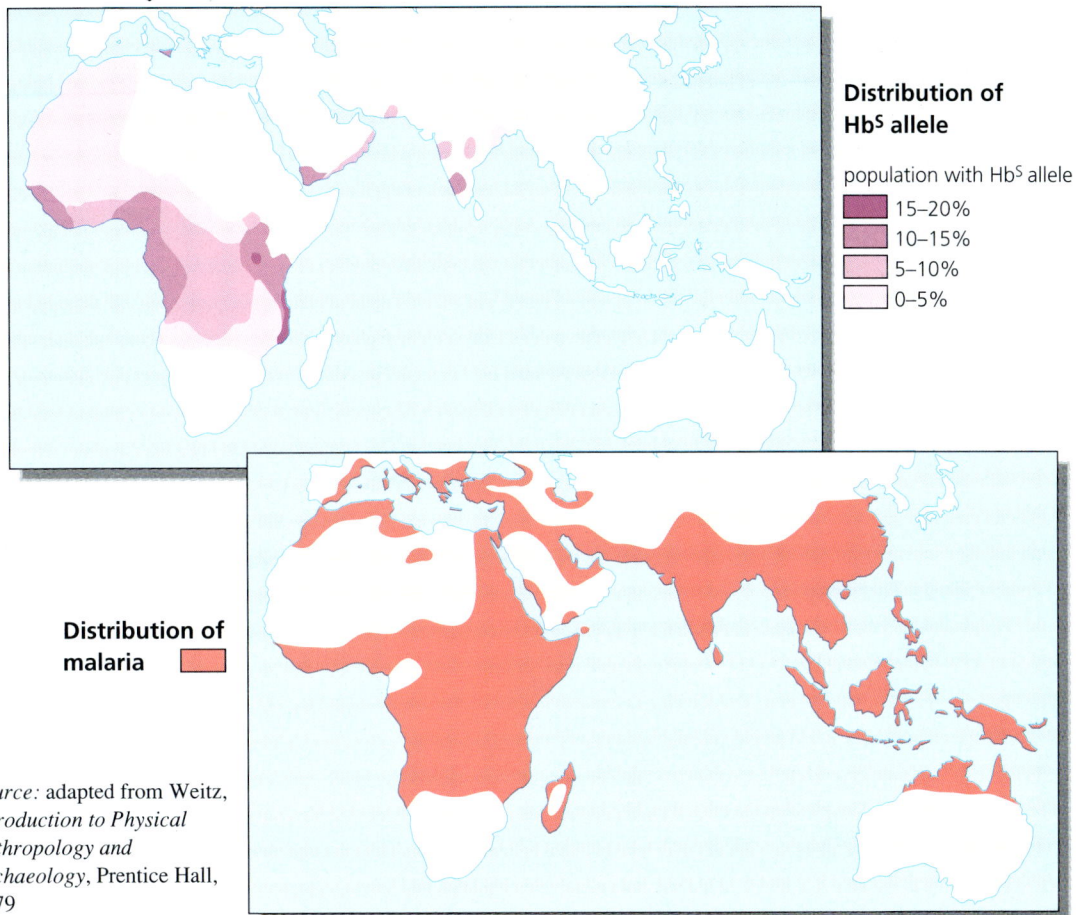

Distribution of Hb^S allele

population with Hb^S allele

- 15–20%
- 10–15%
- 5–10%
- 0–5%

Distribution of malaria

Source: adapted from Weitz, *Introduction to Physical Anthropology and Archaeology*, Prentice Hall, 1979

13 a Study Table 9. How does this information support the hypothesis that having the Hb^{A}Hb^S genotype offers some resistance to malaria?

b Why are there no children with the Hb^{S}Hb^S genotype in this study?

14 Study Figure 7. What is the relationship between the distributions of malaria and the Hb^S allele?

15 Outline one advantage and one disadvantage of having the Hb^{A}Hb^S genotype.

Key ideas

- Haemoglobin A is normal and haemoglobin S is a mutant form.

- A person with the genotype Hb^{S}Hb^S has sickle cell anaemia and is unlikely to survive early childhood.

- An Hb^{A}Hb^S genotype gives some protection against malaria at the expense of having a reduced number of normal red blood cells to carry oxygen.

9.5 The genetic lottery

Each of us is the product of a genetic lottery; on top of that, the way our genes are expressed can be modified by our environment. For example, nutrition and disease affect the way we look. Genetic characteristics can be visible, like eye colour, or hidden, like blood groups. Some are inherited from our parents, others could be the result of mutations in our own cells during our lifetime. If mutations occur in the egg or sperm cells, the features they determine can be inherited by our offspring. Some people inherit conditions such as cystic fibrosis, Huntington's disease and Down's syndrome (*Biology Core*, Chapter 10).

Gene therapy

Cystic fibrosis is caused by a mutant recessive gene. Carriers are people who have one defective allele but do not develop the disease. Cystic fibrosis sufferers must have inherited two defective alleles, one from each parent. People with cystic fibrosis depend on physiotherapy and a strict diet to improve their quality of life. However, advances in **gene therapy**, the use of normal genes to replace or override defective ones, might one day bring relief to those with this disease (Fig. 8).

An inhaler transfers normal genes to the nose and lungs to overcome the effects of the defective gene in the lung cells. If trials are successful, it will cost around £200 million to develop and market the gene transfer inhaler.

16 How does gene therapy for cystic fibrosis work?

Genetic counselling

Many genetic diseases can be detected during pregnancy, although false positives (a fetus is incorrectly diagnosed as having a disease) and false negatives (a fetus is incorrectly thought to be free from the disease) can occur. However, tests are becoming more and more reliable, and they

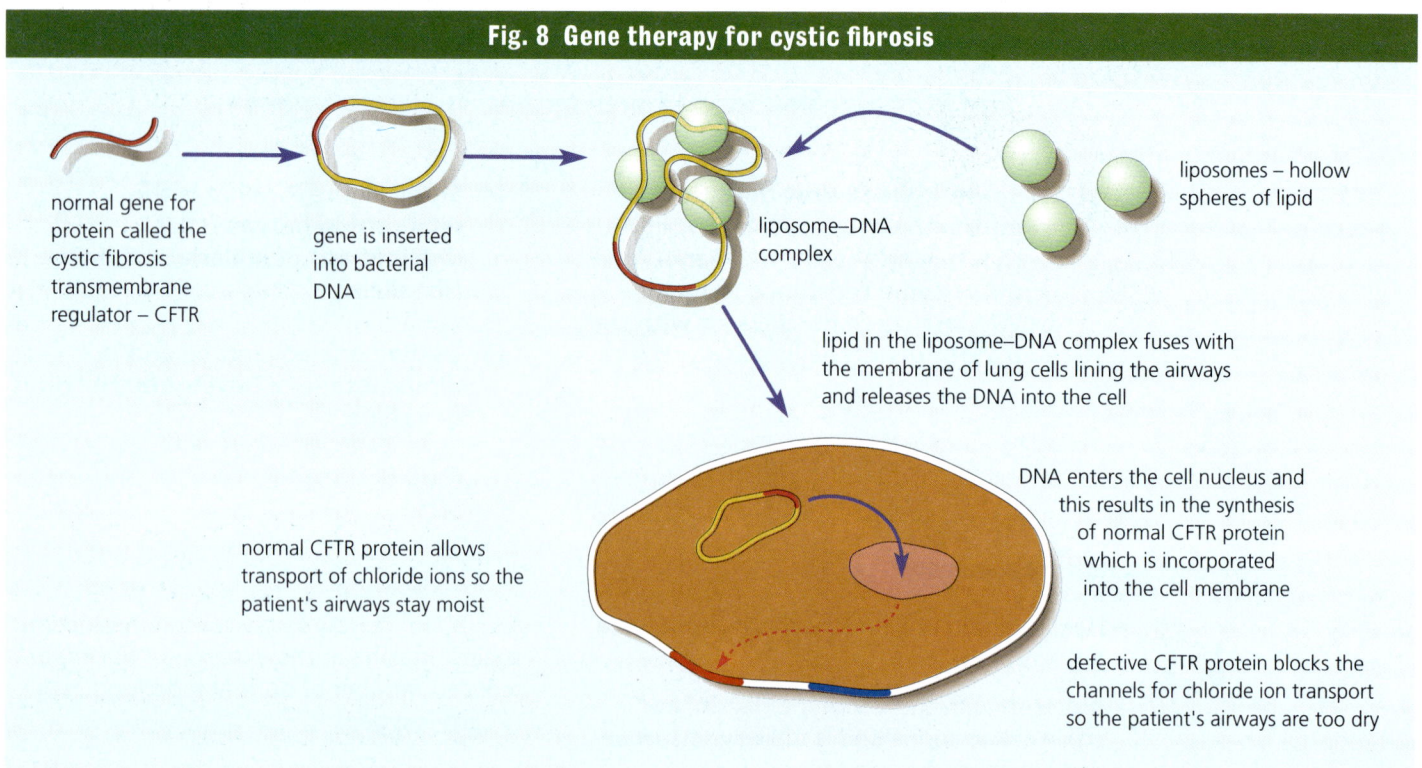

Fig. 8 Gene therapy for cystic fibrosis

normal gene for protein called the cystic fibrosis transmembrane regulator – CFTR

gene is inserted into bacterial DNA

liposome–DNA complex

liposomes – hollow spheres of lipid

lipid in the liposome–DNA complex fuses with the membrane of lung cells lining the airways and releases the DNA into the cell

DNA enters the cell nucleus and this results in the synthesis of normal CFTR protein which is incorporated into the cell membrane

normal CFTR protein allows transport of chloride ions so the patient's airways stay moist

defective CFTR protein blocks the channels for chloride ion transport so the patient's airways are too dry

enable parents-to-be to make informed decisions. Professional help is available for those who wish to take advantage of it. This help is called **genetic counselling** and is very important to those who have to make difficult decisions about whether or not to continue a pregnancy. It is also possible to detect the genes for some diseases in people before they become parents. Genetic counsellors can then talk with potential parents about the likelihood of combinations of genes arising which would lead to difficult questions concerning the termination of a pregnancy.

Medical technology

Down's syndrome is caused by an extra chromosome and is more common in children born to older mothers. Individuals with Down's syndrome have changed facial features, short stature, and learning

Down's syndrome doesn't stop children having fun.

difficulties. With care and support, people with this syndrome can become self-sufficient and lead an active life. Medical technology can correct some of the physical problems. For example, an enlarged tongue can be reduced to make speech easier.

What's the outcome?

We can modify our inherited characteristics by making environmental changes. For example, we can change the way we look, and can surgically correct some physical problems. Even gene therapy is still an environmental change. As yet, we have no control over our genetic inheritance. Natural selection has acted over many generations to increase the frequency in the gene pool of random mutations that are advantageous. But other, disadvantageous, mutations also exist. Will we ever be able to do away with them?

Might the selection pressures on which human evolution is founded be changed in the future? Developments in medicine are not likely to diminish the number of patients with inherited diseases: if a formerly devastating illness can be successfully overcome with treatment, the number of patients could actually increase. Genetic counselling might affect the numbers of patients with certain inherited conditions, but that is a matter of individual decision.

Q 17 What is likely to be the effect on the incidence of genetic diseases of gene therapy and other developments in medicine?

Key ideas

- Gene therapy might soon be available for sufferers of cystic fibrosis. The treatment makes up for genetic defects in the patient's lung cells. It is an environmental change.

- Genetic counselling consists of help for would-be parents to predict the risk of genetic disease in their offspring and consider the implications.

- Developments in medicine and medical technology can help to ease distressing inherited conditions, but they do not change the genes a person has inherited.

- New selection pressures in societies with medical services might change the incidence of genetic disease in the future.

What price our food?

Not a single item in this load of shopping was grown in the UK. The rice, tea and mangetout come from India; the coffee, drinking chocolate, melon, grapes and asparagus are from south American countries. The bananas are from Costa Rica, the lime from Mexico, the grapefruit and tomatoes from Israel, and the celery, lettuce, clementines, lemon, potatoes and broccoli were all grown in southern Europe. The apples, sweetcorn, beans and carrots were grown in western Europe.

In the 1990s food is transported all round the world and prices in the supermarket must reflect the cost of production, packaging and transport. Exotic tropical fruits and other crops that cannot be grown in the UK are imported throughout the year. Consumer choice is not limited by British growing seasons or climate. All this involves an enormous energy cost that is often met by the use of fossil fuels (Fig. 1).

Does this make sense? The world can produce enough food for everyone but can we afford to use so much energy to pack, transport and process food? And what about the environmental and social costs? Can everything be reduced to pounds and pence? How should we plan food production for the future? Is a local or a global approach the best way forward?

Fig. 1 Slicing up the costs

Retail
shop heat and light 1.7 slices
transport 2.5 slices

Bakery
packaging 1.7 slices
baking fuel 4.7 slices
other ingredients 1.9 slices
transport 1 slice

Mill
packaging 0.4 slices
milling fuel 1.5 slices
other 0.4 slices
transport 0.3 slices

Farm
fertiliser 1.5 slices
other 0.1 slices
tractor fuel 2.3 slices

The food energy value in a standard 750g loaf is less than 40% of the energy used to produce and market it.

10.1 Learning objectives

After working through this chapter, you should be able to:

- **describe** the characteristics of cattle which made them suitable for domestication;

- **explain** how selection of features from wild grasses produced cultivated cereals;

- **describe** the features of modern varieties of wheat and breeds of cattle that reflect their use in intensive farming;

- **describe** the biological principles of preservation of grain and meat products;

- **describe** how agriculture led to surplus production, trade and social change;

- **interpret** archaeological evidence relating to domestication;

- **describe** the effects of agriculture on soils and ecosystems;

- **explain** the principles of soil conservation;

- **compare** subsistence and intensive agriculture with reference to energy input, mechanisation, crops, and genetic diversity;

- **explain** the biological principles behind biomass as a renewable energy source.

10.2 Revolution or evolution

Table 1 The Neolithic revolution

	Hunter–gatherer group	Farming community
Lifestyle	• seasonal movement to sources of food • nomadic – move large distances	• permanent home base in favourable farming area • limited movement – stay near to home base
Group features	• small and scattered • slow population growth, restricted by food sources and a nomadic life	• large and settled • more rapid population growth
Skills	• hunting and gathering techniques • knowledge of wild animals and plants and how to obtain food from natural sources	• farming techniques such as cultivation and husbandry • knowledge of how to farm plants and breed animals for food
Social focus	• finding food • hunting cooperation within a group • competition between groups	• supplying food • exchange of skills between groups • trade in ideas, plants, and animals

It's a long way from the hunter–gatherer lifestyle to modern farming methods. Along the way, humans have:
- produced cultivated cereals from wild grasses;
- domesticated a range of animals;
- learned about food preservation;
- developed trade.

The change from hunter–gatherer to farmer is called the **Neolithic revolution** (Table 1). However, it was a very slow process. The change started in a few areas with suitable wild species and climate. It then slowly radiated into other parts of the world. Beginning over 10 000 years ago in the Middle East, the change took almost 6000 years to reach Britain and Scandinavia, a rate of less than 1 km per year. Hunter–gatherers probably went through an intermediate stage of settled hunting and gathering that led to the first permanent farming settlements.

1 What is the Neolithic revolution and where did it start?

Grass and wheat

The wild grass ancestors of our modern cereal crops could have been harvested simply by cutting or breaking the stems and detaching the grains or seeds. Remains of stored wild grains have been found at Jarmo in eastern Iraq (Fig. 2). These finds date from over 9000 years ago. Grain found at archaeological sites shows that by around 6000 BC, the wild grains in our ancestors' diet were being replaced by cultivated varieties.

Wheat grows in dry, sunny conditions and produces rows of grains on the end of the flowering stem or **rachis**. Inside each grain is an embryo plant with a store of food, and around the outside is a tough protective coat. The grains are rich in carbohydrate and protein, have a high energy value, and are easy to store when dry.

Modern wheat differs from its wild ancestors in:
- strength of the rachis;
- size of the **glume** – the scale-like growth around the grain;
- number of sets of chromosomes.

The wild ancestors of wheat had a very fragile rachis that shattered easily to disperse the grains. Modern wheat has a much stronger rachis. What lead to this change? There are two models for the selection of a stronger rachis: deliberate and accidental. The deliberate selection model

---

Fig. 2 Domesticating plants

Centres of plant domestication

1 Central American highlands
2 Northern Andes
3 Abyssinia
4 Mediterranean
5 Middle East
6 Southeast Asia

Source: adapted from Baker, *Plants and Civilisation*, Fundamentals of Botany Series, Macmillan Press, 1972

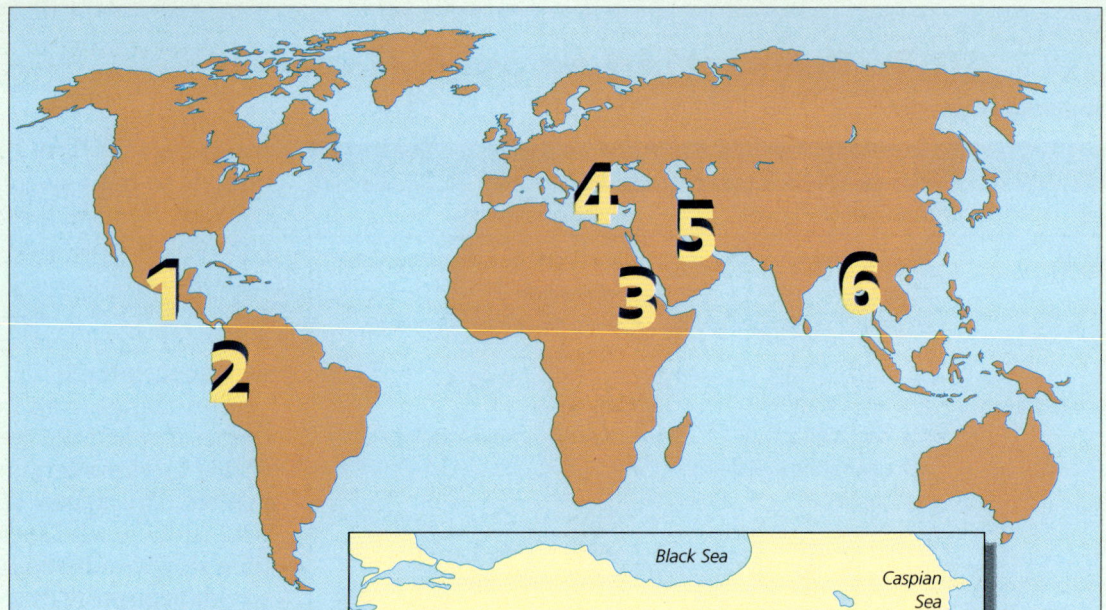

The fertile cresent

The fertile crescent was an area of good soil and sufficient water for simple cultivation.

Source: adapted from Leakey, *The Making of Mankind*, Michael Joseph, 1981

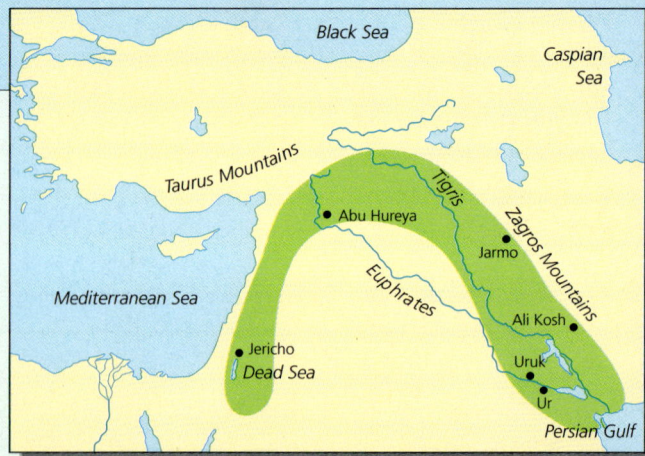

suggests that early farmers deliberately harvested patches of wheat with a stronger rachis and then sowed the grain. They chose these plants because harvesting was easier – fewer grains fell to the ground. The accidental selection model suggests that grains from plants with a weak rachis are accidentally lost because the rachis shatters during harvesting, so the grain that is collected and sown tends to be from plants with a strong rachis. This is thought to be the more likely model.

Small glumes are more likely to have been deliberately selected later. Wild varieties of wheat have such tightly fitting glumes that heat is needed to release the grains – the process of doing this is called **parching**. The cultivated wheats have 'naked grains' that easily detach from the inedible parts without heat.

2 Give two physical differences between cultivated wheat and its wild ancestors.

Modern wheats are **polyploid** – they have more than two sets of chromosomes. Grasses readily crossbreed and can produce fertile polyploid hybrids. **Diploids** have two sets, **tetraploids** four sets and **hexaploids** six sets of chromosomes per cell. Polyploid wheats have larger or more numerous grains, so more food is produced per plant.

From a wild diploid grass with a fragile rachis and large, tight glumes, we now have hexaploid wheats with a strong rachis and four rows of large naked grains (Fig. 3).

3 How does modern wheat differ genetically from its wild ancestors?

Fig. 3 Wheat pedigree

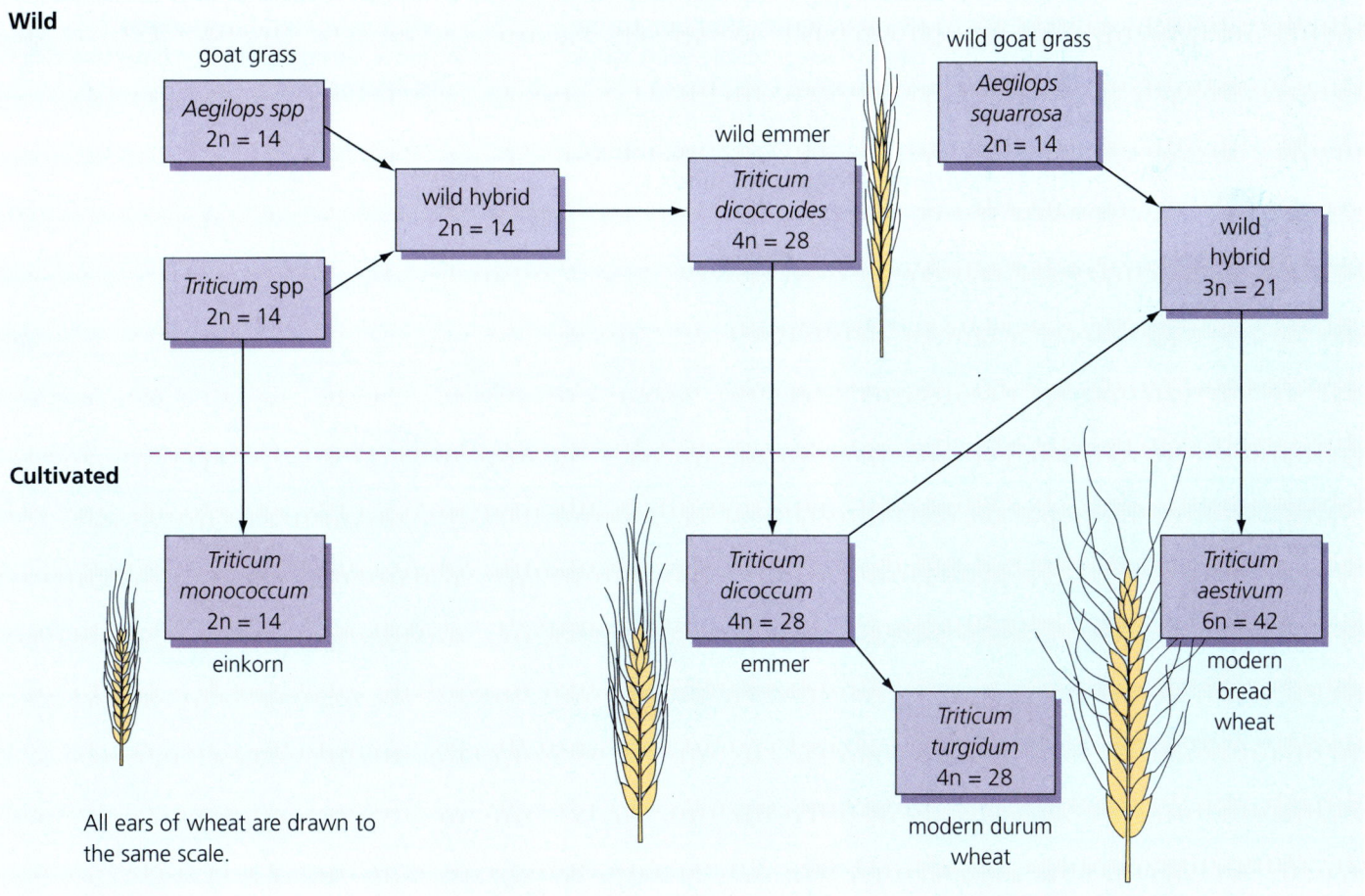

Wild

goat grass
Aegilops spp
2n = 14

Triticum spp
2n = 14

wild hybrid
2n = 14

wild emmer
Triticum dicoccoides
4n = 28

wild goat grass
Aegilops squarrosa
2n = 14

wild hybrid
3n = 21

Cultivated

Triticum monococcum
2n = 14
einkorn

Triticum dicoccum
4n = 28
emmer

Triticum turgidum
4n = 28
modern durum wheat

Triticum aestivum
6n = 42
modern bread wheat

All ears of wheat are drawn to the same scale.

New winter wheats

Modern varieties of wheat are produced by cross breeding and careful selection of characteristics that are suitable for intensive agriculture. A high yield of grain depends on four characteristics:

• short straw – a short stem bearing the grain;
• standing power – wheat does not collapse in the field;
• resistance to shedding – a strong rachis;
• resistance to fungi that cause diseases such as mildew, rust and eyespot.

Between them, these characteristics affect photosynthesis, the amount of food stored in the grain, and how easy or difficult it is to harvest the grain.

4 **Suggest how each of the four features above increases the yield of wheat.**

The National Institute for Agricultural Botany (NIAB) and plant breeders keep seed banks of old wheat varieties that are grown from time to time to keep viable seed stock. This maintains genetic diversity and preserves genes for features that might be important in the future. Farming a limited list of recommended varieties tends to reduce genetic diversity.

Domesticating animals

Of the thousands of animal species, four mammals were domesticated in Europe and Asia: sheep, goats, cattle and pigs. In South America, there was domestication of the llama and the guinea pig.

Three features are important for successful domestication:

• the animals must be herd animals that adapt to handling and do not take flight if startled;
• they are usually herbivores (pigs are an exception, they are omnivores with the advantage of consuming waste);
• the animals must breed in captivity.

Nomadic pastoralism is a way of life in which humans travel with and tend their grazing animals. It is still practised. However, overgrazing can destroy the environment and threaten the existence of the nomadic peoples who keep the animals.

5 a **Which feature of pigs is unusual in a domesticated farm animal?**
  b **How can grazing increase the risk of soil erosion?**

In East Asia, yaks and water buffalo were domesticated. This boy and his water buffalo are in a rice field in Vietnam.

Unrestricted grazing has made huge changes to this deforested landscape.

Cattle story

Cattle are big animals and need careful management to ensure there is enough grazing, to keep the animals away from crops, and to protect them from predators. Besides milk and meat, the skins, horns, and bones of cattle are useful for clothing or making tools. Cattle can pull ploughs and carts. This increases the ability of humans to farm the land and to transport food.

The huge number of modern cattle breeds are all descended from *Bos primigenius*, a wild species that was not easily managed (Fig. 4). An important goal for early cattle farmers must have been to breed smaller, more docile animals.

Domestication of cattle began in the Middle East around 6500 BC, and then spread outward from this area. There was then a long period during which individual communities selected characteristics that suited their particular farming conditions. Eventually, this resulted in the evolution of a large number of extremely localised land-races of cattle.

6 a Which features of cattle made them suitable for domestication?

b How would early farming communities have affected the evolution of cattle?

Fig. 4 Cattle domestication

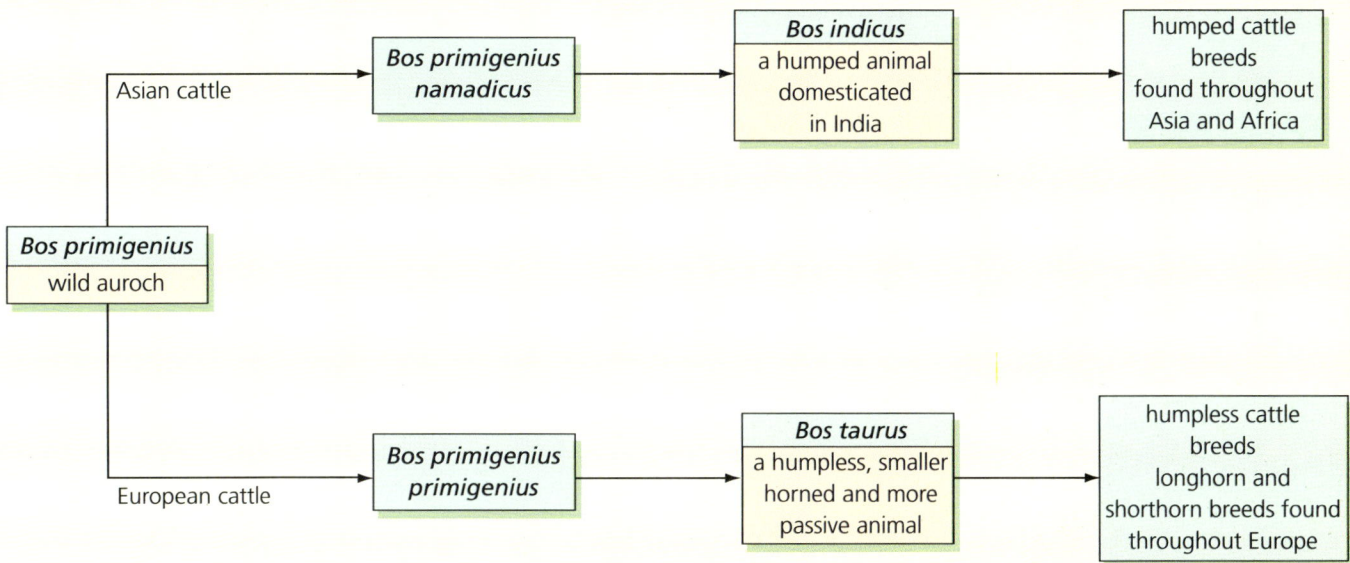

Asian cattle → *Bos primigenius namadicus* → **Bos indicus** a humped animal domesticated in India → humped cattle breeds found throughout Asia and Africa

Bos primigenius wild auroch

European cattle → *Bos primigenius primigenius* → **Bos taurus** a humpless, smaller horned and more passive animal → humpless cattle breeds longhorn and shorthorn breeds found throughout Europe

Humped cattle are tolerant of hot dry conditions.

This longhorn bull is a humpless rare breed.

Belted Galloway cattle are a distinctive humpless rare breed.

Modern cattle are very large and specialised to suit modern farming. In the 1990s, we use a limited number of breeds in the UK, for example, Herefords for meat and Friesians for milk (Table 2).

At first, the gene pool of farmed cattle would have shown good genetic diversity. However, modern farming has shown a progressive selection from a smaller and smaller number of specialised breeds. This means that many less obviously useful or desirable features are being lost – unfortunately, this reduces the gene pool available for future breeding programmes. The Rare Breeds Trust aims to conserve Britain's farm animals, including cattle, and to increase the numbers of minority breeds (Fig. 5).

7 a Why is there a loss of genetic diversity in cattle?
b Why is this loss a cause for concern?

Table 2 Modern European cattle

	Dairy breed	Beef breed
Body shape	• angular bony body with good udder support	• stocky muscular body with strong legs to support a large body mass
Product	• milk	• meat (muscle)
Product quantity	• cattle bred for high milk yield per day and long period of lactation (milk production)	• cattle bred for good growth rate as food efficiently converted into body mass (a high 'conversion rate')
Product quality	• different breeds produce milk with different fat and protein content	• newer breeds produce leaner, less fatty meat in greater quantity

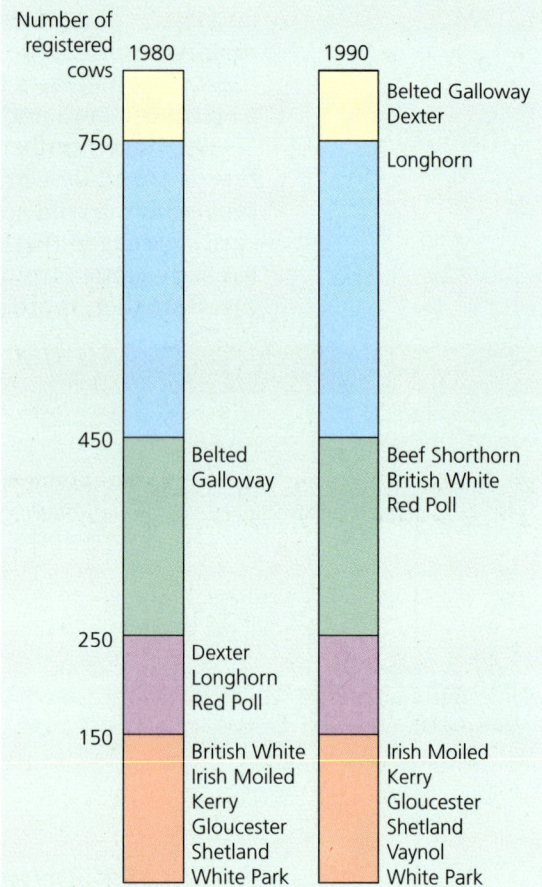

Fig. 5 Rare breeds of cattle

Ladder of success

Number of registered cows

1980 — Belted Galloway; Dexter Longhorn Red Poll; British White Irish Moiled Kerry Gloucester Shetland White Park

1990 — Belted Galloway Dexter; Longhorn; Beef Shorthorn British White Red Poll; Irish Moiled Kerry Gloucester Shetland Vaynol White Park

(scale: 750, 450, 250, 150)

Key ideas

• A long transitional period occurred between hunter–gathering and agriculture.

• Wheat has been produced by a mix of chance hybridisation and deliberate selection of characteristics from wild species. Modern species reflect the need for high-yield plants.

• Agriculture requires domestication of wild species, the time to tend animals, and the energy to cultivate crops.

• Domestication of animals is most successful in herding herbivores that breed in captivity. Modern cattle show a high food conversion rate with lean meat for beef cattle, and a high milk yield for dairy cattle.

10.3 Save and trade

Fungi and bacteria break down and destroy plant and animal products stored for food. They do this by releasing enzymes that digest the foodstuff. This decay is most rapid in moist, warm conditions and can quickly make the food inedible. Some enzymes within the foodstuff itself can remain active after the animal is killed or the plant collected. The action of these enzymes can also spoil the food (Fig. 6).

To preserve foods, techniques must be found that prevent or restrict microbial growth, or that inactive enzymes within the food (Table 3). Salt dehydrates cells by osmosis; its presence prevents the action of microbes and reduces enzyme activity in the foodstuff cells.

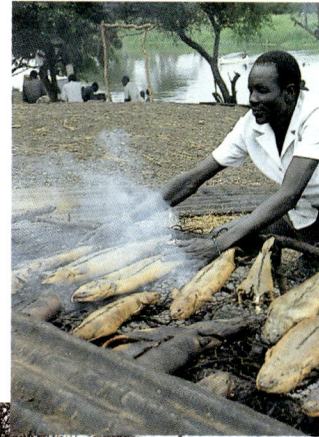

Burning wood releases tars and oils released that inhibit microbial growth. The heat from the fire dries and seals the surface of the food.

Table 3 The treatment and storage of food

Food	Treatment	Storage conditions
grain	ripen on the plant	in a cool dry place
	dry to reduce moisture content	in circulating air
meat and fish	smoke by hanging over a fire	in a cool airy place
meat and fish	cover with salt	in cool dry place
meat and fish	dry small pieces in full sun or next to a fire	in cool dry place

The !Kung dry strips of meat. Drying food reduces enzyme activity in the food cells and in any microbial cells.

Fig. 6 Food decay

Agent	enzymes in foodstuff	bacteria and fungi
Action	enzymes cause changes in cells of foodstuff	microorganisms release enzymes to digest food so it can be absorbed	microorganisms release toxic substances

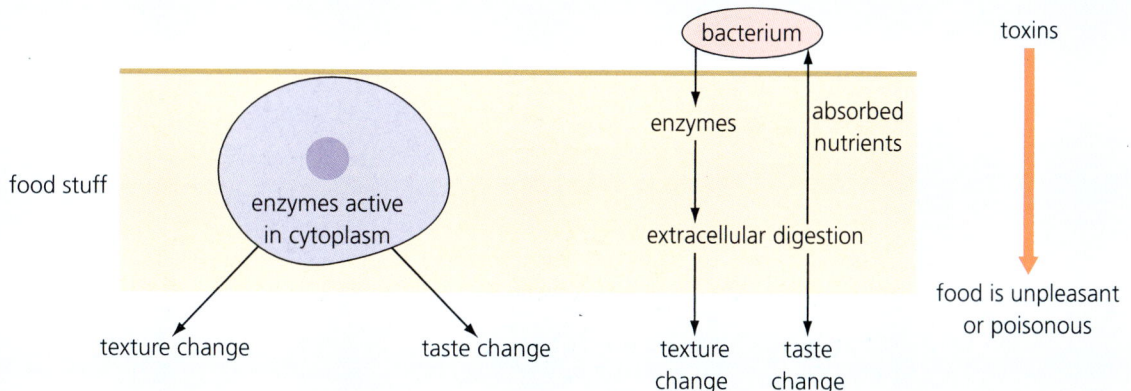

food stuff — enzymes active in cytoplasm

bacterium

enzymes → extracellular digestion | absorbed nutrients

toxins

food is unpleasant or poisonous

| **Result** | texture change | taste change | texture change | taste change | |

Fig. 7 Agriculture and social change through time

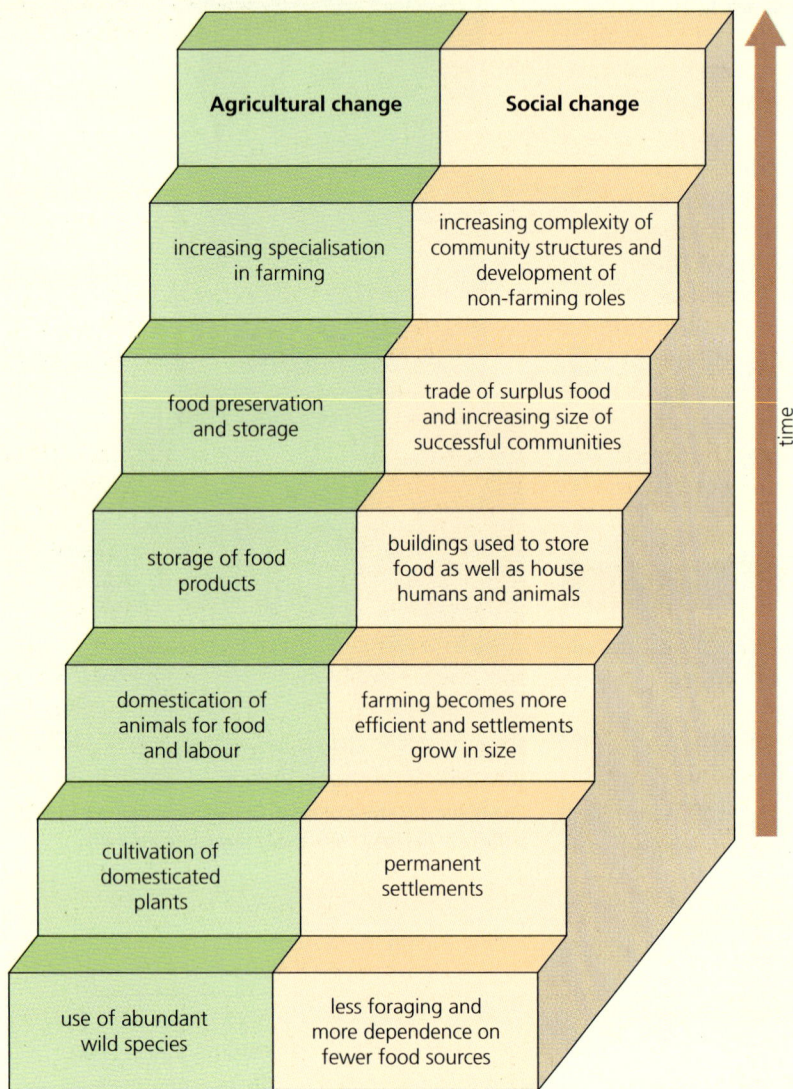

Agricultural change	Social change
increasing specialisation in farming	increasing complexity of community structures and development of non-farming roles
food preservation and storage	trade of surplus food and increasing size of successful communities
storage of food products	buildings used to store food as well as house humans and animals
domestication of animals for food and labour	farming becomes more efficient and settlements grow in size
cultivation of domesticated plants	permanent settlements
use of abundant wild species	less foraging and more dependence on fewer food sources

(vertical arrow labelled "time")

Sometimes the changes brought about by microorganisms are desirable because they improve the taste or help to preserve food. For example, milk keeps longer if it has been turned into cheese, fruit juice keeps longer as wine. These processes are called fermentations.

Agriculture reduces choice and variety in the diet because it brings increasing dependence on fewer species of both plants and animals. It also brings social change. Cooperation between individuals in farming villages would be important in providing enough food to last the community between growing seasons. This in turn would accelerate learning and develop specialist knowledge of cultivation, animal husbandry and food storage. Archaeological evidence suggests that the human population began to increase in the Neolithic period as communities became more settled. In better years, there would be a surplus of crops and animals. Surpluses, and the ability to store food, could create the opportunity to trade (Fig. 7).

8 Looking back to the begining of Section 10.3, give one method of preserving grain and one of preserving meat in a primitive farming community that had fire.

9 Describe four ways in which communities changed during the transition from hunter–gatherer groups to farming communities.

Key ideas

- Food can become inedible through the action of microorganisms or through the action of enzymes in the food itself.

- Preservation methods usually prevent or restrict microbial growth or inhibit enzyme activity or both.

- Fermentation is a type of microbial activity that is beneficial in food preservation.

- Agriculture is a stimulus for a change in social structure. Communities are more settled, increase in size, learn cultivation and husbandry skills and begin to trade.

10.4 An archaeological jigsaw

Evidence for agricultural development comes from a range of scattered and patchy sources. Only the most recent agricultural history is recorded in words or pictures.

In the areas where we think farming first developed, we still find the wild ancestors of cereals growing. Studying these might tell us how modern cereals developed.

Agricultural implements provide information about farming methods, while utensils for preparing or storing food can indicate how our ancestors milled grain and butchered animals.

The layering of remains at archaeological sites can provide a record of changes with time. Table 4 shows the number of animal remains at two sites in Jericho, 1000 years apart. However, a lot of this type of evidence has been destroyed because of later activities that disturb the ground, for instance, cultivating, rebuilding, exploring for minerals, and so on.

This Nyangatom woman is grinding millet using a hollowed-out stone called a quern. Querns have been used for thousands of years and are found around the world.

Table 4 1000 years of change

Date of site	Number of bones/teeth			
	gazelle	sheep/goat	cattle	fox
8000 BC	294	20	34	128
7000 BC	109	388	91	64

10 **What does the data in Table 4 suggest about the change in diet of people in Jericho from 8000 BC to 7000 BC?**

Although direct evidence such as animal bones and plant remains of domesticated species is sparse, the remains of bones of animals can provide information about domestication. The presence of many incomplete skeletons of many different species is typical of hunter–gatherer communities. Large, hunted animals were butchered where they were killed and the useful parts of the animals carried back to the settlement.

More systematic killing and butchering of a few species, in or very near the settlement, indicate that domestication has taken place. Large numbers of butchering tools are commonly found in such settlements. The preserved cattle skeletons are smaller than the remains of wild animals found in the same area. This suggests that early farming communities selected smaller animals that were easier to handle.

Another sign of domestication is the presence of the remains of many young or immature animals. Bones from an immature animal can be recognised by their structure and physical proportions. When a mature breeding herd is maintained, large numbers of young animals can be culled.

As in the domestication of cereals, accidental selection is likely to have played a part in the early stages of domestication of animals. For instance, unmanageable cattle are likely to have been slaughtered thereby leaving a more passive breeding stock. Over many generations, passive features would become more common.

Q11 **Give three pieces of evidence that would indicate that a community had begun to domesticate cattle. Explain how this evidence can be interpreted.**

Mapping archaeological sites can show settlement patterns and give some clues about activities and social structure. For example, agricultural settlements are much larger than hunter–gatherer settlements, and the buildings include small internal rooms with no hearths that might have been storage areas. New sorts of tools for cultivation and harvesting are found, and land around the settlements might show evidence of enclosures for animals or irrigation channels to water crops.

Q12 **Give four types of evidence that have been used to assemble our agricultural past.**

Key ideas

- Evidence for the development of agriculture is patchy and has to be carefully interpreted.
- Wild wheats and their relatives still alive today can be compared with modern wheats.
- The remains of farmed plants and animals are scarce, but tools and equipment can provide information about farming communities.
- Farming settlements get bigger and more complex with time.

10.5 Modern subsistence farming

Some farmers still depend on human power and produce only enough food for their immediate family. This is **subsistence farming** and is very common in the developing world. Subsistence farmers are often forced to farm in ecologically sensitive areas because the best land has been given over to the production of **cash crops** such as tea, coffee and cocoa for export to the developed world. In some countries rain forests are being cleared to provide land for subsistence farmers. Farming in newly cleared areas can quickly lead to soil exhaustion and erosion and has serious effects on the natural plants and animals in these delicate ecosystems.

Conservation in Costa Rica
The upland subsistence farmers of Costa Rica (Fig. 8) are encouraged by the Costa Rican government to use cultivation techniques that minimise environmental damage, exploit native species, and increase productivity.

Winds of 100 km per hour and 9 metres of rain per year have been measured in exposed upland sites. Such conditions quickly erode soil from cleared sites. Soil conservation is essential and several techniques are employed:
- planting immediately after clearance;
- planting windbreaks of native species;
- contour-ploughing;
- replanting immediately after harvest;
- ensuring there is vegetation cover;
- improving soil fertility by using vegetable waste and manures;
- installing drainage systems of pipes and channels to carry away surface water;
- conserving areas of natural rain forest.

Q13 **Explain how each of these measures conserves soil.**

Fig. 8 Costa Rica

Costa Rica is only 445km by 120km but must support a population of around 3 million people. The economy is dependent on agriculture and the country earns foreign currency by selling coffee, sugar and bananas that are grown on large plantations in the lowlands. Small subsistence farmers are found in the uplands.

This tree nursery is growing native species for use as windbreaks to divide fields, offer shelter and anchor the soil. They also become part of the local ecosystem unlike imported 'exotic' species.

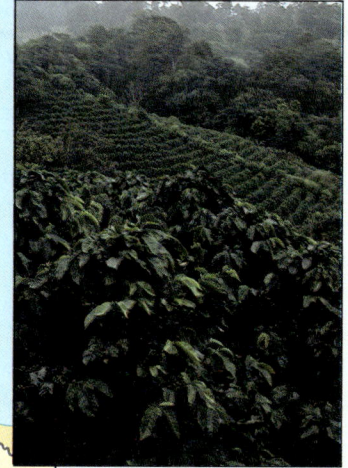

Cleared areas of rain forest are often steep and soil erosion is a problem. Contour-ploughing around the hills (rather than up and down) prevents soil from being washed away.

Manuel's farm

Manuel is a subsistence farmer in Costa Rica. He and his family live in a wooden farmhouse in a clearing in the forest. Fresh water is piped from a spring, cooking is done on a wood-burning stove, and paraffin is used for lighting. There are six other farms in the area.

Manuel grows native vegetables and beans in a small area surrounded by native trees and banana plants. He grazes a few cows on rough pasture, and chickens provide eggs. The farm produces enough food for a simple, mainly vegetarian, diet for Manuel and his family. Surplus milk is sent to a cheese factory and sugar extracted from sugar cane is sold through a farm cooperative. After putting aside a little for clothing and things needed in the home, any profit is put back into the farm.

The home of a subsistance farmer in Costa Rica.

Table 5 Time and energy required to plough a hectare				
Agricultural method	Required time/hours	Human energy consumption/kJ	Ox energy consumption/kJ	Total energy consumption/kJ
human power alone	400	814 800	0	814 800
using oxen	65	132 405	1 066 800	1 199 205

Manuel and his two eldest sons do all the farm work. They have two oxen to pull a cart and their plough. Using oxen substantially reduces the time and human energy consumption needed for some tasks (Table 5).

14a Study Table 5. Why is human energy still needed when using the oxen to plough?

b From Table 5, what is the advantage of using oxen to plough the land?

Role of multinational companies

Agriculture in many developing countries is modified by transnational companies to meet the needs of business and the developed world, rather than provide food for local people.

Governments of developing countries need to export cash crops to earn foreign currency. But the transnational companies who buy these crops often process them, adding considerably to the cost, before selling them on to customers in the developed world. In order to keep costs down and provide cheap food for the developed world, the companies pay very low rates for the original raw materials. So, the producers of bananas in the developing world may get less than 10% of the price paid for the same bananas in the UK.

Many developing countries are deeply in debt to international moneylenders who encouraged them to borrow in the 1970s and 1980s. This debt must be repaid, which means more cash crops must be exported. However, since the return is so poor, developing countries are often forced to produce more and more, using less and less suitable land to make the same amount of money. Subsistence farmers like Manuel often have no choice but to farm the less fertile and environmentally more fragile areas to grow food for their own use.

15 What is meant by the term 'cash crops'?

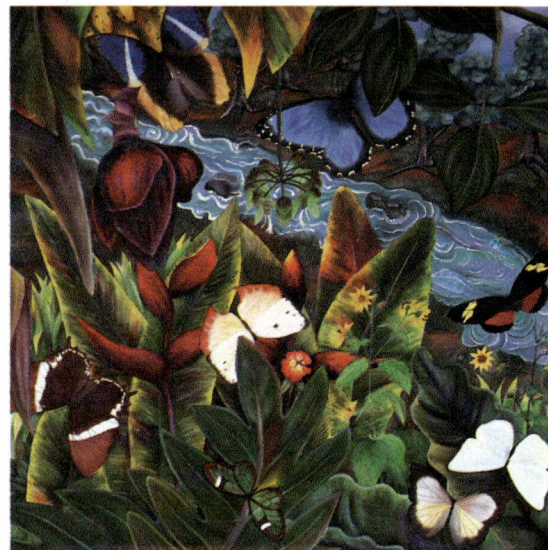

This wall-painting of biodiversity in the rain forest decorates a subsistence farmers' cooperative in Costa Rica.

Key ideas

- Subsistence farmers rely on human power and usually produce only enough food for their immediate family.

- Clearing sites for subsistence agriculture is depleting the tropical rain forest in many areas. Soil conservation is a major concern in these parts of the world.

- Transnational food companies can affect the agriculture and economics of developing countries in ways that are damaging to the environment.

10.6 Farming here and now

Modern, conventional intensive farms have a small workforce. They achieve high yields by mechanisation, the use of **crop protection products** such as artificial fertilisers and pest control chemicals, and the use of drugs such as antibiotics. The level of mechanisation means that intensive production has hidden energy costs in the use of fossil fuels for farm equipment.

Organic farming is often seen as a healthier option, but many organic farmers are also intensive producers even though they do not use the same methods as conventional farmers. Organic farmers cannot use chemical fertilisers or pest control. They rely on rotation and natural manures to provide nutrients, and cultivation to control pests or diseases. Plant remains and animal manure add humus to the soil. This contributes to soil structure as well as being a rich source of nutrients such as nitrates, phosphates (potash) and potassium.

Round in circles

Rotation is the basis of modern crop farming (Fig. 9).

16 a Study Figure 9. How many years does one rotation take in each system?

b What are the main differences between an organic rotation and a conventional rotation?

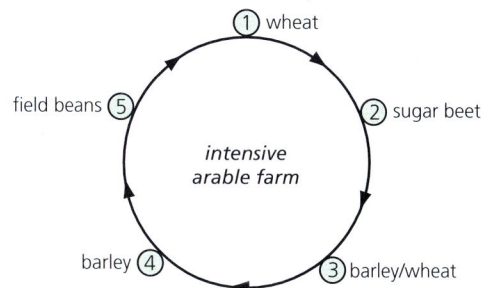

c How do peas contribute to soil fertility?

17 Again from Figure 9, give two reasons why silage is not produced on an organic farm.

Fig. 9 Organic and conventional rotation

Organic rotation

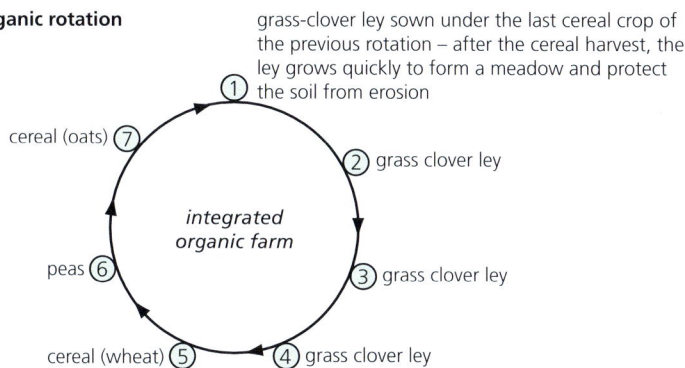

grass-clover ley sown under the last cereal crop of the previous rotation – after the cereal harvest, the ley grows quickly to form a meadow and protect the soil from erosion

1
2 grass clover ley
3 grass clover ley
4 grass clover ley
5 cereal (wheat)
6 peas
7 cereal (oats)

integrated organic farm

Conventional rotation

1 wheat
2 sugar beet
3 barley/wheat
4 barley
5 field beans

intensive arable farm

Grass-clover ley can be grazed by animals or cut as a hay crop. The hay is collected and stored to provide a nutritious winter feed. Droppings from grazing animals and dung from the winter quarters provide natural manure to maintain the soil structure and fertility.

Cereals require a lot of soil nutrients. Nitrates are particularly important for growth and eventually become part of the stored food in the cereal grains. Nitrates are added to the soil by including peas in the rotation system.

Artificial fertilisers provide nutrients for cereals, herbicides kill weeds, and fungicides kill disease-causing fungi. Post-harvest treatments reduce storage losses. Conventional farms can produce silage, a highly nutritious animal food. New grass is cut in spring and put in plastic bags or a concrete pit. Without air, chemical changes convert the grass into silage. Artificial fertilisers are needed to get good grass growth in spring. Specialist cutting and storage equipment is needed and has a high energy consumption.

Peas/beans are cultivated for three reasons:
• their root nodules contain bacteria that produce nitrate for the pea plant and eventually improve the nitrate content of the soil;
• they can be made into protein-rich animal feed;
• the break between cereal crops stops disease build-up in the soil.

Table 6 Estimated annual losses in conventional intensive agriculture

Cause of loss	Estimated loss %
insects	6
crop diseases	8
weeds	12
post-harvest losses	18

Even with crop protection products, there are still losses (Table 6).

The situation is most severe when a single crop is grown in the same ground for successive years. This is called **monoculture**

Fig. 10 Cow routines

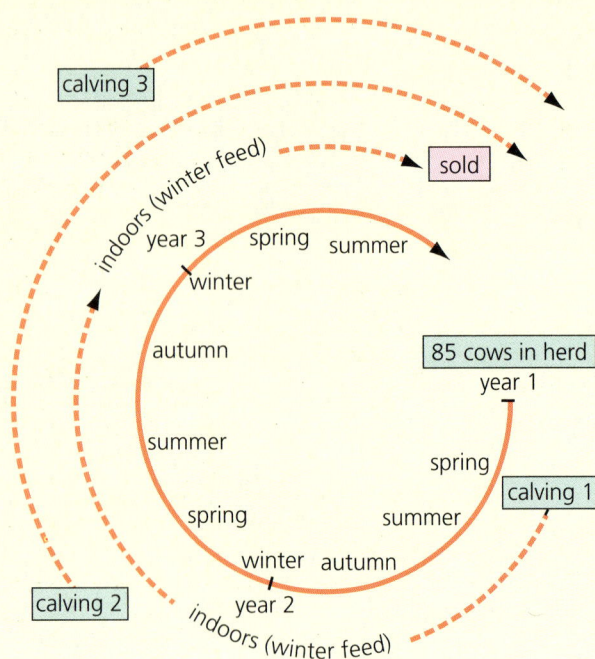

and high yields can only be maintained with the use of fertilisers and crop protection products.

18 What does Table 6 show about losses even with crop protection products?

Looking after livestock

In organic meat production, the animals graze organic meadows in summer and are fed on organic products in winter. Cattle housed indoors in winter are sheltered and keep each other warm, and food can be supplied efficiently. However, diseases are more readily transmitted between animals kept indoors but routine drug use is not allowed on organic farms. Any medication is strictly monitored, and a specified minimum time must pass before treated animals can be marketed.

In Figure 10, the nominal herd size is 85 cows but there could be 250–300 cows and calves on the farm because it takes 2–3 years for calves to reach a marketable size on an organic farm.

On conventional farms, the winter feed is silage plus a variety of manufactured concentrates and there may be routine drug treatment, such as the use of antibiotics to prevent disease. On this sort of farm, cattle are ready to go to market in 2 years or less.

19 What is the big disadvantage of housing cattle indoors?

20 Describe three differences between subsistence farming and intensive farming.

Key ideas

- Modern conventional intensive farms have a small workforce and high capital investment in farm equipment.

- Most intensive agriculture achieves high yields though mechanisation, and use of crop protection products and drugs such as antibiotics.

- Intensive production has hidden energy costs in the use of fossil fuels.

- Modern organic farming is intensive, but uses less energy and reduces environmental impact by not using synthetic fertilisers and pesticides, or the routine administration of drugs.

10.7 Energy crisis

In developed countries:
- populations are increasing slowly;
- intensive farming is supported by fossil fuels;
- there is a surplus of cheap food;
- there is high technology storage, processing and transport.

In developing countries:
- populations are rapidly expanding;
- subsistence farmers cannot afford high-yield crops, expensive fertilisers, and crop protection products;
- it is a struggle to produce enough food;
- cash crops are exported;
- the economic power of the developed world ensures that prices paid for cash crops are kept artificially low.

Only about 10% of the world's land will support agriculture. Of the land used for agriculture, about 50% is suffering from overproduction or bad agricultural practice.

Looking ahead

An energy crisis is forecast within the next 50 years as non-renewable fossil fuels are used up. Sustainable agriculture is possible, but only with changes in both what we farm and how we do it. One change might be increased use of **biological fuels**, fuels produced from renewable sources.

21 Why will continued use of fossil fuels will lead to an energy crisis?

Biomass is plant material produced by photosynthesis. It offers an alternative source of energy. For instance, sugar cane can be fermented to produce industrial ethanol, which can power vehicle engines. Sugar cane waste is burned to provide heat for the distillation process.

Another alternative is **biogas**. This is methane produced by fermenting plant and animal wastes in a generator. The gas can be used as a fuel for vehicles, for heating, and to generate electricity. The fermented waste can be used as animal food.

22a What is biomass?
   b Name two biological fuels.

To feed the world's population successfully, we have to take account of the social and environmental costs as well as the more obvious costs of production and transport.

This car runs on alcohol rather than petrol.

Key ideas

- Agriculture tends to fall into two camps, intensive in the developed world and subsistence farming in developing countries.

- Feeding the world requires sustainable methods that rely on renewable energy sources.

- Biological fuels are an alternative to fossil fuels; biomass can produce ethanol and methane which are both renewable sources of energy.

Life story

We are very lucky to have a baby son. When my husband and I wanted to start a family, we found I just didn't conceive. Our GP eventually referred us to a specialist clinic. The decision to use IVF (in vitro fertilisation) was not easy, and we were told that success was not guaranteed. We really had to learn a lot about the biology of reproduction.

IVF and other reproductive technologies now make it possible to offer women the chance to become mothers long after they have stopped producing eggs naturally. Sperm can be stored indefinitely, so dead men can become fathers. What safeguards, if any, can or should be applied to these new technologies?

11.1 Learning objectives

After working through this chapter, you should be able to:

- **describe** the structure of the male and female reproductive systems and how fertilisation occurs;

- **describe** how hormones control the female reproductive cycle;

- **recall** the biological principles of birth control;

- **distinguish** between innate reflexes and learned behaviour;

- **recall** that development and learning are related to an extended period of dependency;

- **interpret** graphs showing growth data and calculate growth rate;

- **recall** the main physical changes which occur at puberty;

- **explain** how growth hormone, thyroxine, and the sex hormones control growth during childhood and adolescence;

- **distinguish** between sex differences and gender appropriate behaviour;

- **explain** the meaning of ageing;

- **interpret** physiological data relating to the ageing process;

- **explain** the process of ageing in terms of genetic error, tissue degeneration, and malfunction of the immune system.

11.2 Life begins

In humans, like other mammals, the fertilisation of **ova** (egg cells) and the development of the baby are internal. Without a breeding season or a visible sign that an ovum is ready to be fertilised, the chances of a sperm and ovum meeting at the right time are reduced.

For fertilisation to occur, the sperm must reach the ovum in the oviduct at the right time. Male mammals have a penis to ensure that semen, a mixture of sperm and fluids, is transferred to inside the female's vagina. Once in the vagina, the millions of ejaculated sperm swim the remaining distance to the ovum. Because fertilisation is inside the oviduct, the chances of sperm and ovum meeting are high (Fig. 1).

However, the female system is hostile to sperm in two ways:
- the inside of the system is coated in mucus – this slows down the swimming movements of the sperm;
- the vagina is acidic – this shortens the life of the sperm.

It is thought that sperms are assisted on their journey by miniature waves of muscle contraction from the uterus towards the oviduct. This allows sperms to reach the oviduct 0.5–3 hours after coitus. Some substances in seminal fluid stimulate these contractions while others neutralise the acidity. The sperm can remain active for up to 2 days in the female system.

1 Where does fertilisation occur?

Fig. 1 Coitus

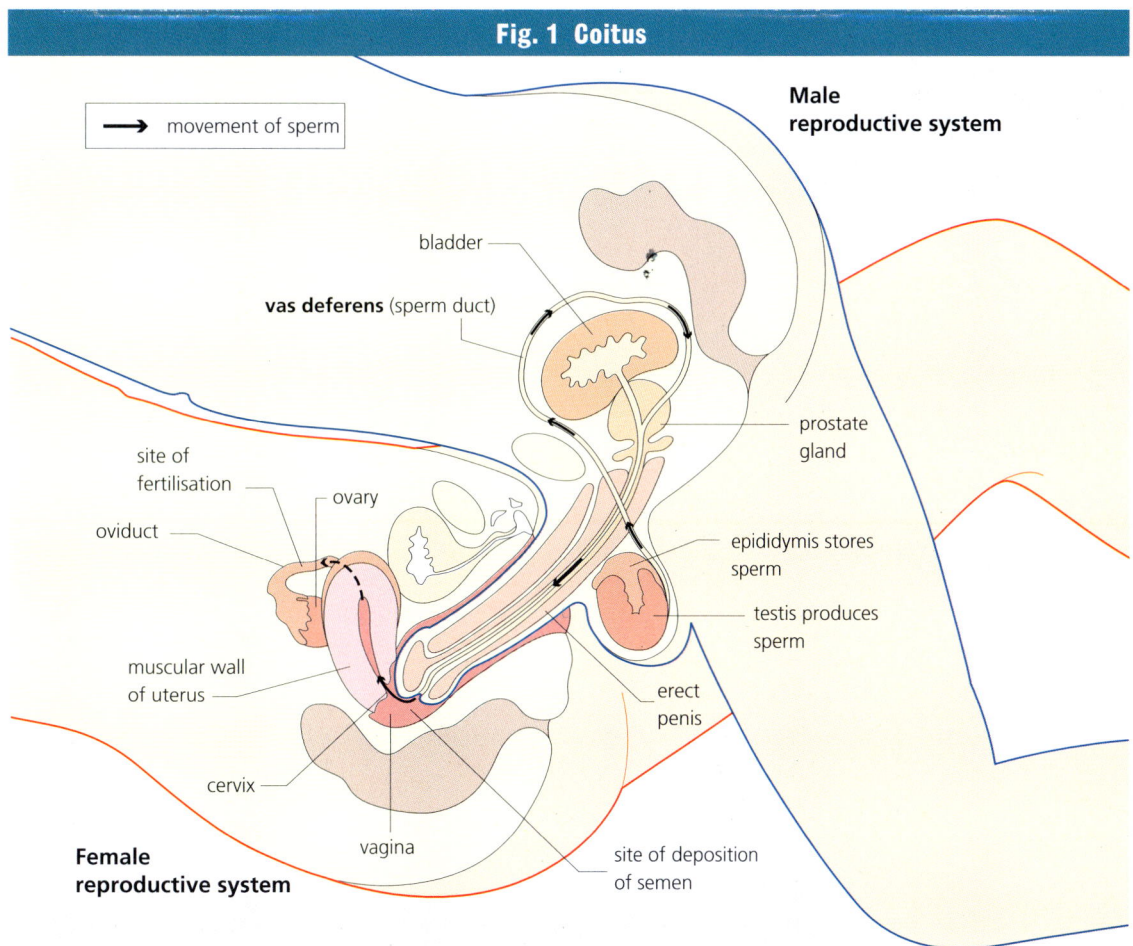

→ movement of sperm

Male reproductive system

bladder

vas deferens (sperm duct)

prostate gland

site of fertilisation

ovary

oviduct

epididymis stores sperm

testis produces sperm

muscular wall of uterus

erect penis

cervix

vagina

site of deposition of semen

Female reproductive system

Usually, a single ovum is released by an ovary each month. Only one sperm is needed to fertilise the ovum and form a new cell or zygote. The zygote divides and develops into a ball of cells called the **morula** that develops into an embryo and implants in the **endometrium** or uterus lining (Fig. 2).

Neither an unfertilised ovum nor a developing morula can move by itself, so both rely on muscle contractions and the cilia lining the oviduct to carry them into the uterus.

The embryo will continue to grow in the uterus for the next 40 weeks. After 2 months, the embryo has developed the main organ systems of the adult body and from this time it is called a **fetus**. The exchange of materials between the blood systems of the developing fetus and the mother is achieved by the **placenta**.

2 Explain how a blocked oviduct could affect the likelihood of fertilisation.

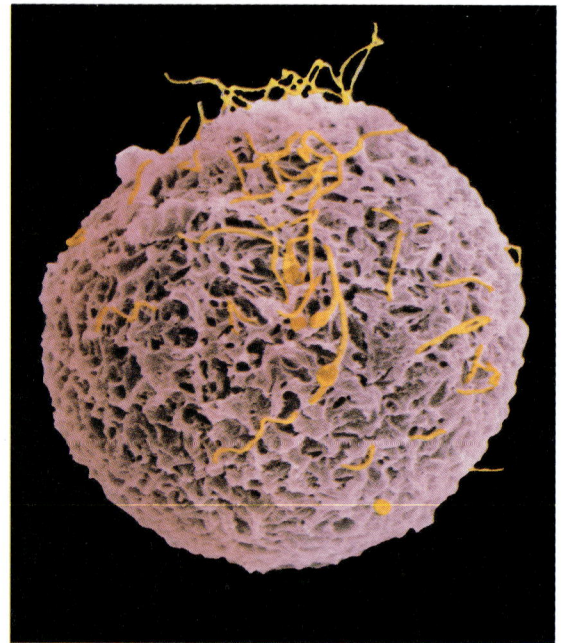

False-colour scanning electron micrograph of a human ovum (pink) with sperm (yellow) attached. Only one sperm will succeed in fertilising the ovum.

Fig. 2 Fertilisation

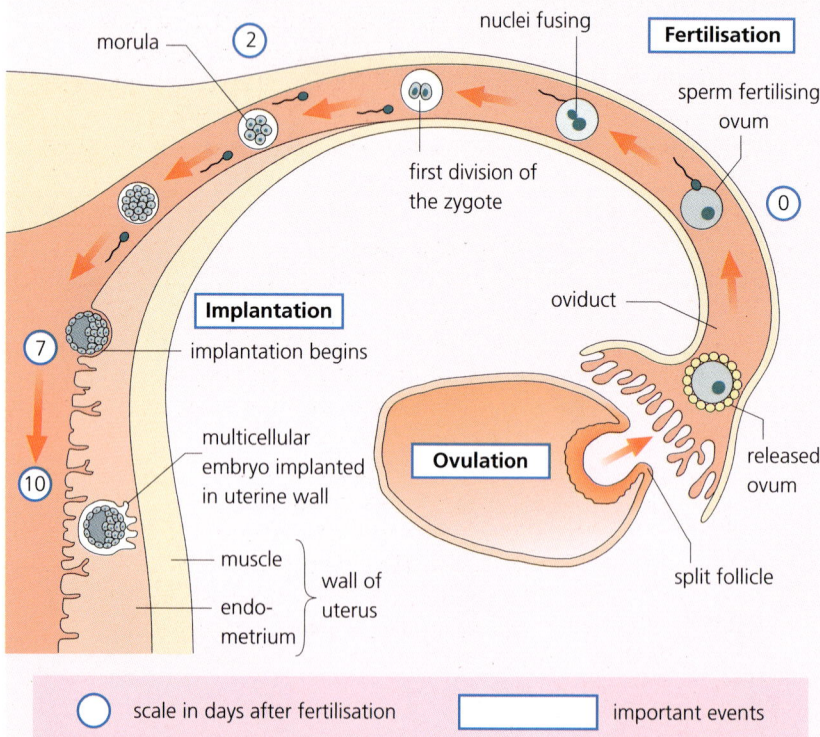

morula — ②
nuclei fusing
Fertilisation
sperm fertilising ovum
first division of the zygote
⓪
Implantation
oviduct
⑦ — implantation begins
multicellular embryo implanted in uterine wall
Ovulation
released ovum
⑩
muscle
wall of uterus
endo-metrium
split follicle

○ scale in days after fertilisation   ☐ important events

Source: adapted from Tudor and Tudor, *Understanding the Human Body*, Pitman, 1981

A question of timing

Ova are usually released singly from the ovaries. This is called **ovulation** (Fig. 3).

Before an ovum is released, the uterus or womb develops a new lining in case the ovum is fertilised. This stage of the ovarian cycle is called **repair**. It is important that the ovum is released when the lining of the uterus is in the right condition for a fertilised ovum to develop. Ovulation is marked by a slight rise in body temperature and a change in the mucus in the vagina. After ovulation, the endometrium begins a process called **proliferation**. It continues to thicken slightly and to develop a rich blood supply.

If fertilisation does not occur, the lining breaks down and is lost through the vagina. This is called **menstruation** or a menstrual period. The first day of a new menstrual period is taken as day 1 of a new reproductive cycle. The cycle usually lasts about 28 days, but can be anywhere between 24 and 32 days. Ovulation usually occurs about halfway between periods, around day 14 in a 28-day cycle (Fig. 4). Getting pregnant is most likely just after ovulation.

Fig. 3 Ovarian cycle

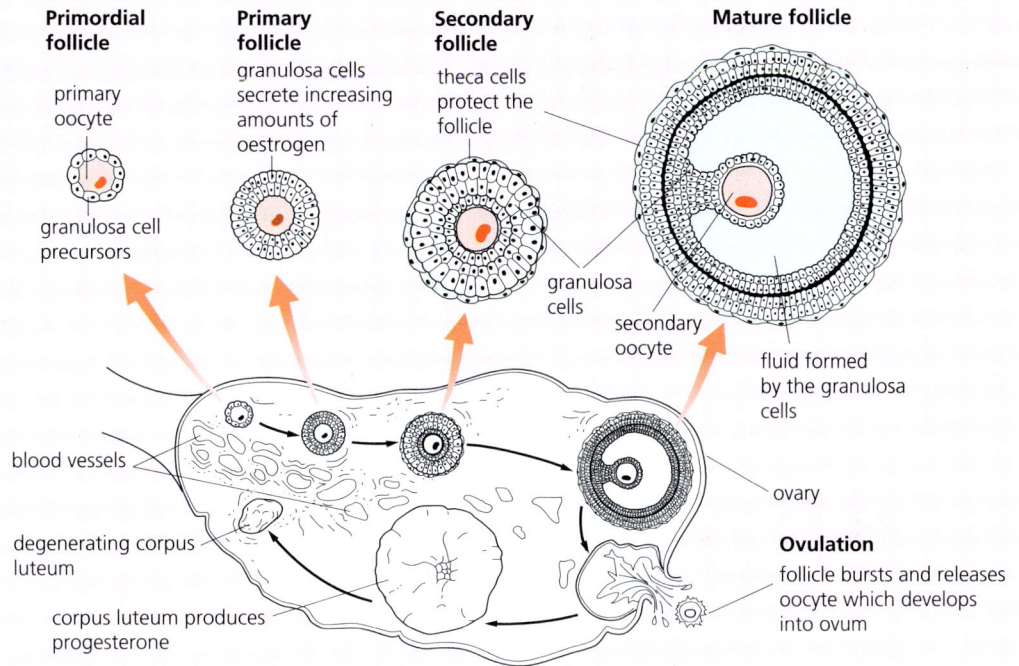

Primordial follicle

primary oocyte

granulosa cell precursors

Primary follicle

granulosa cells secrete increasing amounts of oestrogen

Secondary follicle

theca cells protect the follicle

granulosa cells

secondary oocyte

Mature follicle

fluid formed by the granulosa cells

blood vessels

degenerating corpus luteum

corpus luteum produces progesterone

ovary

Ovulation

follicle bursts and releases oocyte which develops into ovum

Source: adapted from Berne and Levy, *Principles of Physiology*, Wolfe, 1990

Fig. 4 Changes in the uterus

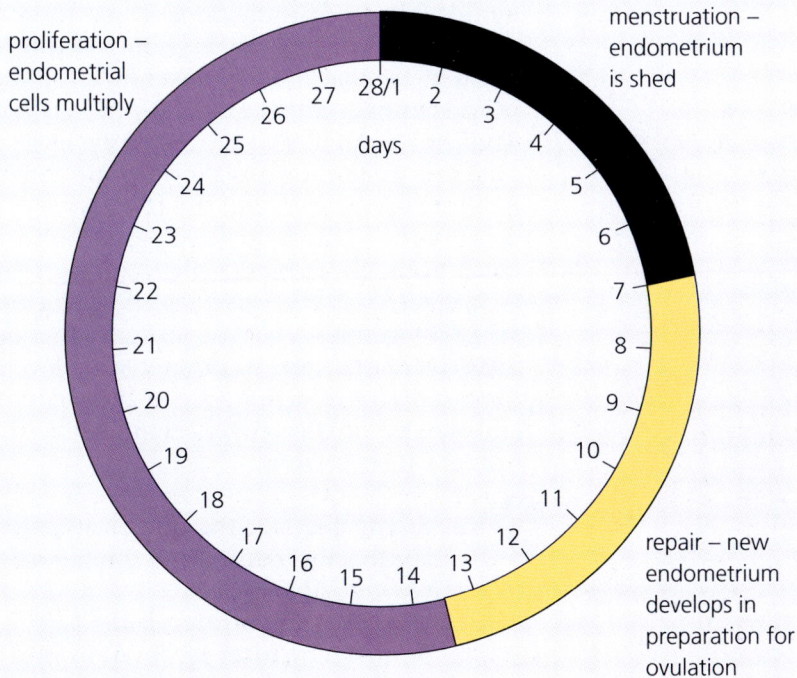

proliferation – endometrial cells multiply

menstruation – endometrium is shed

days

repair – new endometrium develops in preparation for ovulation

3 a What marks the start of a new reproductive cycle in women?

b At what stage in the human female cycle does ovulation occur?

c How often does ovulation occur in women?

Controlling events

The female reproductive cycle is controlled by two hormones produced in the pituitary gland. This gland is located on the base of the mid-brain and is formed from specialised neuro-secretory tissue. The two hormones it produces are:

• follicle stimulating hormone (FSH);
• luteinising hormone (LH).

Once secreted into the blood, both hormones cause changes in the ovaries. For this reason they are called **gonadotrophic hormones**. The ovary also makes two hormones:

• oestrogen;
• progesterone.

These are the **ovarian hormones**. Hormones are chemicals transported in

157

solution in the plasma of the blood. The body depends on hormones to stimulate (switch on) or inhibit (switch off) various processes. The gonadotrophic and ovarian hormones work together to give fine control over the female reproductive cycle (Fig. 5).

4a **4a Name the two gonadotrophic hormones and say where they are produced.**
  b Name the two ovarian hormones.

Fig. 5 Hormonal control of the cycle

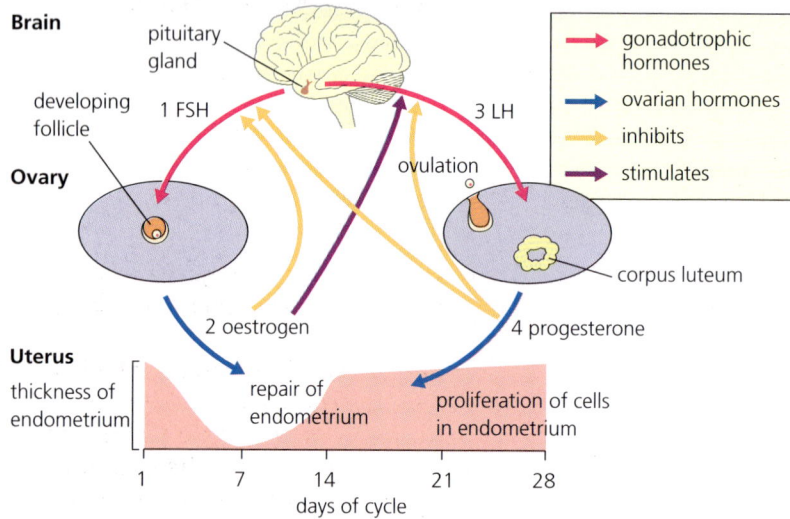

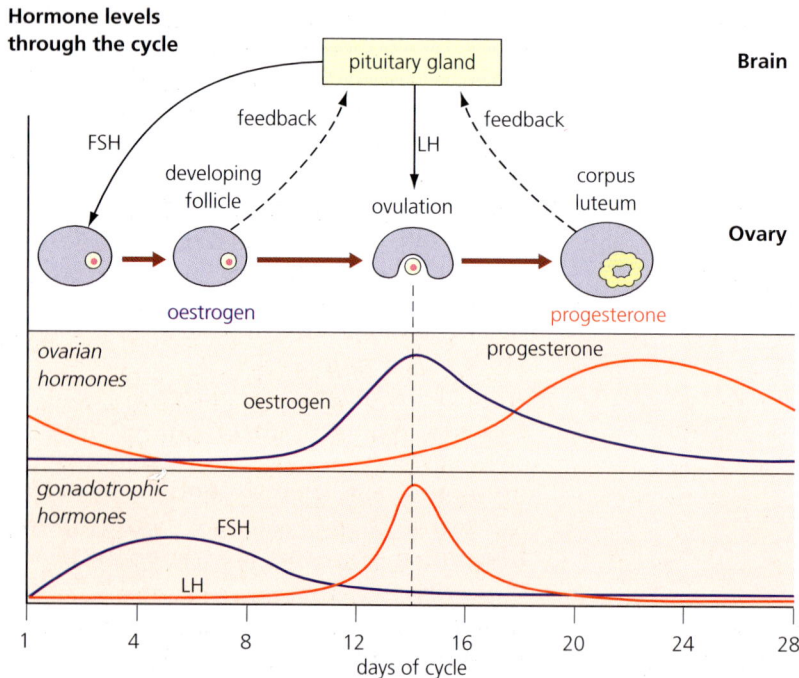

FSH stimulates the production of oestrogen by the ovary as the ovum develops. Oestrogen stimulates the repair of the endometrium. So, by the end of the first half of the cycle, the uterus lining is prepared, and an ovum is ready to be released from the ovary. Oestrogen also has two effects on the pituitary gland:
• it inhibits FSH production;
• it stimulates LH production.

The inhibition of FSH by an increase in the oestrogen level leads to a reduction in the level of oestrogen. This type of regulation of the level of oestrogen is called **negative feedback**.

The second half of the cycle begins with ovulation, triggered by LH from the pituitary. After the ovum is released, the empty follicle develops into the corpus luteum and produces progesterone. Progesterone stimulates the proliferation of the endometrium cells. Progesterone also has two effects on the pituitary gland:
• it inhibits the production of LH;
• it inhibits FSH production.

Progesterone level is also controlled by negative feedback: a high level of progesterone inhibits production of LH and leads to a reduction in progesterone level.

If fertilisation does not occur, the corpus luteum degenerates, and the level of progesterone falls. Without the inhibitory effect of progesterone on the pituitary, FSH is produced and the cycle begins again. If fertilisation does occur, the corpus luteum does not degenerate, so progesterone levels remain high in the presence of a developing embryo. Eventually, the placenta will also produce progesterone.

5a Briefly describe the role of negative feedback in controlling the female reproductive cycle.
  b Why does the progesterone level remain high after fertilisation, and what effect does this have on the reproductive cycle?

Key ideas

- Fertilisation takes place in the oviduct when one sperm cell joins with a single ovum.

- The female reproductive system has a regular monthly cycle in which an ovum develops in preparation for ovulation on or about day 14.

- Four hormones control the female reproductive cycle in humans.

- The two gonadotrophic hormones, FSH and LH, are produced by the pituitary gland.

- The two ovarian hormones, oestrogen and progesterone, are made by the ovary.

- Oestrogen and progesterone levels are controlled by negative feedback.

11.3 Controlling fertility

For many couples, controlling fertility is about **contraception**, or preventing conception, rather than assisting conception. **Birth control** is another name for the many ways of preventing conception. The effect of birth control on populations is discussed in Chapter 12.

Hormonal methods of contraception

Women can take additional female hormones in tablet form to control their fertility. This is called **oral contraception**. Hormones are the most effective and reliable type of birth control other than sterilisation by surgery.

There are several varieties of oral contraceptive for women. Most contain a relatively high level of oestrogen. This inhibits the production of FSH. Without FSH, no ovum develops and ovulation does not occur, so conception is not possible. Some contraceptive pills, called combined pills, also contain progesterone. Progesterone inhibits the production of LH (thereby further reducing the likelihood of ovulation), but the role of progesterone in hormonal contraception is not as clear as that of oestrogen. Combined pills are taken once a day for 21 days of the 28-day cycle. During the 7 days when the pill is not taken, the levels of oestrogen and progesterone fall, and the endometrium breaks down. This produces a lighter than usual menstruation.

Q 6 Explain how oestrogen in oral contraceptives prevents ovulation.

Hormonal 'patches' can now be inserted under the skin to give contraceptive protection to women. These patches are active for the several months and slowly release hormones into the bloodstream.

Current research includes work on a hormone pill for men. The male hormone testosterone is made in the testes. Derivatives of this hormone drastically alter the development of sperm, and can be used to control male fertility. However, effectiveness has been found to be very variable in trials.

Barriers, sterilisation and IUDs

Barriers can be physical (for example, the condom and the diaphragm) or chemical (for example, spermicide creams and pessary tablets that are placed in the vagina before intercourse to kill or immobilise sperm). These methods all aim to prevent the sperm and ova from meeting.

Sterilisation is considered permanent, because the effects of the techniques used are very difficult to reverse. Sterilisation for women usually involves putting a clip on the oviducts so that there is no passage from the ovary to the uterus. This is called **tubal ligation**. For men, sterilisation is by removing a small section of the vas deferens or sperm duct. This is called a **vasectomy**.

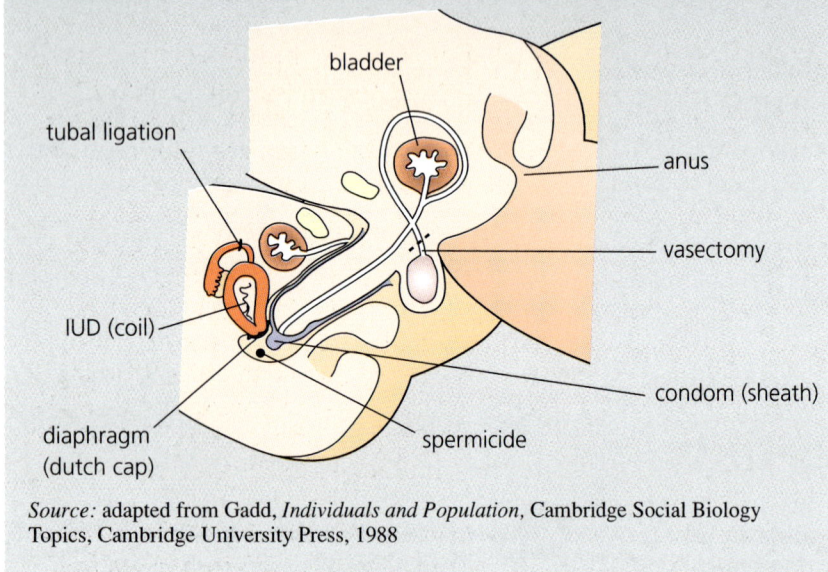

Fig. 6 Non-hormonal contraceptive methods

bladder
tubal ligation
anus
vasectomy
IUD (coil)
condom (sheath)
diaphragm (dutch cap)
spermicide

Source: adapted from Gadd, *Individuals and Population*, Cambridge Social Biology Topics, Cambridge University Press, 1988

The intrauterine device (IUD) or 'coil' is a modern development of the ancient discovery that the presence of a small object in the uterus can prevent conception. A modern IUD is kept permanently in the uterus. There are two main types of IUD available – those that contain copper and those that contain progesterone. We know that IUDs prevent implantation, and that non-hormonal ones are more effective when they contain copper, but exactly how they work is not known.

Figure 6 shows the placing of these methods of birth control in men and women.

Natural methods

Natural methods of birth control are based on the practice of avoiding intercourse during the most fertile time of the monthly cycle – immediately after ovulation. A woman's body temperature increases at the time of ovulation, so it is possible to identify when ovulation has occurred by keeping accurate body temperature records.

Lactation, or breast feeding, can have a contraceptive effect because it causes changes in the levels of the hormones that control ovulation and thereby reduces the chances of pregnancy. Lengthy lactation is common in some communities, for example, the !Kung hunter–gatherers of Africa where babies may continue to suckle for up to 2 years after birth.

Comparison of methods

The effectiveness of various birth control methods can be compared using the number of pregnancies to occur in a year for every 100 women using the method (Table 1).

Table 1 Effectiveness of some birth control methods	
Method	Pregnancies per year per 100 women
no method	50
natural methods	17–22
spermicide alone	20–30
diaphragm with spermicide	2–9
condom	2–7
IUD	2–4
combined pill	0.1
sterilisation	less than 0.05

Q 7 Study Table 1. Briefly explain the effectiveness of each method in terms of the way it works.

Key ideas

- Stopping conception is called contraception.
- Contraception can be achieved by a number of artificial methods. These include hormonal control, physical and chemical barriers, IUDs, and surgery.
- Natural methods of contraception are based on identifying the female's most fertile time and avoiding intercourse at that time.
- The effectiveness of each method of birth control is related to how it works.

11.4 Growth and development

During **gestation** (the period between conception and birth) the fetus grows very rapidly. However, not all parts of the fetus or its organs grow at the same rate. This means that the proportions of the body change during gestation (Fig. 7). The growth of body parts at varying rates at different times is called **allometric growth**.

The mass of the brain increases rapidly during gestation. Brain growth continues after birth and the brain reaches its final mass around puberty. The evolutionary advantages that a large brain gives humans are discussed in Chapter 7. The relationship between brain mass and body mass changes dramatically during gestation (Fig. 8). However, at birth, brain mass is about 10% total body mass. So, in human evolution, the advantages of developing a large brain before birth must be balanced against the disadvantages of a large head during birth.

At birth, other parts of the body are relatively small when compared with the head, but they then grow at a relatively faster rate. Newborn human infants are helpless and totally dependent on their mothers; as children grow, they rely on their parents for food and protection for many years. During this period of extended dependency, skills and abilities that will be useful throughout life are learned.

Q 8 Which part of the fetus grows most rapidly in the first 3 months of life?

Fig. 7 Allometric growth of the fetus

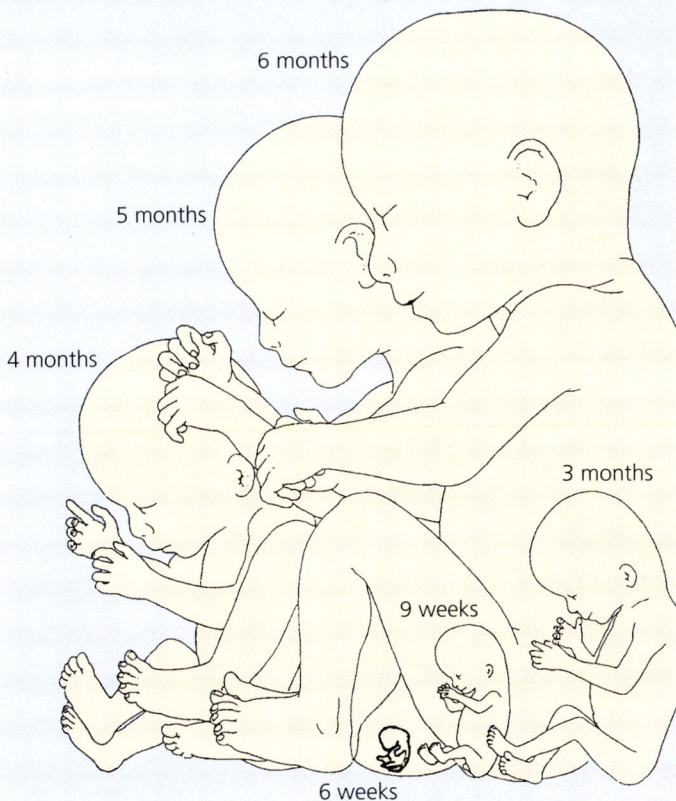

6 months
5 months
4 months
3 months
9 weeks
6 weeks

Source: adapted from *Science: Human Physiology and Health*, SEG

Fig. 8 Brain growth before birth

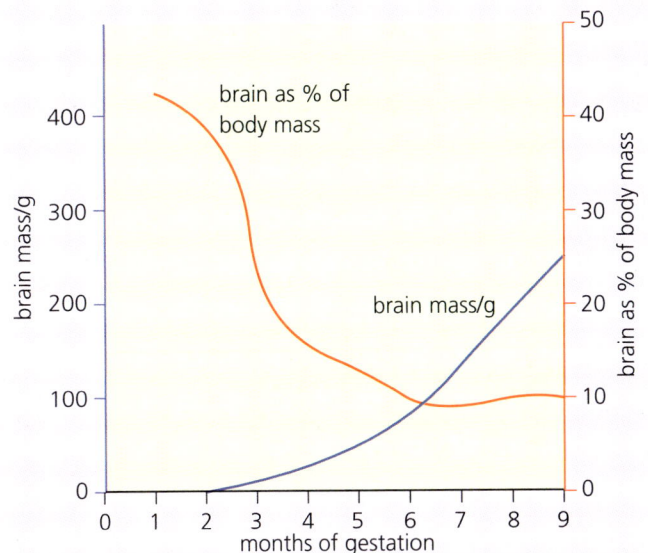

Source: based on data from Young, *Introduction to the Study of Man*, Oxford University Press, 1971

Table 2 Some developmental firsts	
Action	Age
rolls over	about 3 months
sits with support	about 5 months
sits alone	about $5\frac{1}{2}$ months
stands alone	about $9\frac{1}{2}$ months
walks holding on	about 10 months
walks alone	about 12 months

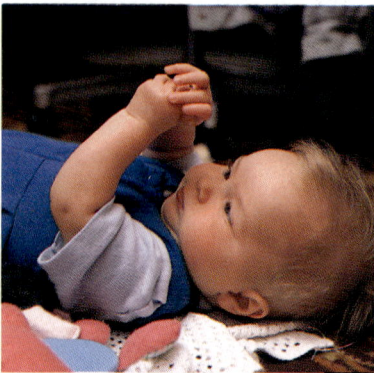
At 2–5 months, a baby can clasp its hands.

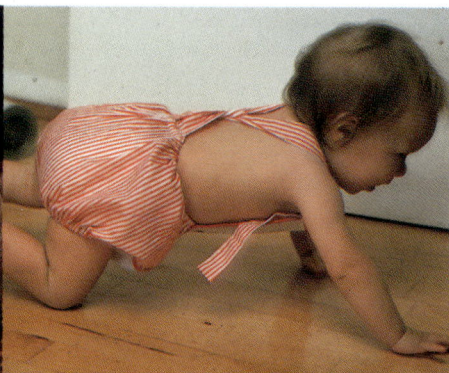
Crawling starts at 6–9 months.

Drinking from a cup is achieved at 10–15 months.

Scribbling develops at 1.5–2 years old.

At 2–3 years old, a child can ride a tricycle.

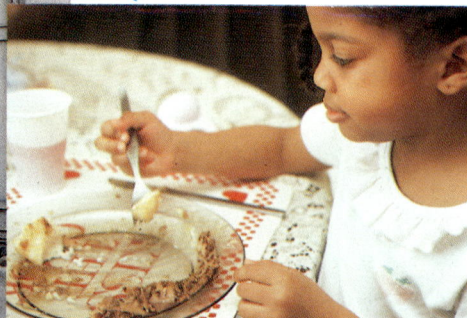
How to use a knife and fork is learned at 3–5 years.

If you touch the palm of a newborn baby with a finger, the baby will grasp your finger so tightly that you could lift the child up. This is innate behaviour, not learned behaviour, it is instinctive and is due to a reflex action (Chapters 4 and 5). Many innate reflexes in humans disappear soon after birth. Most of a child's behaviour and abilities are learned. Although all children are individuals, it is easy to identify some milestones in their development from being born to learning to walk (Table 2).

Child development is a result of interaction between genetic and environmental effects. Many features and attributes appear as the child ages, but they can be grouped into four important areas:
• physical maturation;
• social development;
• intellectual development;
• emotional development.

Physical maturation

Physical changes such as growth are called **maturation**. The simplest measures of growth are changes in height or mass. Special charts called **centile charts** are available for boys and girls. These charts show the expected pattern of change based on the national average. They also show the expected spread of results for 94% of the population (Fig. 9). A very erratic pattern of growth with plots outside the expected range would be cause for concern. Control of growth is dealt with in Section 7.5.

9 Study Figure 9. What does the graph suggest about (a) the baby boy's mass at birth and (b) his growth during the first year?

Social development

Social development is very important in humans. As social animals we have many rules regarding what is and what is not acceptable. These rules apply to everyday situations and often change according to the child's age. For example, babies are very demanding, and are totally dependent on adults. A similar level of demanding behaviour is likely to be seen as inappropriate in an older child. Social

Fig. 9 Centile chart for boys

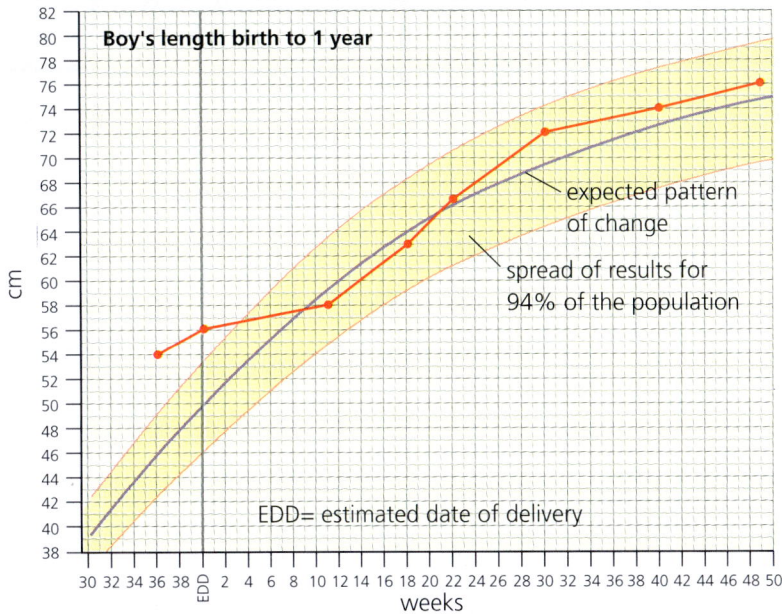

Boy's length birth to 1 year

expected pattern of change

spread of results for 94% of the population

EDD= estimated date of delivery

cm (y-axis): 38, 40, 42, 44, 46, 48, 50, 52, 54, 56, 58, 60, 62, 64, 66, 68, 70, 72, 74, 76, 78, 80, 82

weeks (x-axis): 30 32 34 36 38 EDD 2 4 6 8 10 12 14 16 18 20 22 24 26 28 30 32 34 36 38 40 42 44 46 48 50

parents towards independence and self-reliance. It includes learning different ways to display affection towards others, and learning how to respond to displays of affection.

Intellectual development

The changes in our abilities, ideas, comprehension and communication skills are all part of our intellectual development. The speed of learning and ability to grasp the significance of what we learn is a measure of our intelligence.

Q 10 Use examples to distinguish between an innate reflex and learning.

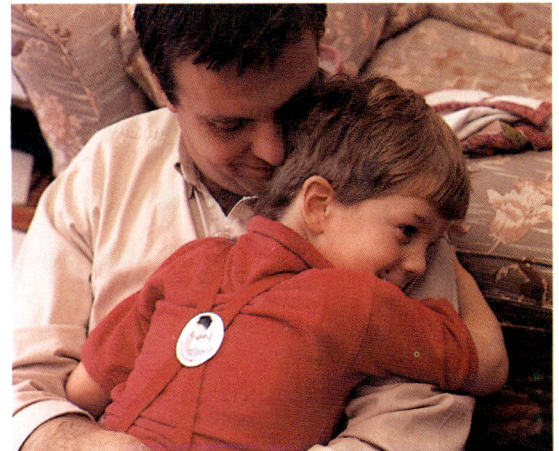

Hugging is one of the ways we learn to show affection for each other.

development and behaviour are influenced by the interactions between a child and other humans. For example, a baby mimics its mother's smiles and sounds. When older children play together, they learn to share toys and take turns in activities; sharing and turn-taking are important social attributes for adults.

Emotional development

Emotional development in children is the progress from complete dependence on the

Key ideas

- Growth can be measured by an increase in size or mass. Relative growth indicates the change in proportions of an individual.

- Humans are born with relatively large brains and heads. Other parts of the body grow more rapidly after birth.

- Babies are born with innate reflexes such as grasping. These reflexes might relate to primate ancestry. Innate reflexes soon disappear.

- Milestones in the development of babies and children can be identified.

- Children have an extended period of dependency. This offers time for the learning associated with social, emotional and intellectual development.

11.5 Growing up

Puberty marks the physical change from child to adult. It is controlled by hormone levels, and results in a period of both physical and emotional change. Humans have a long **pre-pubertal** stage when they are physically smaller and sexually immature. This coincides with the period of extended dependency.

Growth rate

There are four distinct phases in human life (Table 3). Each phase of life is characterised by a change in **growth rate** (Fig. 10).

Table 3 Phases of life	
Life phase	Approximate age/years
infancy	up to 4
childhood	4–11
adolescence	11–18
adulthood	18+

Q 11 Study Figure 10b. Give the two ages when the growth rate is greatest (a) in males and (b) in females.

Fig. 10 Mass, age and growth rates for girls and boys

(a) Growth and age

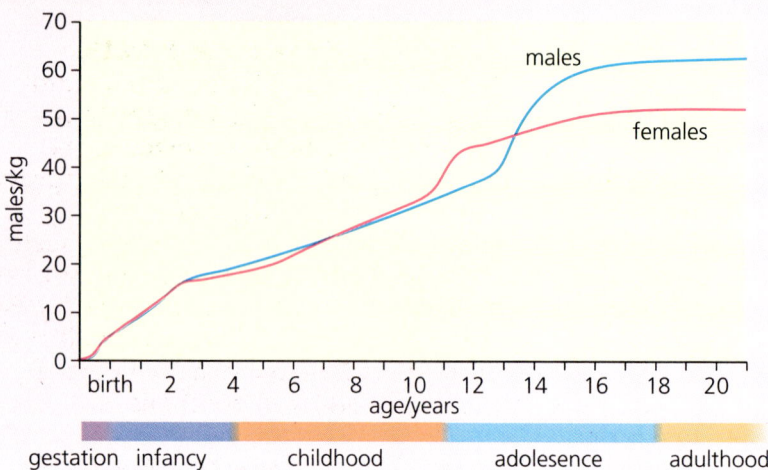

(b) Growth rate and age

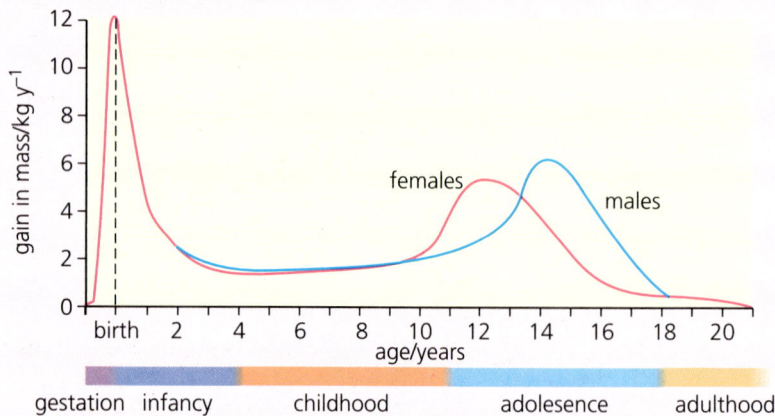

Source: adapted from Gadd, *Individuals and Population*, Cambridge Social Biology Topics, Cambridge University Press, 1988

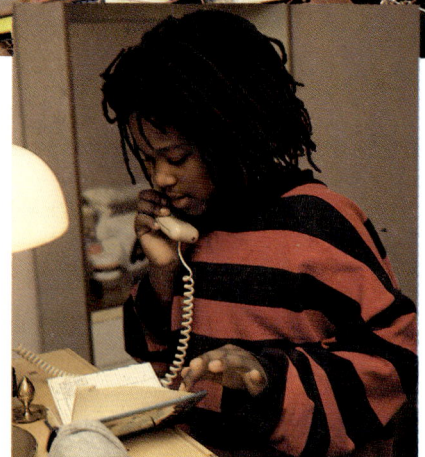

Puberty occurs during the phase of an individual's life that is called adolescence and usually happens sooner in girls than in boys.

Fig. 11 Age and body proportions in girls and boys

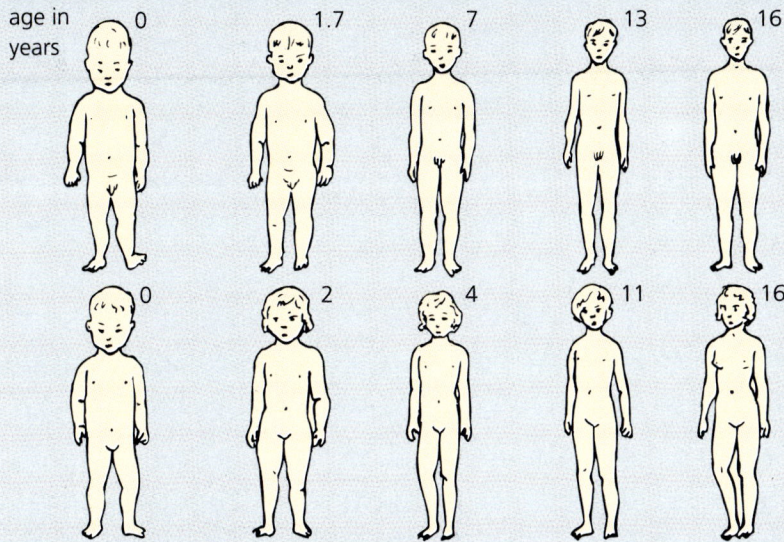

age in years

Source: adapted from Forma and Linstead (eds), *Human Biology Laboratory Manual*, Science Teachers Association of Western Australia, 1984

Fig. 12 Growth curves

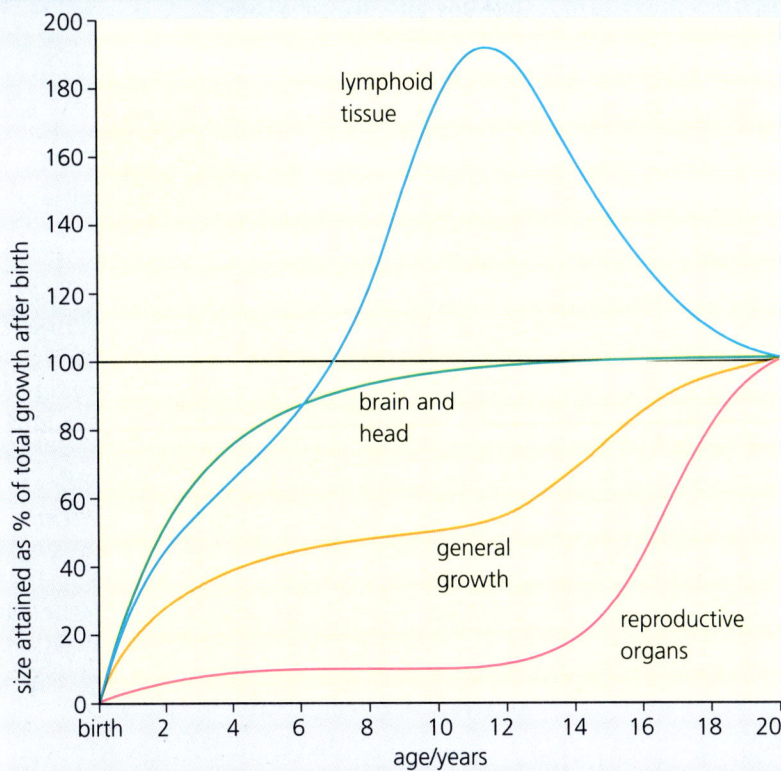

Source: adapted from Young, *Introduction to the Study of Man*, Oxford University Press, 1971 (after Tanner, 1962)

Relative growth

Although humans grow from birth until they are adults, the growth is not even. Many parts of the body change at different times and girls and boys grow at different rates (Fig. 11).

12 Study Figure 11. Describe how the *relative* lengths of the head, body and legs change as humans grow.

Throughout childhood and adolescence, internal organs also grow at very different rates. Lymphoid tissue is an important part of the immune system and grows most rapidly during childhood. Brain growth is most rapid in early life. The growth of the reproductive organs is most rapid during adolescence, and is accompanied by a general growth spurt (Fig. 12).

Hormones and control of growth

The pituitary gland is in overall control of growth. During childhood, growth is controlled by a hormone made in the pituitary gland and called **pituitary growth hormone (PGH)**. PGH stimulates the growth of body tissues, and the elongation of the long bones. This leads to an increase in height. PGH also stimulates the thyroid gland, which is in the neck, to produce a hormone called **thyroxine**. Thyroxine controls growth and metabolism.

13 How is growth controlled during childhood?

At puberty, environmental and internal genetic triggers cause the release of large amounts of the sex hormones. Oestrogen is the main sex hormone in females; in males, the sex hormone is testosterone. Both these hormones stimulate further growth, but testosterone is much more powerful than oestrogen.

14a How is growth controlled during puberty in males?
** b Describe how the timing of puberty and the growth rate differ in males and females.**

165

Fig. 13 Hormones and growth

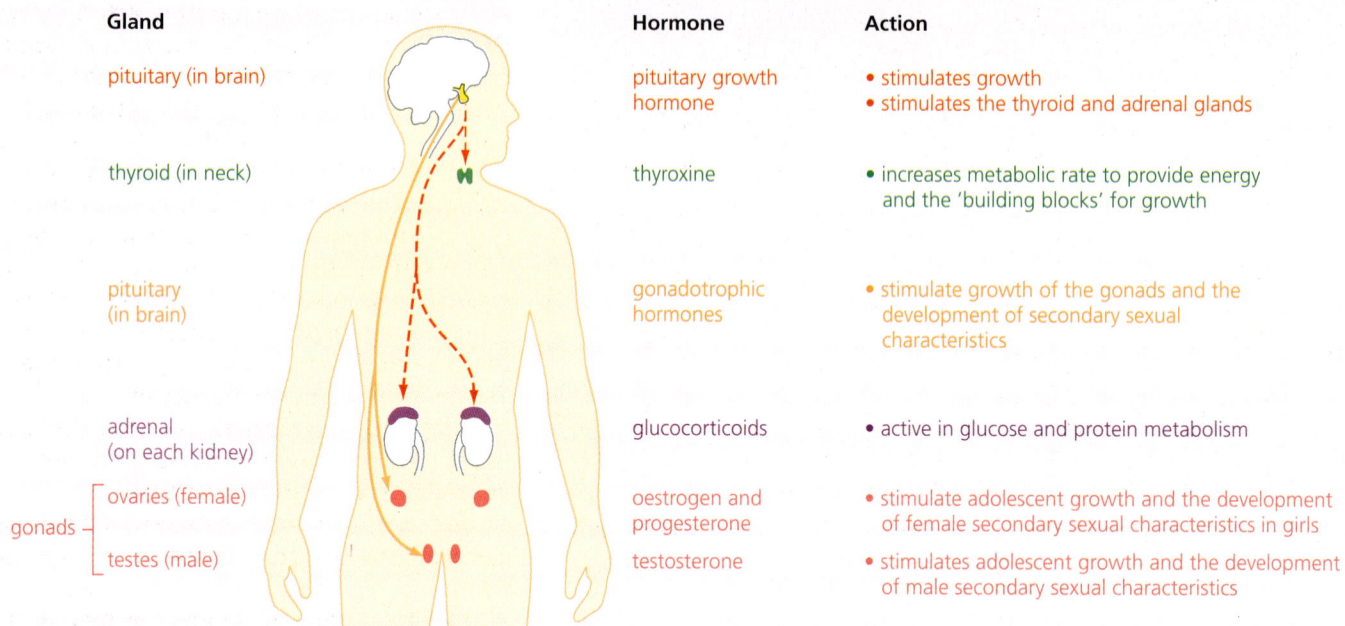

Gland	Hormone	Action
pituitary (in brain)	pituitary growth hormone	• stimulates growth • stimulates the thyroid and adrenal glands
thyroid (in neck)	thyroxine	• increases metabolic rate to provide energy and the 'building blocks' for growth
pituitary (in brain)	gonadotrophic hormones	• stimulate growth of the gonads and the development of secondary sexual characteristics
adrenal (on each kidney)	glucocorticoids	• active in glucose and protein metabolism
gonads — ovaries (female)	oestrogen and progesterone	• stimulate adolescent growth and the development of female secondary sexual characteristics in girls
gonads — testes (male)	testosterone	• stimulates adolescent growth and the development of male secondary sexual characteristics

Figure 13 summarises the roles played by hormones in the control of growth in humans.

Table 4 Secondary sexual characteristics

Secondary sexual characteristic	Male	Female
Hair growth	• on face, chest, in groin and armpits	• in groin and armpits
Voice	• larynx enlarges greatly and voice deepens	• larynx enlarges slightly
Body shape	• muscles and bones grow, shape of body changes	• pelvis becomes broader, fat deposited on hips and thighs changes body shape • breasts develop
Secondary sex organs	• penis, scrotum, and prostate enlarge • sperm formation begins	• uterine tubes, uterus and vagina enlarge; uterine and vaginal linings thicken • ovulation and menstruation begin
Psychological changes	• feelings and sexual drives associated with adulthood begin to develop	• feelings and sexual drives associated with adulthood begin to develop

Sexual characteristics

The **primary sexual characteristics** are the sex organs. At puberty, the sex organs begin their reproductive function and release sex hormones. These hormones produce the physical changes that are associated with puberty and called **secondary sexual characteristics** (Table 4).

An increased self-awareness develops during puberty and individuals become more conscious of how they look and how they relate to others. There is a lot of variation within each sex but, on average, males tend to have a larger body mass and greater physical strength. Studies show that girls develop better verbal abilities than boys from the age of 11. Boys' behaviour tends to be more aggressive, both physically and verbally. However, boys are, on average, better at mathematics from the age of 12 and tend to have better spatial skills. Spatial ability enables us to visualise objects in relation to each other in space.

Gender roles

The biological differences between males and females have a bearing on what each does in the society in which they live. The

The Naxi (pronounced Nashi) people live in Yunnan in south-west China. They have a matriarchal society where the women trade and control business, and inheritance follows the female line.

The men look after the children.

type of activity or job generally done by each sex is called a **gender role**. The fact that women produce children often means that their gender role is related to caring for offspring and the home environment. In many societies, women are more likely to be in caring or child-related jobs. Men tend to be physically stronger and often have the role of provider for the family. So, hard physical work and jobs that require long periods away from the home are often done by males. However, gender roles are closely related to the particular society in which they are found. For example, among the !Kung, one of the few hunter–gatherer societies of the twentieth century, the men hunt while the women gather and look after the children (Chapter 8).

Gender appropriate behaviour is linked to a society's understanding of appropriate gender roles. Every culture has certain expected codes of behaviour for men and women during courtship, in family life, and in the workplace. Although it is linked to sex differences, most gender appropriate behaviour is learned.

15 **Distinguish briefly between sex differences and gender appropriate behaviour.**

Key ideas

- Puberty marks the change from child to adult.

- There are four distinct phases in human life: infancy, childhood, adolescence, and adulthood.

- Children grow from birth to adulthood. The pattern of growth is not even.

- In children, is growth controlled by PGH, and metabolism by thyroxine.

- The growth spurt during adolescence is due to testosterone in males, and oestrogens in females.

- Puberty happens sooner in girls than in boys. The timing of sex hormone release differs in males and females. Sex hormone production results in the development of secondary sexual characteristics.

- Sex differences include both primary and secondary sexual characteristics.

- Gender roles and gender appropriate behaviour for both males and females are products of biological and social factors.

11.6 Growing old

From a biologist's point of view, ageing begins as soon as we are born and all the changes in a lifetime are part of the ageing process. As the years go by, physical appearance changes due to the ageing of external features. Internally, ageing may alter the way that the organs function, and their efficiency declines. **Senescence** is the term for the later stages of the ageing process. The rate of change varies in different organs of the body (Fig. 14).

16 Biologically, what is meant by ageing and when does it start?

17a Study Figure 14. Which physiological function declines most with increasing age?
b What will be the effect of this decline?

Elderly people often say they still feel like they did when they were young even though their bodies have changed substantially (Fig. 15).

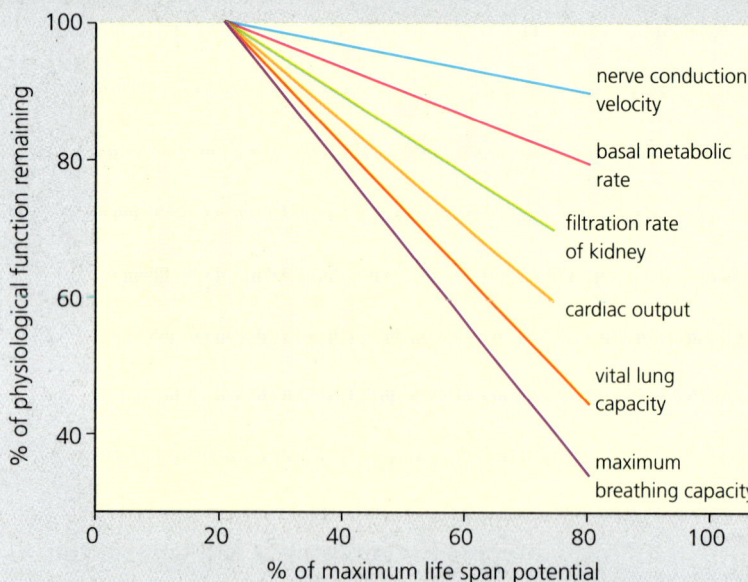

Fig. 14 Physiological function and age

% of physiological function remaining (y-axis: 40, 60, 80, 100)
% of maximum life span potential (x-axis: 0, 20, 40, 60, 80, 100)

nerve conduction velocity
basal metabolic rate
filtration rate of kidney
cardiac output
vital lung capacity
maximum breathing capacity

Source: adapted from SATIS 61, *Why do we grow old*, Association for Science Education

18 Study Figure 15. Which physical changes due to ageing are sex-related?

Osteoporosis

Osteoporosis is a disease associated with ageing, in particular the ageing of women. It is often called 'brittle bone disease'. After the **menopause**, the end of the female reproductive cycle, the levels of the ovarian hormones drop and osteoporosis can develop. There is a loss of bony tissue, so the bones become less dense and more likely to fracture. Fractures, especially those of the hip and femur, can be serious and life-threatening in the elderly.

Investigating ageing

There are two types of study that could be used to assess the ageing process:
- a **longitudinal study** follows the same group of people through their lives, as they age;
- a **cross-sectional study** looks at a groups of people of different numerical ages.

There are advantages and disadvantages to each kind of study. For instance, a longitudinal study takes very much longer to do than a cross-sectional one, but the data gathered from a cross-sectional study might be less reliable because it is not known what other factors, such as diet, might have influenced the apparent effects of ageing.

Cross-sectional studies have a particularly important disadvantage. These studies involve taking means of numbers of people of the same numerical age. However, these people might not be ageing at the same rate. So, they might have different physiological ages. This can make the rapid changes that occur in individuals at certain physiological ages, such as puberty, less obvious.

19 Give a disadvantage of each type of study used to investigate ageing.

Fig. 15 Key changes of ageing

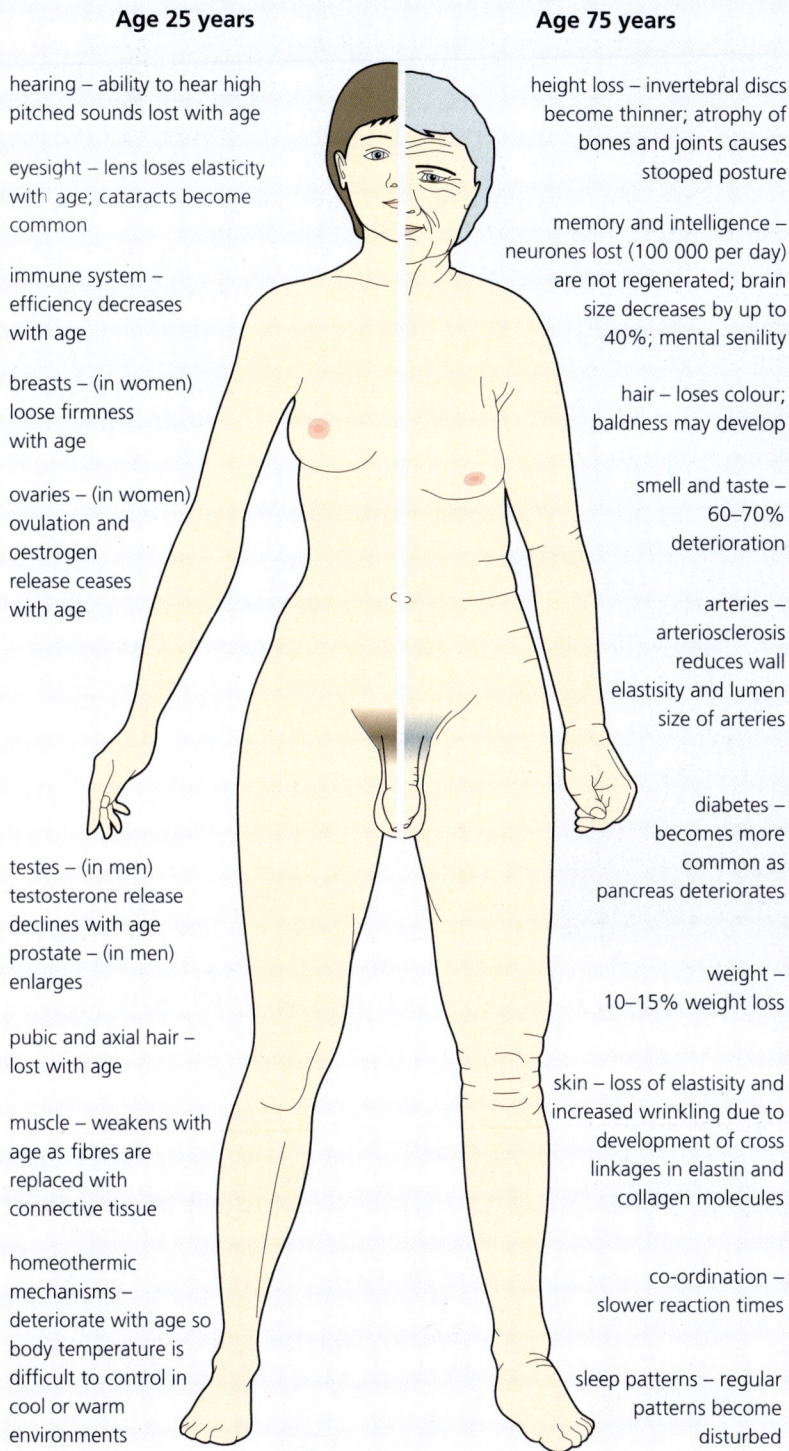

Age 25 years

Age 75 years

hearing – ability to hear high pitched sounds lost with age

eyesight – lens loses elasticity with age; cataracts become common

immune system – efficiency decreases with age

breasts – (in women) loose firmness with age

ovaries – (in women) ovulation and oestrogen release ceases with age

testes – (in men) testosterone release declines with age
prostate – (in men) enlarges

pubic and axial hair – lost with age

muscle – weakens with age as fibres are replaced with connective tissue

homeothermic mechanisms – deteriorate with age so body temperature is difficult to control in cool or warm environments

height loss – invertebral discs become thinner; atrophy of bones and joints causes stooped posture

memory and intelligence – neurones lost (100 000 per day) are not regenerated; brain size decreases by up to 40%; mental senility

hair – loses colour; baldness may develop

smell and taste – 60–70% deterioration

arteries – arteriosclerosis reduces wall elastisity and lumen size of arteries

diabetes – becomes more common as pancreas deteriorates

weight – 10–15% weight loss

skin – loss of elasticity and increased wrinkling due to development of cross linkages in elastin and collagen molecules

co-ordination – slower reaction times

sleep patterns – regular patterns become disturbed

Source: adapted from Gadd, *Individuals and Population*, Cambridge Social Biology Topics, Cambridge University Press, 1988

Why do we age?

Ageing consists of many complex body changes caused by the interaction of many factors. The two main groups of factors are:
• genetic;
• environmental.

Comparison of the patterns of physical change, such as hair going grey, and the life span of parents and their offspring shows that the main cause of ageing is not inherited. However, genetic changes can occur to the DNA in body cells during a person's lifetime. These changes are called **somatic mutations** and they do contribute to ageing.

There are a variety of environmental factors that can affect body tissues. Pollutants in the air, ultraviolet light, diet, and disease, are just a few aspects that contribute to the ageing of our external appearance. Such factors can also contribute to changes in our internal body processes.

There are three ways to explain how genetic and environmental factors might interact in the ageing process:
• accumulation of genetic error;
• degeneration of tissues;
• malfunction of the immune system.

Studying ageing presents problems as it is difficult to measure ageing even in the features that we can see changing.

169

Somatic mutations are passed to daughter cells at cell division when new cells are produced for growth and repair of tissues. Errors in the DNA can lead to incorrect cell function. Eventually, the accumulation of genetic errors leads to more and more malfunctioning tissue and results in some of the features of ageing.

Degeneration of tissue is largely due to 'wear and tear' and incorrect repair of damaged tissue. For example, changes in tissue elasticity lead to changes in our outward appearance. Similarly, changes in the structure of internal organs lead to reduced efficiency. Exposure to substances in the environment (for example, air pollutants) might accelerate this degeneration.

The immune system recognises 'self' (the body's own proteins) as different from 'non-self' (other proteins such as invading bacteria or viruses). So, the immune system can protect us from harmful microorganisms by destroying them. It also protects us from our own tissue when things have gone wrong, for instance, when a cell has developed cancerous features. As we age, our immune system becomes less efficient, and less able to destroy invading pathogens or cancerous cells. The immune system might also destroy the body's own healthy tissue, this is called **autoimmunity**.

Pollutants such as smoke or gases in the air affect the lungs and impair gaseous exchange. This is Mexico City.

20 Briefly describe the part played by genes and by the environment in
(a) accumulation of genetic error
(b) degeneration of tissues
(c) malfunction of the immune system.

Will understanding more and more about the processes of ageing enable us to delay its effects? Like the advances in reproductive technology, it could lead to children being born to older and older parents. Will we ever be able to halt ageing? Would you want to live for ever?

Key ideas

- Ageing is an on-going process from birth throughout life.

- Ageing causes changes in external appearance and a decline in the functioning of body organs.

- Ageing can be studied by longitudinal and cross-sectional studies.

- Ageing can be understood in terms of the interactions between genetic and environmental factors that lead to accumulation of genetic error, degeneration of tissue, and malfunction of the immune system.

Enough space?

After gaining her place on a 3-month Raleigh International expedition, Emily has a tough time ahead of her. She must raise funds to go to Chile, and buy equipment for projects based in coasts, deserts, and the high mountains.

Visiting another country brings health risks. Emily will camp and stay in villages and farming settlements. Before travelling to Chile, she has to have a series of immunisations to reduce her chances of catching some diseases that can be passed from one person to another. Illness could be particularly dangerous in remote areas where healthcare is not available.

12.1 Learning objectives

After working through this chapter, you should be able to:

- **recall** how antibodies are produced in response to antigens and explain the principle of vaccination and immunisation;

- **explain** the influence of vaccination and immunisation on population growth;

- **interpret** age pyramids, survival curves and population growth data;

- **calculate** population growth rate from data on birth rate, death rate, emigration and immigration;

- **relate** changes in population size and structure to the demographic transition model;

- **describe** how food supply and disease affect populations;

- **describe** the effect of clean water and sewage treatment on mortality patterns;

- **evaluate** the effect of different methods of birth control on the growth of populations.

12.2 Infection and immunisation

In the UK, there is an immunisation programme for children against poliomyelitis, tetanus and tuberculosis. It is important that immunisations are kept up to date as some become less effective with time. Before going to Chile, Emily must have additional immunisation against hepatitis A, cholera and typhoid fever as these diseases are more common in South America. Organisms able to cause disease are called **pathogens** (Fig. 1).

Fig. 1 Effect of a bacterial pathogen

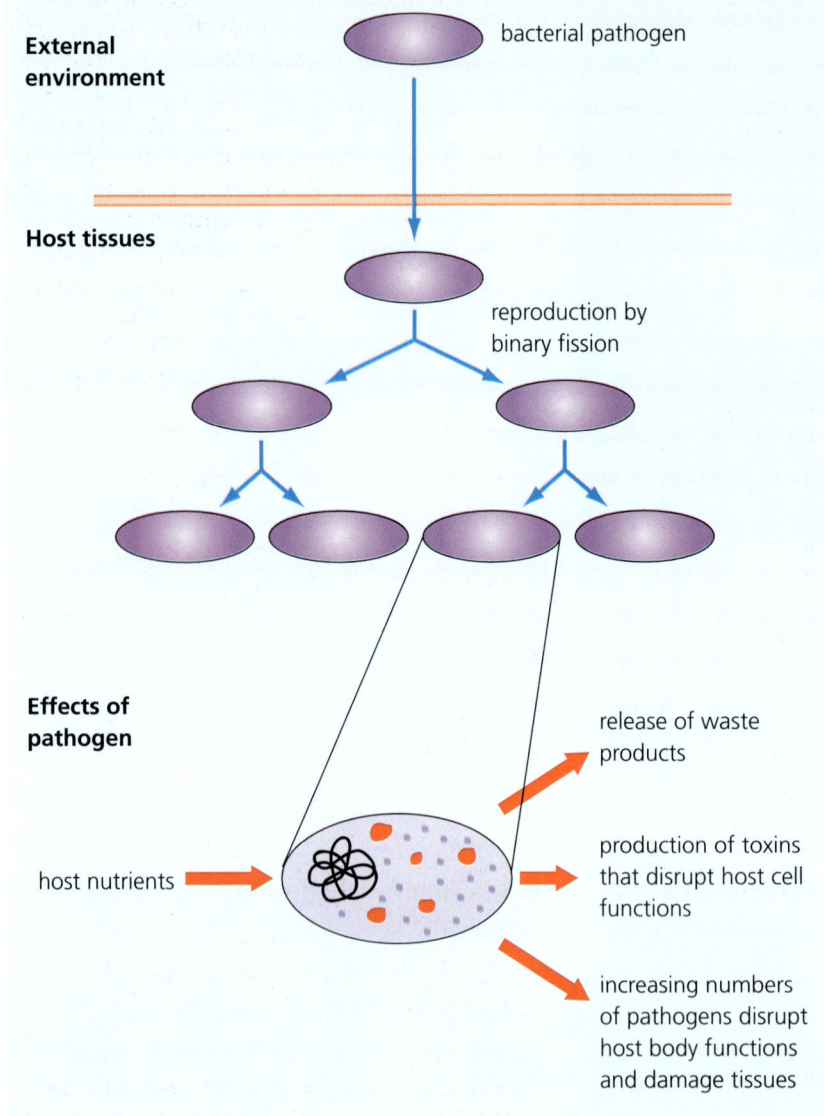

External environment

bacterial pathogen

Host tissues

reproduction by binary fission

Effects of pathogen

host nutrients

release of waste products

production of toxins that disrupt host cell functions

increasing numbers of pathogens disrupt host body functions and damage tissues

1 Study Figure 1. Describe three ways that pathogens affect the body.

Preventing infection

The human body has several ways to keep itself free of pathogens. The first line of defence is the presence of physical and chemical barriers that stop pathogens entering body tissues (Table 1).

Table 1 Preventing the entry of pathogens

Possible entry site	Nature of barrier
skin	• tough layer of keratin is a physical barrier to pathogens • sweat contains salt and has antimicrobial properties
airways and lungs	• hair-like cilia remove dirt and microorganisms • mucus is a physical barrier to pathogens
gut	• saliva contains lysozyme – an enzyme that destroys bacteria • stomach acids destroy pathogens • mucus is a physical barrier to pathogens

The second part of the body's defences is to attack pathogens that have entered the body tissues. All cells are covered with chemical substances called **antigens**. If a pathogen enters the body tissues, the **immune system** can distinguish the foreign antigens from the body's own antigens. The immune system responds to foreign antigens by making specific substances called **antibodies**. Each antibody combines with its specific antigen, and enables one of range of responses that destroy or inactivate the pathogen (Fig. 2).

2 Give two physical and one chemical barrier to pathogens.

3 Describe how the immune system combats a pathogen.

Fig. 2 Pathogens and the immune system

Entry

pathogen

via wound in skin via gut via lung

Reproduction of pathogens in body

antibodies antitoxins

Immune system

foreign antigens stimulate immune system → memory cells

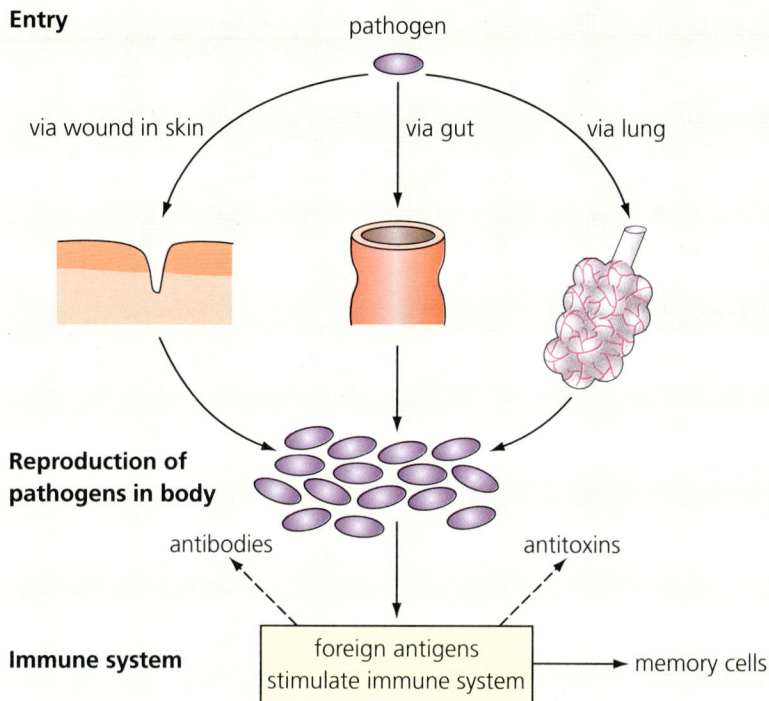

The first time a pathogen invades the body, the immune response is quite slow and there is often time for the pathogen to multiply and cause illness. After an immune response, **memory cells** retain the ability to recognise the pathogen and to make antibodies against it. So, if the same pathogen invades a second time, the memory cells enable large amounts of antibody to be produced very quickly. This **secondary response** usually destroys the pathogen before any symptoms of the disease are shown – the individual has developed **immunity** to the disease.

Immunisation is the term for giving individuals **artificial immunity** to a disease so that they are protected without having been infected.

There are two types of artificial immunity:
* **active immunity** – the body is stimulated to make its own antibodies;
* **passive immunity** – ready-made antibodies are injected.

Passive immunity does not produce an immune response, and the immunisation only gives protection for as long as the antibodies are active. For example, passive immunity can be used against Hepatitis A. The injection contains a concentrated dose of antibodies, but it is only effective for a few weeks. A preparation of ready-made antibodies is called an **antiserum**.

In active immunity, a weakened or dead form of the pathogen is injected and stimulates the immune system to produce antibodies and memory cells. These preparations are called **vaccines**.

Pathogens can be killed with heat or chemicals to make a **dead vaccine**. The immune system can still recognise the foreign antigens, so antibodies and memory cells are made in response to the vaccine. The new Havrix vaccine is an example of a dead vaccine against the disease Hepatitis A. It is more effective than using an antiserum because it causes an immune response. **Live vaccines** (also called **attenuated vaccines**) contain a weakened pathogen that cannot reproduce quickly enough to cause disease. Live vaccines are usually more effective than dead vaccines in stimulating the immune system to produce antibodies and memory cells.

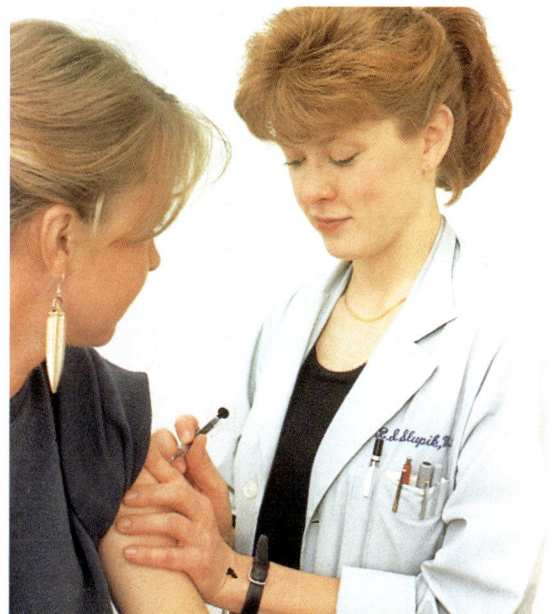

Gaining and maintaining immunity to a range of diseases usually involves only an injection or two.

Tetanus bacteria produce a toxin that causes muscle cramps and can cause death by heart failure and breathing failure. Immunisation against tetanus uses a **toxoid** or harmless form of the toxin. When it is injected, the immune system produces antibodies to the toxoid but these antibodies also protect against the more harmful tetanus toxin.

Even active immunity does not always last a lifetime (Table 2).

Table 2 Disease and active immunisation

Disease	Source of immunity	Extent of immunity
poliomyelitis	live vaccine	10 years
tetanus	toxoid	10 years
tuberculosis	BCG live vaccine	lifelong
hepatitis A	Havrix dead vaccine	10 years
cholera	dead vaccine	3–6 months
typhoid fever	TyhimVi dead vaccine	3 years

4 List four ways of providing artificial immunity.

5 Account for the difference in effectiveness of vaccination against poliomyelitis and vaccination against cholera.

A clean water supply is essential for good health.

Transmission of disease

Infectious diseases are passed or **transmitted** from a person who has the disease to another person. Droplets from coughing or sneezing spread common diseases such as influenza, whereas direct contact is necessary to transmit venereal diseases. **Transmissible diseases** affect all human populations.

A disease spreads when sufferers infect others before recovery. So, the greater the number of people in a population who have immunity, the less the chance of transmission of the disease. However, it is not necessary to immunise 100% of the population to prevent widespread outbreaks of disease. Immunising a sufficiently large number of people to protect to the whole population is called the **herd immunity effect.**

People who are fortunate enough to live and eat in hygienic surroundings, drink clean water and have good sanitation, tend to take such conditions for granted. When drinking water is contaminated by faeces, or food contains harmful organisms, or people

Table 3 Cause and transmission of some diseases

Disease	Type of pathogen	Transmission method
poliomyelitis	virus	• water contaminated with faeces of an infected person • droplets coughed or sneezed by an infected person.
tetanus	bacterium	• dirt in cuts or wounds (this bacterium is common in soil)
tuberculosis (TB)	bacterium	• droplet infection from an infected person (risk is higher in overcrowded conditions)
hepatitis A	virus	• water contaminated with faeces of an infected person
cholera	bacterium	• water contaminated by faeces of an infected person or carrier (a carrier doesn't show symptoms of the disease)
typhoid fever	bacterium	• water contaminated with faeces of an infected person or carrier

live in overcrowded conditions, many life-threatening diseases can easily be passed from person to person (Table 3).

When visiting some countries it is not sufficient to rely on immunisation to prevent infection (Fig. 3).

6 Explain briefly how each precaution in Figure 3 reduces the danger of catching an infectious disease.

Fig. 3 Extra precautions

- wash your hands before eating
- fully cook food and eat it hot
- don't buy food from roadside stalls
- drink bottled water or purify water by boiling or using iodine–resin filters
- avoid ice in drinks
- peel uncooked fruit and vegetables before eating
- carry a sterile syringe and needle kit for any medical injections

Key ideas

- Physical and chemical barriers protect the body against invasion by pathogens.
- The immune system attacks pathogens that have invaded the body.
- Antibodies are produced in response to foreign antigens.
- Artificial immunity can be given by vaccination and immunisation.

12.3 The numbers game

Life is physically demanding and living conditions simple in remote areas of Chile. This community of a few farms and a medical centre has no electricity, sewage treatment, or running water.

Spatial distribution of a population is the way the population is distributed. It depends on climate, geography, and the economic state of the country (Table 4).

Table 4 Some population facts

Country	Population 1995	Area/km^2	People per km^2
Chile	14.3 million	748 795	19
UK	58.5 million	241 595	242

Emily will find the spatial distribution of the population in Chile very different from that of the UK. Chile is three times larger but has less than a quarter of the population of the UK. However, 92% of the UK population live in various urban areas, compared with 82% of the Chilean population, most of whom live in the capital, Santiago. So, the rural population (18% of the total in Chile) are spread over a massive area.

Most countries keep accurate records so that the total population can be determined. These records include:
- births;
- deaths;
- immigration (people entering the country);
- emigration (people leaving the country):
- census data.

There are several useful ways of examining population data (Fig. 4).

Fig. 4 Manipulating population data

Growth = (births − deaths) + (immigration − emigration)

Percentage growth rate in a given period

$$= \frac{\text{population increase during the period}}{\text{population at the start of the period}} \times 100\%$$

Birth rate $= \dfrac{\text{number of births per year}}{\text{total population that year}} \times 1000$

Death rate $= \dfrac{\text{number of deaths per year}}{\text{total population that year}} \times 1000$

Fertility rate = average number of children produced per woman of child-bearing age (15–49 years) in the population

7 From Figure 4, what information do you need in order to calculate the growth rate of a population?

Table 5 lists the total population figures for Chile since 1950.

Table 5 Chile population data 1950–1995

Year	Total population thousands
1950	6 082
1955	6 766
1960	7 614
1965	8 579
1970	9 504
1975	10 380
1980	11 145
1985	12 122
1990	13 173
1995	14 237

Source: Population Concern, Population Reference Bureau, June 1995

8 a Using Table 5, calculate the increase in the population (in thousands) for each 5-year period and tabulate your results.

b Now, calculate the growth rate for each 5-year period and tabulate your results.

c Describe how the percentage increase in population has changed between 1955 and 1995.

Further information about populations round the world is provided in Table 6. A high fertility rate is typical of populations that are growing rapidly. Populations with a fertility rate higher than 2.0 tend to increase. So, fertility rate can be used to predict future trends in population growth. However, death rate will also have an effect on the actual population.

Table 6 Population data 1995

Country	Total population thousands	Fertility rate
Sub-Saharan Africa	16 200	6.2
North America	293 000	2.0
South America	319 000	3.0
Asia	345 100	2.9
Europe	729 000	1.5

Shape of populations

Population pyramids show the percentage of males and females in each age group in a country (Fig. 5).

Conditions in developed countries with high living standards, plenty of food and good medical services contrast sharply with conditions in developing countries. These factors affect the number of births and the pattern of deaths in each age group. So, the shape of the population pyramids of a developing country and a developed country show differences in:
- the angle of the sides of the pyramid;
- the height of the pyramid;
- the width of the base of the pyramid.

Developing countries usually have both a high birth rate and high **infant mortality rate** (death rate for infants). Over 35% of the population is under 15 years of age, so

Fig. 5 Features of population pyramids

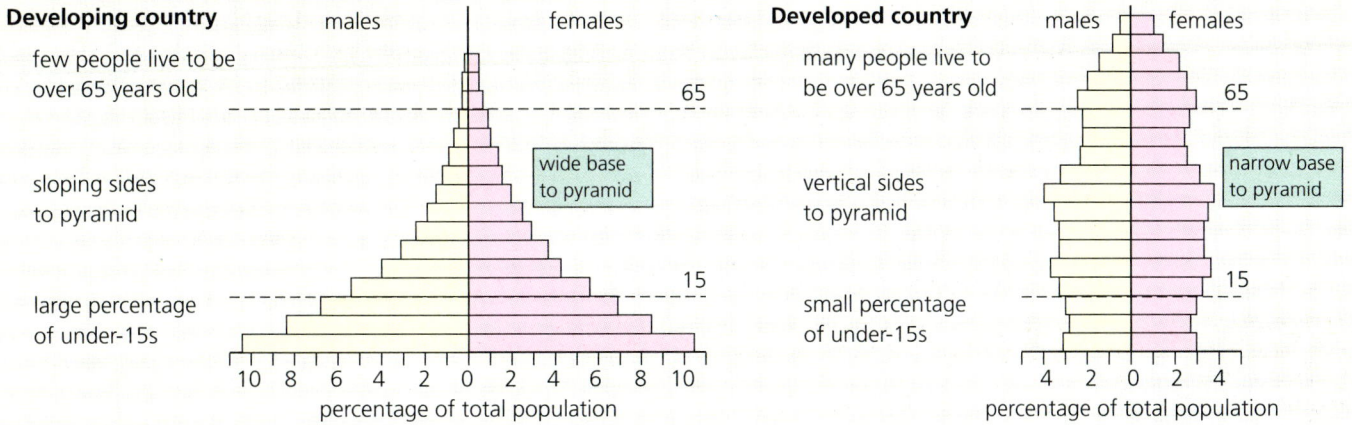

Developing country

few people live to be
over 65 years old

sloping sides
to pyramid

large percentage
of under-15s

males | females

wide base
to pyramid

65

15

10 8 6 4 2 0 2 4 6 8 10

percentage of total population

Developed country

many people live to
be over 65 years old

vertical sides
to pyramid

small percentage
of under-15s

males | females

narrow base
to pyramid

65

15

4 2 0 2 4

percentage of total population

Source: adapted from Hornby and Jones, *Introduction to Population Geography,* Cambridge University Press, 1993

the base of the population pyramid is broad. Death rates for adults are high, and few survive to old age.

A **survival curve** can be plotted from records of how long each individual in a group of 10 000 lives (Fig. 6). The **average life expectancy** is the age at which 50% of

the people in the sample are still alive. So, average life expectancy can be calculated from a survival curve.

9 Using Figure 6, calculate the average life expectancy in each type of country.

Fig. 6 Survival curves for developing and developed countries

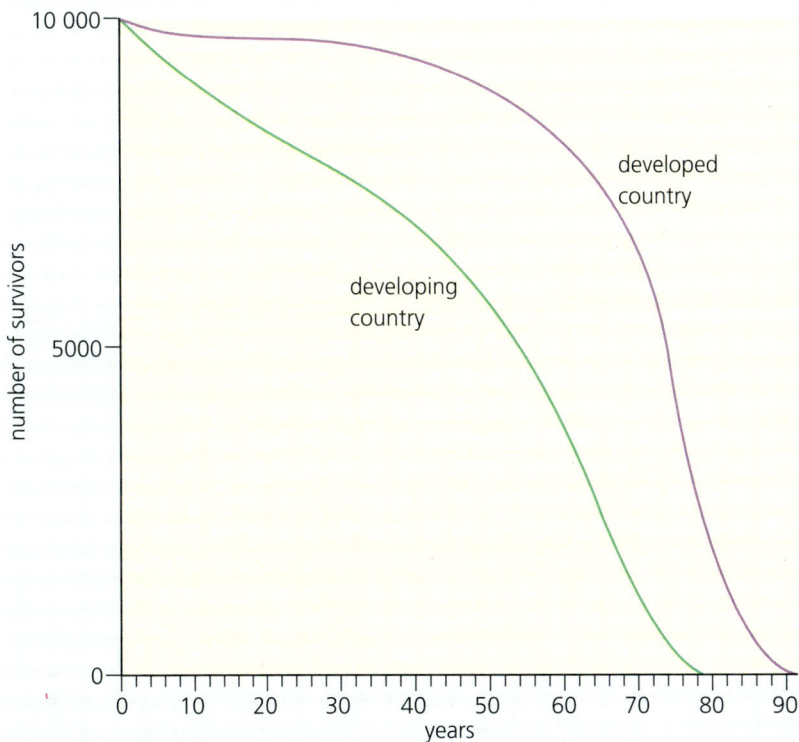

number of survivors

10 000

5000

0

developed
country

developing
country

0 10 20 30 40 50 60 70 80 90

years

Santiago is much like any other large city – it has high rise buildings, good roads and many shops.

	Table 7 Population structure – Chile				
	1950		2000 (projected)		
Age group years	males % total population	females % total population	males % total population	females % total population	
0–4	7.2	7.0	4.9	5.0	
5–9	6.1	6.0	5.0	4.9	
10–14	5.1	5.0	5.0	4.9	
15–19	4.7	4.7	4.9	4.7	
20–24	4.3	4.3	4.5	4.3	
25–29	3.6	3.7	3.9	3.8	
30–34	3.3	3.4	4.0	3.9	
35–39	2.9	2.8	3.9	3.9	
40–44	2.7	2.8	3.3	3.4	
45–49	2.3	2.4	2.8	2.9	
50–54	2.0	2.1	2.2	2.4	
55–59	1.6	1.7	1.8	2.0	
60–64	1.2	1.3	1.3	1.6	
65–69	0.83	0.95	1.1	1.4	
70–74	0.57	0.7	0.8	1.2	
75–79	0.31	0.41	0.48	0.75	
80+	0.18	0.27	0.42	0.80	

Chile has seen enormous changes since 1955, especially in the growing towns. By 1995, the total population was more than double the 1945 figure, and there has been rapid urban expansion. The population structure has also changed (Table 7).

10 a Plot two population pyramids, one for 1950 and one for 2000.
  b Describe three differences between the 1950 and 2000 population pyramids.

Key ideas

- Population growth is the difference between the birth rate and the death rate *plus* the difference between immigration and emigration.

- Growth rate can be shown as percentage increase per year in a population.

- The age structure of a population is shown as a population pyramid. The shape of the pyramid reflects social and economic differences between populations.

- A survival curve shows the number of survivors of a group of 10 000 people plotted against time.

- Average life expectancy is the age at which 50% of the population sample used for the survival curve is still alive.

- Developed countries and developing countries have distinctive growth rates, population pyramids, survival curves and average life expectancy.

12.4 One world – many populations

The average world rate of population growth is 1.5% per year. However, there is considerable variation between growth rates in different countries (Table 8). So, although the total human population will double in less than 45 years, the doubling in some countries will occur in less than 30 years, whereas in others it will take more than a century.

Table 8 Growth rates and doubling times		
Location	Growth rate % y^{-1}	Doubling time /y
World	1.5	45
Africa	2.8	24
Central and South America	1.9	36
Asia	1.7	42
Northern Europe	0.2	443
North America	0.7	105
Australia	0.8	91

11 Study Table 8. Describe the relationship between doubling time and growth rate.

The growth and structure of a population is the result of a complex interaction of many social and economic factors. However, as countries become more economically developed, it is possible to see general trends in birth rate, death rate, and total population growth. The pattern of these trends has four stages and is known as the **demographic transition model** (Fig. 7).

Chile is economically successful and has a large urban population. The current birth and death rates put Chile in stage 3 of the demographic transition model.

12 Study Figure 7. Briefly summarise the trends of the birth rate, death rate, and the total population in each of the four stages of the demographic transition model.

Fig. 7 The demographic transition model

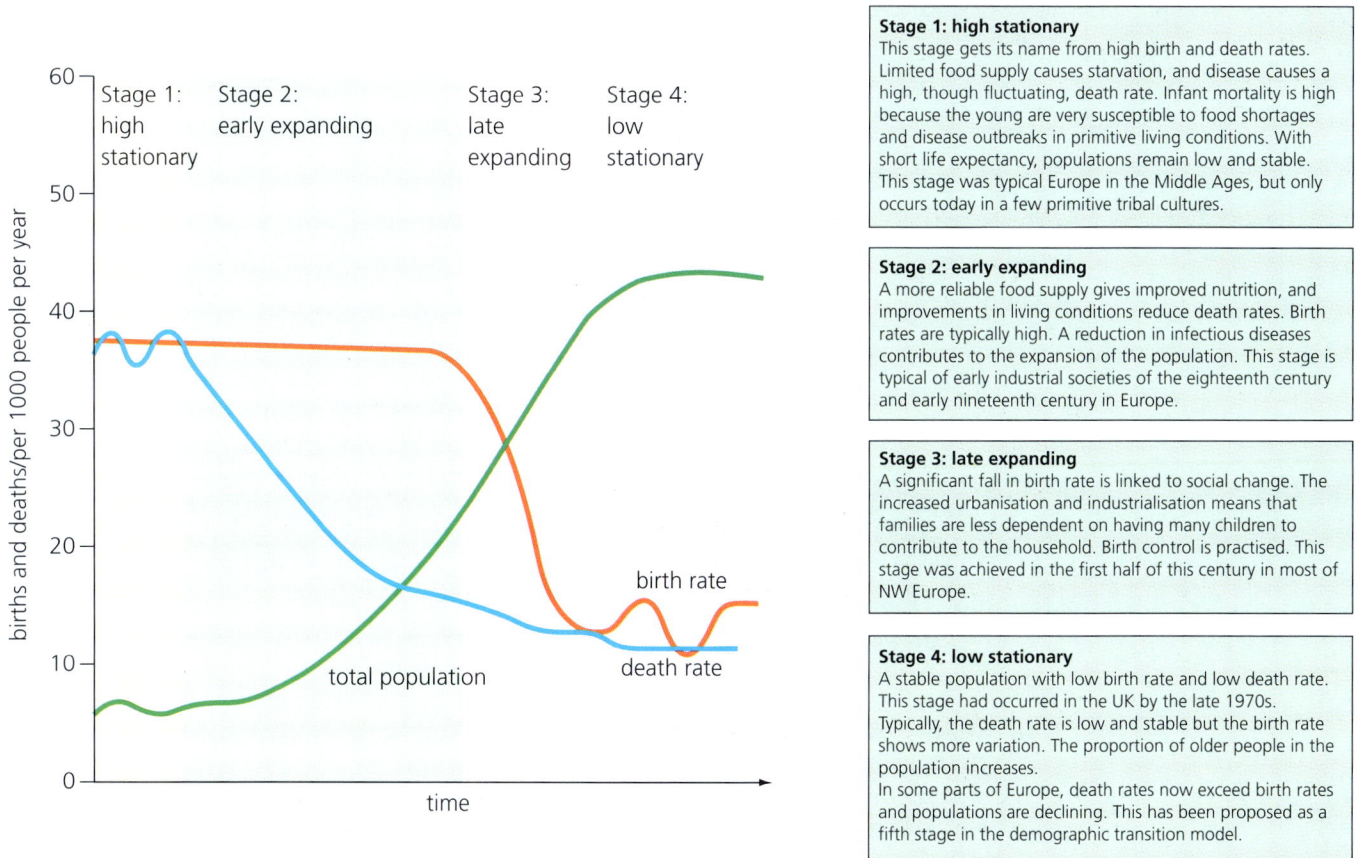

Stage 1: high stationary
This stage gets its name from high birth and death rates. Limited food supply causes starvation, and disease causes a high, though fluctuating, death rate. Infant mortality is high because the young are very susceptible to food shortages and disease outbreaks in primitive living conditions. With short life expectancy, populations remain low and stable. This stage was typical Europe in the Middle Ages, but only occurs today in a few primitive tribal cultures.

Stage 2: early expanding
A more reliable food supply gives improved nutrition, and improvements in living conditions reduce death rates. Birth rates are typically high. A reduction in infectious diseases contributes to the expansion of the population. This stage is typical of early industrial societies of the eighteenth century and early nineteenth century in Europe.

Stage 3: late expanding
A significant fall in birth rate is linked to social change. The increased urbanisation and industrialisation means that families are less dependent on having many children to contribute to the household. Birth control is practised. This stage was achieved in the first half of this century in most of NW Europe.

Stage 4: low stationary
A stable population with low birth rate and low death rate. This stage had occurred in the UK by the late 1970s. Typically, the death rate is low and stable but the birth rate shows more variation. The proportion of older people in the population increases.
In some parts of Europe, death rates now exceed birth rates and populations are declining. This has been proposed as a fifth stage in the demographic transition model.

Source: adapted from Hornby and Jones, *Introduction to Population Geography*, Cambridge University Press, 1993

179

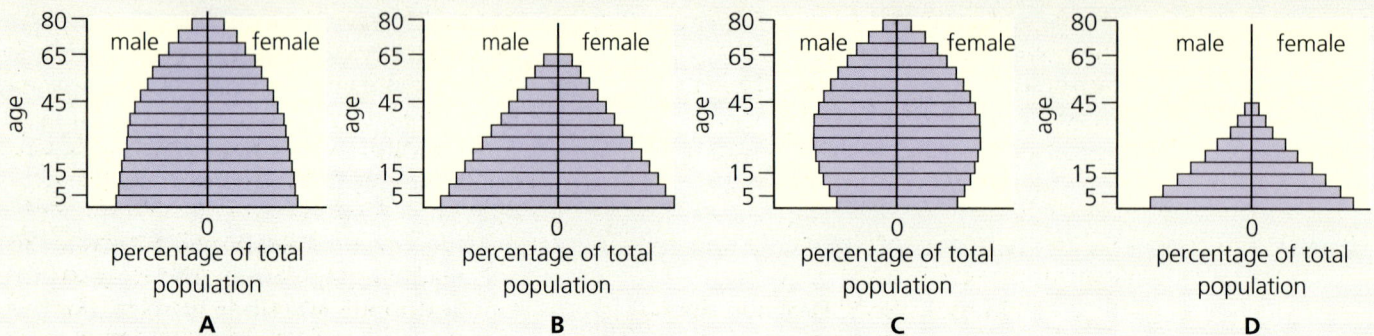

Fig. 8 Population pyramids for the demographic transition model

Source: adapted from Musgrove, *Data Response for GCSE Geography*, Hutchinson, 1988

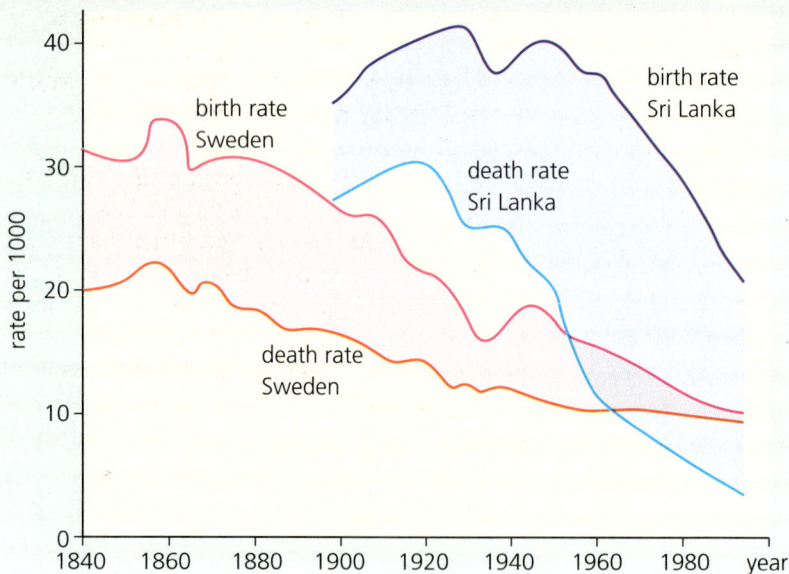

Fig. 9 Population change in Sweden and Sri Lanka

A tale of two factors

The growth of a population rarely follows the demographic transition model exactly. There are many social factors that interact and are, in turn, affected by environmental factors. However, there are two particularly important factors that affect fertility rate, birth rate, and death rate:

• food supply;
• disease.

Individual growth and general health depend on sufficient food. Lack of food increases child mortality and decreases the birth rate. In times of plenty, when there is surplus food, the effect is reversed.

In overcrowded conditions, infectious diseases spread rapidly and contribute to a high death rate. Reducing overcrowding significantly reduces death rates, but infectious diseases are present in all populations. Widespread outbreaks of disease are called **epidemics**. Epidemics of some infectious diseases have noticeably increased death rates and slowed population growth. The 'black death' in pre-industrial Britain reduced the population by 30–50% in the fourteenth century. Epidemics which spread internationally are called **pandemics**.

13 Study Figure 8. Which stage of the demographic transition model does each pyramid represent?

14a Study Figure 9. Describe the changes in birth rate, death rate and population growth in Sweden and Sri Lanka since 1900.

b Which phase of the demographic transition model is each country in?

Case study 1: the Irish potato famine

After a long period of very slow growth, the human population rose throughout western Europe between 1750 and 1850. In Ireland, the population doubled to over 8 million during this period and then declined for almost a hundred years (Fig. 10).

Population data is often unreliable for the eighteenth century and early nineteenth century. During this time, Ireland had large but scattered rural communities that kept only crude local records of births, marriages, and deaths. There were no central records. The 'hearth taxes' for each family do not record the total family size. Movement of labour, typical in farming communities, meant many people were not recorded in census data, making it inaccurate. Social factors can be inferred from employment, estate, and rent records as these provide useful information about working and living conditions. The first reliable census in Ireland was in 1821.

Before 1730, the rural population of Ireland depended on wheat as the staple food, but yields were low and unpredictable. Dairy farming provided milk, and a good deal of the land was pasture for grazing. The spread of the potato as a staple crop coincides with a rapid population expansion in Ireland. This population growth is attributed to four factors:
- productivity;
- land use;
- fertility rate;
- death rate.

Productivity – In the second half of the eighteenth century, potatoes became the staple food. A family could grow sufficient food for their needs on only 0.8 hectare of land, a fraction of the area necessary to subsist on wheat and dairy produce. An intake of 4–7 kg of potatoes per day plus half a litre of milk provides a nutritious though very boring diet.

Land use – Irish farms were divided between the sons of the household. By growing potatoes, it was possible to support all the sons on smaller areas. So, many sons married and set up home on these smallholdings. Previously, a mixed farm would have been much bigger, even to support one or two sons from a large family. The rest would have moved away or not married.

Fertility rate – The number of children born to a woman in her lifetime is a measure of fertility rate. More marriages, an earlier age of marriage, and longer-lasting marriages increase fertility rate. A reliable food supply, in the form of potatoes, provided the economic and nutritional security for more and larger farming families.

Death rate – A potato-based diet provided everyone with enough to eat, so resistance to disease improved and infant mortality was reduced.

We cannot be sure which of these aspects had the biggest effect on the large growth in population in Ireland. Estimates give a birth rate of over 35 per 1000 in the early 1800s.

And then, in 1845, the potato crop failed. Potato blight caused by a fungus, *Phytophthora infestans*, destroyed plants and reduced stored tubers to a stinking inedible pulp. Between 1845 and 1851 about a million people died from starvation and infectious diseases. The squalid living conditions and poor nutrition allowed

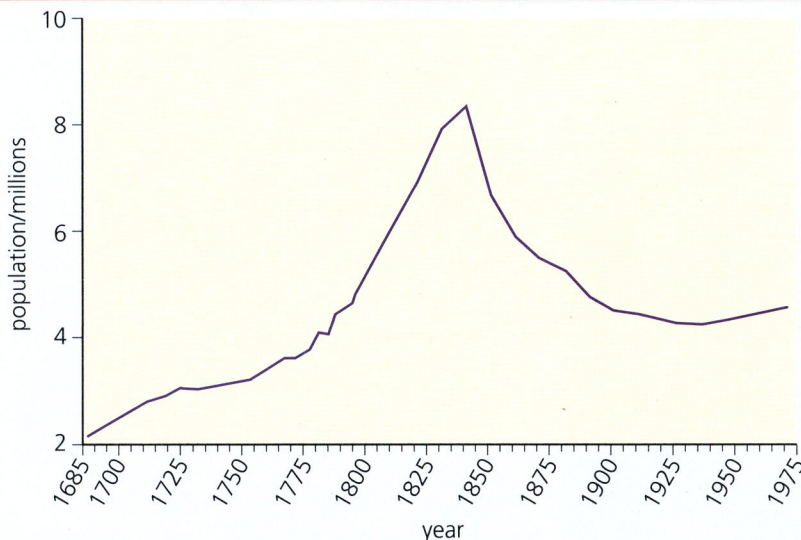

Fig. 10 Population in Ireland 1687–1971

Source: based on data in Grigg, *Population Growth and Agrarian Change*, Cambridge University Press, 1980

typhoid, dysentery and other diseases to flourish.

At the same time, the big estates in Ireland were growing good quality corn. However, this was not used to feed the starving Irish because the landowners were exporting it to Britain.

The decline in the Irish population following the potato famine lasted for almost 100 years. Fertility was reduced because couples could not afford to marry and have children at as young an age as they had previously.

During and after the potato famine, millions of Irish emigrated to America and other countries to seek a better life. St Patrick's Day is celebrated with enthusiasm in New York City.

15 Explain how population growth is affected by (a) an increase in food supply and (b) a decrease in food supply.

16 How is fertility rate affected when the age at which couples marry increases?

Case study 2: cholera in the UK

The relative number of people working on farms and living in rural locations in England and Wales began to decline in the mid-1700s and the population became increasingly town-based (Fig. 11). This is known as **urbanisation**.

The rapid development of urban communities was linked to industrialisation and world trade. Towns offered better opportunities for employment, and working people could earn more money. Imported food reduced the dependence of town-dwellers on the farming communities and farm wages dropped. This accelerated the rural decline.

Water-borne diseases such as cholera are often associated with developing countries and refugee camps, but outbreaks of cholera were frequent in the UK in the mid-1800s. A cholera pandemic started in Bengal in 1817, reached Moscow by 1830, and England by 1831. It was probably brought into Sunderland by infected passengers on a ship from Europe. Once here, the disease spread rapidly through the UK claiming 50 000–70 000 lives.

Cholera is caused by the bacterium *Vibrio cholerae*. Sufferers develop diarrhoea and sickness and eventually become severely dehydrated. Cholera is usually fatal within 2–5 days.

Brighton's Victorian sewers are still in use. In the 1800s, many large sewerage systems were installed in the UK. When clean water is provided and efficient sewage treatment is achieved, the reduction in many water-borne diseases results in rapid human population growth.

Fig. 11 Rural population in England and Wales

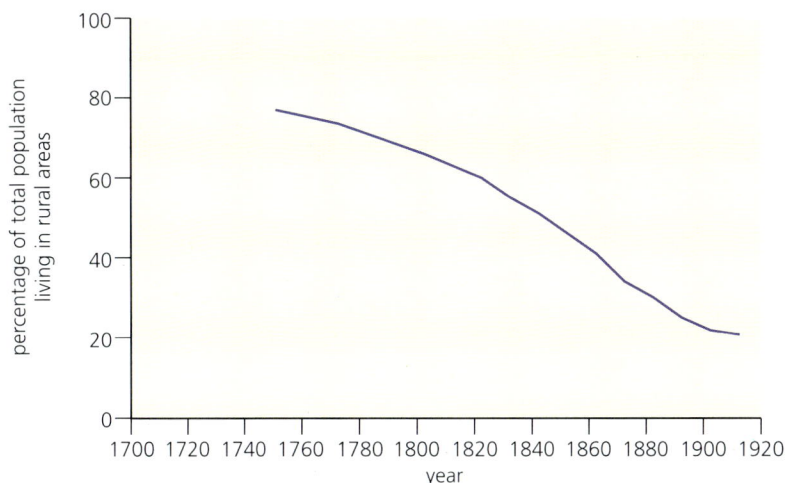

Fig. 11 Rural population in England and Wales

Source: based on data in Grigg, *Population Growth and Agrarian Change*, Cambridge University Press, 1980

John Snow, a medical apprentice at Newcastle, identified the main methods of transmission of cholera. He noticed that the disease did not just strike the poor and undernourished, although many more of them died than the wealthy. Snow carried out detailed studies and concluded that:

- the sick passed the disease to their carers;
- cholera spread quickly in overcrowded houses where up to 15 families shared facilities;
- the source of their drinking water was the main link between sufferers;
- fresh water from street pumps or wells was contaminated by waste from cracked sewer pipes and this was the method of transmission.

Cholera was eradicated from the UK largely because of improvements in water treatment and distribution. Sewage systems were rebuilt in all large towns in the late 1800s and many of these sewers are still in use today. Better housing with sanitary provision, a reduction in overcrowding, and simple personal hygiene were also important.

The latest cholera pandemic started in China in 1963, spread to Iran in 1966 and to the Black Sea area in 1970. This 'el Tor' strain caused many thousands of deaths.

17a **Explain why cholera epidemics have their greatest effects on urban populations.**

b **Describe two contributions to the eradication of cholera from the UK.**

Success story

Vaccination has helped to control the spread of cholera and other infectious diseases. The smallpox virus was officially declared extinct in 1979 after a worldwide vaccination program began by the World Health Organisation in 1967. A similar programme is now underway to eradicate poliomyelitis.

Diphtheria, another disease that is frequently fatal in children, is spread by droplet infection. Routine vaccination of UK schoolchildren against diphtheria began in the 1940s. At that time there were over 100 reported cases per year in UK, and many of them led to death in under-5-year-old children. In the decade from 1974 to 1984, only 10 cases of diphtheria in total were reported in the UK.

Key ideas

- The changes in a population's size and structure can be represented by the demographic transition model.

- Food supply had a major effect on the growth of the population in Ireland in the eighteenth century.

- The control of water-borne diseases, such a cholera, reduced mortality rates in early industrial societies.

- Vaccination and Immunisation have reduced the number of deaths due to infectious diseases.

12.5 Children for the future

Developing countries face difficult decisions. They have taken large loans to stimulate development and become part of the industrial developed world. But industrialisation is very expensive and takes time. An improved standard of living depends on investment in technology and gaining a significant share of world markets.

The biggest population growth is in Africa and Asia. There are problems of food supply, and infectious diseases are common in both poor rural areas and overcrowded urban areas. Many of the African and Asian nations are not well equipped to cope with crop failure or natural disasters. In these countries, government initiatives try to address the problems with international support from other governments and charities.

A family planning poster in Vietnam.

Government initiatives in China

In 1973, the Chinese government began campaigns with the aim of achieving a population of no more than 1. 2 billion by the year 2000. This requires substantial reductions in the birth rate (Table 9).

Table 9 Birth rates in China (estimated)	
Year	Births per 1000
1949	45
1969	36
1975	26
1995	18

The policy for reducing the birth rate is rigorously applied and includes rewards and penalties:

- a 'one-child pledge' signed by parents brings an extra month's salary per year until the child is 14;
- there are housing incentives for one-child families;
- paid leave from work is available for abortions;
- the minimum legal age for marriage is 22 for men, 20 for women;
- larger land areas are available for rural couples with only one child;
- free birth control is provided for all;
- demotion, loss of earnings, and education costs are applied if a couple have a second child.

The most common methods of birth control in China are sterilisation, the IUD and abortion.

18 a Current growth of population in China is 1.1% per year. The 1995 population was 1.218 billion (1 218 000 000). Calculate the population by the year 2000.
   b How is it possible that China could achieve 1.2 billion target by the year 2000?

This 80-year-old man has seen many changes in China in his lifetime. What changes will his 4-year-old granddaughter see in hers?

Table 10 Methods of birth control (contraception)

Method of birth control	Comments
pill	• very effective way of controlling fertility • reduces blood loss during menstrual period
sterilisation	• permanent and effective • requires surgery • not easily reversible
IUD	• more suitable for women who have had a child • may increase menstrual flow • fitted at a clinic and IUD remains inside woman
condom/diaphragm	• easy to use • less effective than the pill, sterilisation or the IUD

19 Despite the one-child policy, Chinese family sizes continue to be larger in rural areas than in urban areas. Suggest two reasons for this.

Not all forms of birth control are equally suitable for all situations (Table 10). There might be religious objections to some forms, others might be unsuitable for social, economic, personal or medical reasons.

Sub-Saharan Africa has a very high birth rate and unpredictable food supply. People might be malnourished and lack essential minerals such as iron. Lack of iron can cause anaemia, because iron is essential for making haemoglobin. Medical facilities, such as clinics, are few and far between in rural areas. For these reasons, the IUD is not a sensible birth control choice for women in sub-Saharan Africa.

20 Give one advantage for using the pill in communities that might have a low iron intake.

Emily found that Chile is a country of great contrasts. Santiago has the amenities and way of life you would expect to find in a modern city. But beyond Santiago is a vast rural area with few people – and even unexplored mountains and deserts. The world's population is expanding rapidly, yet despite increasing urbanisation there are still many isolated and uninhabited places. Will people one day move into these remote areas? Would that mean health risks? What would be the effect on the environment? Could we afford the economic costs? Is keeping the population down a sensible alternative?

Key ideas

• Birth control or contraception can be used to control fertility.

• The best method for fertility control in any situation depends on a range of social, personal, and medical factors.

Data section

1 Time lines and evolution

Fig. 1 Hominid time line

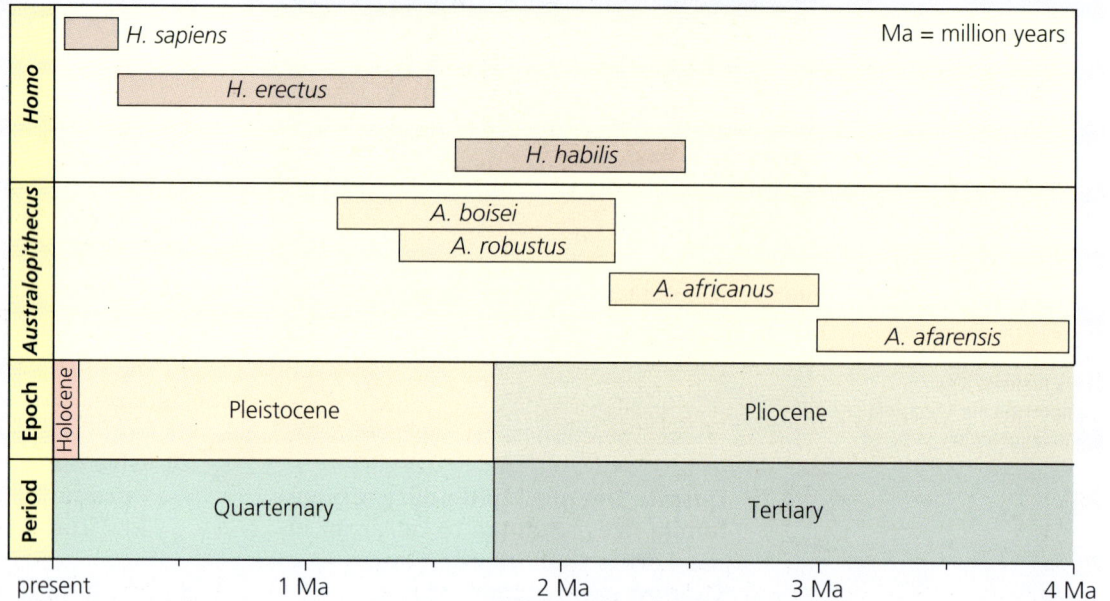

	H. sapiens						Ma = million years
Homo	H. erectus		H. habilis				
Australopithecus			A. boisei				
			A. robustus				
				A. africanus			
					A. afarensis		
Epoch / Holocene	Pleistocene			Pliocene			
Period	Quarternary			Tertiary			
present		1 Ma		2 Ma	3 Ma		4 Ma

Fig. 2 Tool culture time line

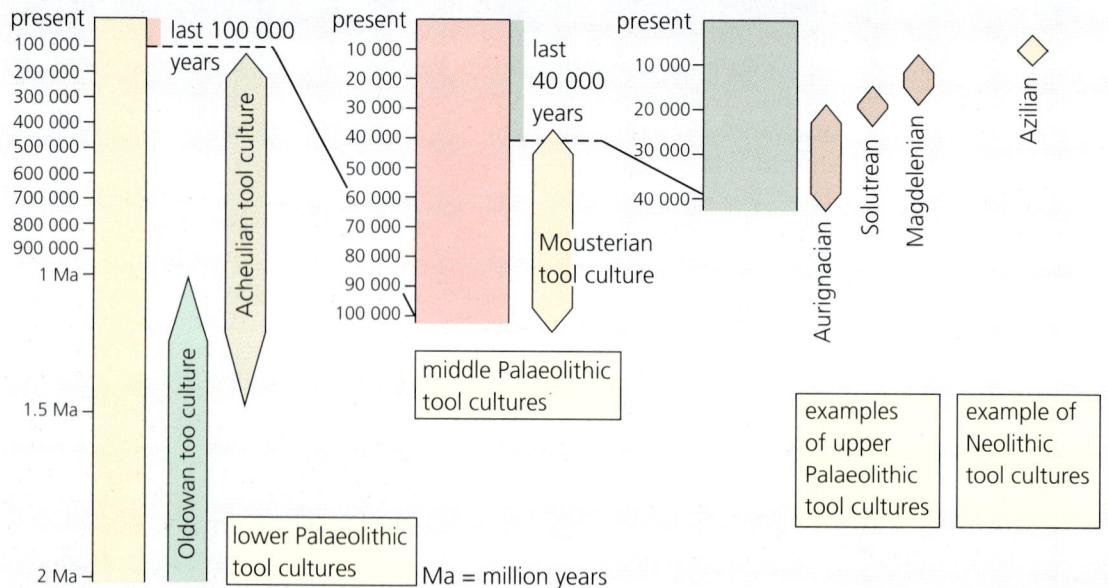

present
100 000
200 000
300 000
400 000
500 000
600 000
700 000
800 000
900 000
1 Ma

1.5 Ma

2 Ma

last 100 000 years

Acheulian tool culture

Oldowan too culture

lower Palaeolithic tool cultures

present
10 000
20 000
30 000
40 000
50 000
60 000
70 000
80 000
90 000
100 000

last 40 000 years

Mousterian tool culture

middle Palaeolithic tool cultures

present
10 000
20 000
30 000
40 000

Aurignacian

Solutrean

Magdelenian

Azilian

examples of upper Palaeolithic tool cultures

example of Neolithic tool cultures

Ma = million years

Fig. 3 Earth's time line

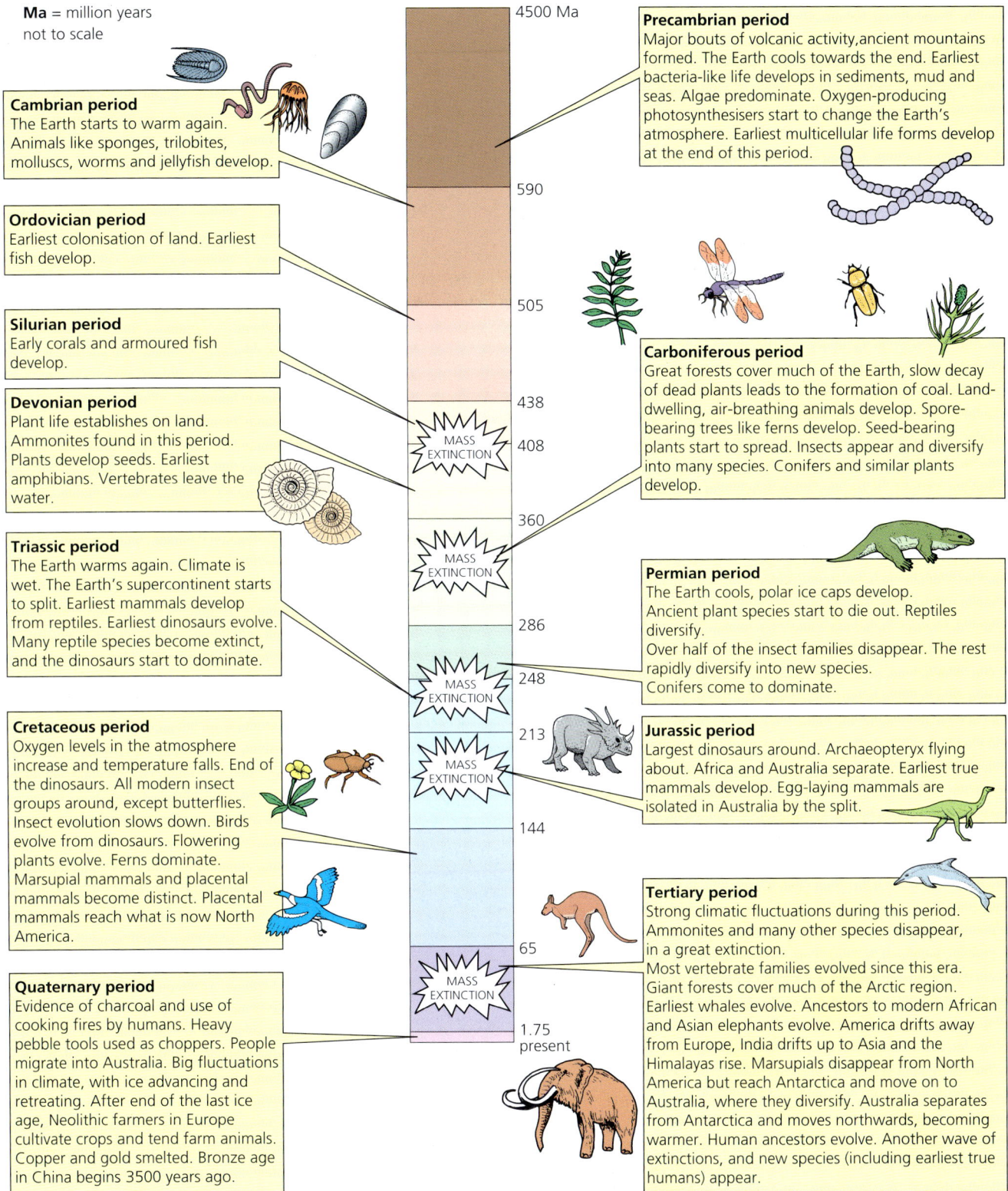

Ma = million years
not to scale

4500 Ma

Precambrian period
Major bouts of volcanic activity, ancient mountains formed. The Earth cools towards the end. Earliest bacteria-like life develops in sediments, mud and seas. Algae predominate. Oxygen-producing photosynthesisers start to change the Earth's atmosphere. Earliest multicellular life forms develop at the end of this period.

Cambrian period
The Earth starts to warm again. Animals like sponges, trilobites, molluscs, worms and jellyfish develop.

590

Ordovician period
Earliest colonisation of land. Earliest fish develop.

505

Silurian period
Early corals and armoured fish develop.

Carboniferous period
Great forests cover much of the Earth, slow decay of dead plants leads to the formation of coal. Land-dwelling, air-breathing animals develop. Spore-bearing trees like ferns develop. Seed-bearing plants start to spread. Insects appear and diversify into many species. Conifers and similar plants develop.

438

Devonian period
Plant life establishes on land. Ammonites found in this period. Plants develop seeds. Earliest amphibians. Vertebrates leave the water.

MASS EXTINCTION **408**

360

MASS EXTINCTION

Permian period
The Earth cools, polar ice caps develop. Ancient plant species start to die out. Reptiles diversify.
Over half of the insect families disappear. The rest rapidly diversify into new species.
Conifers come to dominate.

Triassic period
The Earth warms again. Climate is wet. The Earth's supercontinent starts to split. Earliest mammals develop from reptiles. Earliest dinosaurs evolve. Many reptile species become extinct, and the dinosaurs start to dominate.

286

MASS EXTINCTION **248**

213

Jurassic period
Largest dinosaurs around. Archaeopteryx flying about. Africa and Australia separate. Earliest true mammals develop. Egg-laying mammals are isolated in Australia by the split.

Cretaceous period
Oxygen levels in the atmosphere increase and temperature falls. End of the dinosaurs. All modern insect groups around, except butterflies. Insect evolution slows down. Birds evolve from dinosaurs. Flowering plants evolve. Ferns dominate. Marsupial mammals and placental mammals become distinct. Placental mammals reach what is now North America.

MASS EXTINCTION

144

Tertiary period
Strong climatic fluctuations during this period. Ammonites and many other species disappear, in a great extinction.
Most vertebrate families evolved since this era. Giant forests cover much of the Arctic region. Earliest whales evolve. Ancestors to modern African and Asian elephants evolve. America drifts away from Europe, India drifts up to Asia and the Himalayas rise. Marsupials disappear from North America but reach Antarctica and move on to Australia, where they diversify. Australia separates from Antarctica and moves northwards, becoming warmer. Human ancestors evolve. Another wave of extinctions, and new species (including earliest true humans) appear.

65

MASS EXTINCTION

Quaternary period
Evidence of charcoal and use of cooking fires by humans. Heavy pebble tools used as choppers. People migrate into Australia. Big fluctuations in climate, with ice advancing and retreating. After end of the last ice age, Neolithic farmers in Europe cultivate crops and tend farm animals. Copper and gold smelted. Bronze age in China begins 3500 years ago.

1.75 present

Fig. 4 Hominids move out of Africa

AFRICA

Hadar

Omo
Koobi Fora
Peninj
Lake Victoria
Olduvai
Laetoli

Sites of *Australopithecus* remains

These remains are 1.1–4 million years old

Makapansgat
Taung
Sterkfontein
Swartkrans and Kromdraai

AFRICA

Lake Victoria

Omo
Lake Turkana
Olduvai

Sites of *Homo habilis* remains

These remains are 1.6–2.5 million years old

Swanscombe (England)
Stenheim (Germany)
Arago (France)
Vertessöllös (Hungary)
Petralona (Greece)
Ternifine (Algeria)
Sale (Morocco)
Choukoutieu (China)
Lautian (China)
Awash (Ethiopia)
Olduvai (Tanzania)
Koobi Fora (Kenya)
Laetoli (Tanzania)
Saugiran and Trinil (S.E. Asia)
Modjokerto (S.E. Asia)
Swartkrans (South Africa)

Sites of *Homo erectus* remains
These remains are 0.25–1.6 million years old

Fig. 5 A traditional evolutionary tree

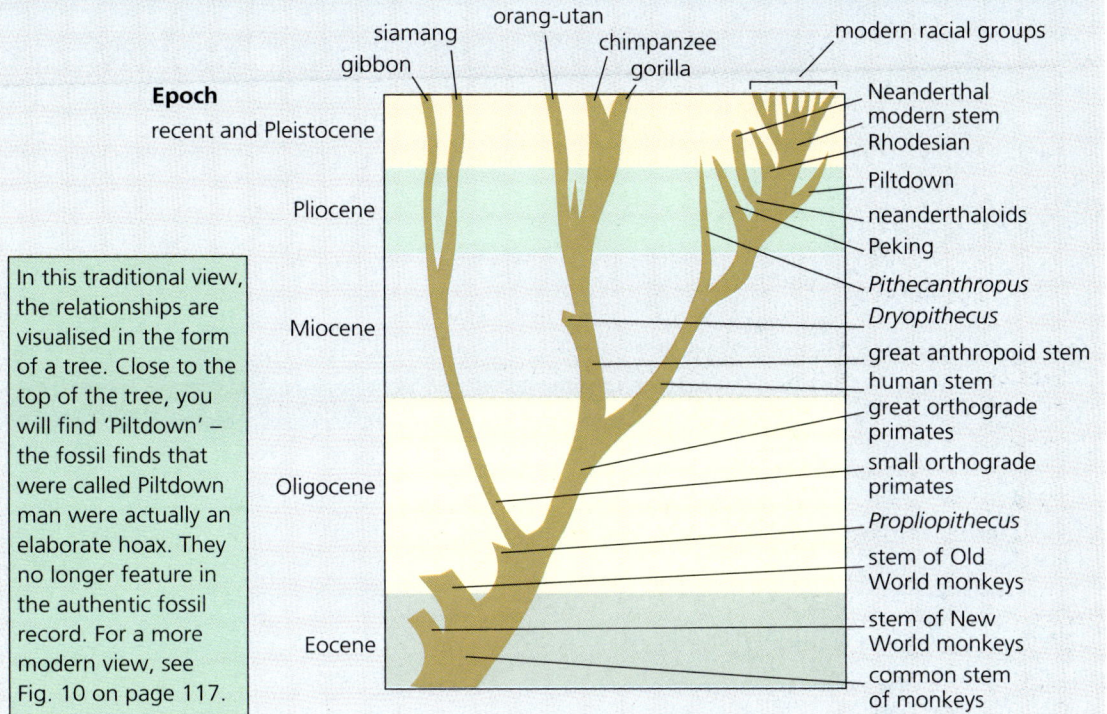

In this traditional view, the relationships are visualised in the form of a tree. Close to the top of the tree, you will find 'Piltdown' – the fossil finds that were called Piltdown man were actually an elaborate hoax. They no longer feature in the authentic fossil record. For a more modern view, see Fig. 10 on page 117.

Labels (top to bottom, left to right): siamang, gibbon, orang-utan, chimpanzee, gorilla, modern racial groups

Epoch:
- recent and Pleistocene
- Pliocene
- Miocene
- Oligocene
- Eocene

Right-hand labels: Neanderthal, modern stem, Rhodesian, Piltdown, neanderthaloids, Peking, *Pithecanthropus*, *Dryopithecus*, great anthropoid stem, human stem, great orthograde primates, small orthograde primates, *Propliopithecus*, stem of Old World monkeys, stem of New World monkeys, common stem of monkeys

Source: Lewin, *Human Evolution, an Illustrated Introduction*, 2nd edn, Blackwell Scientific, 1989

Fig. 6 Two evolutionary models

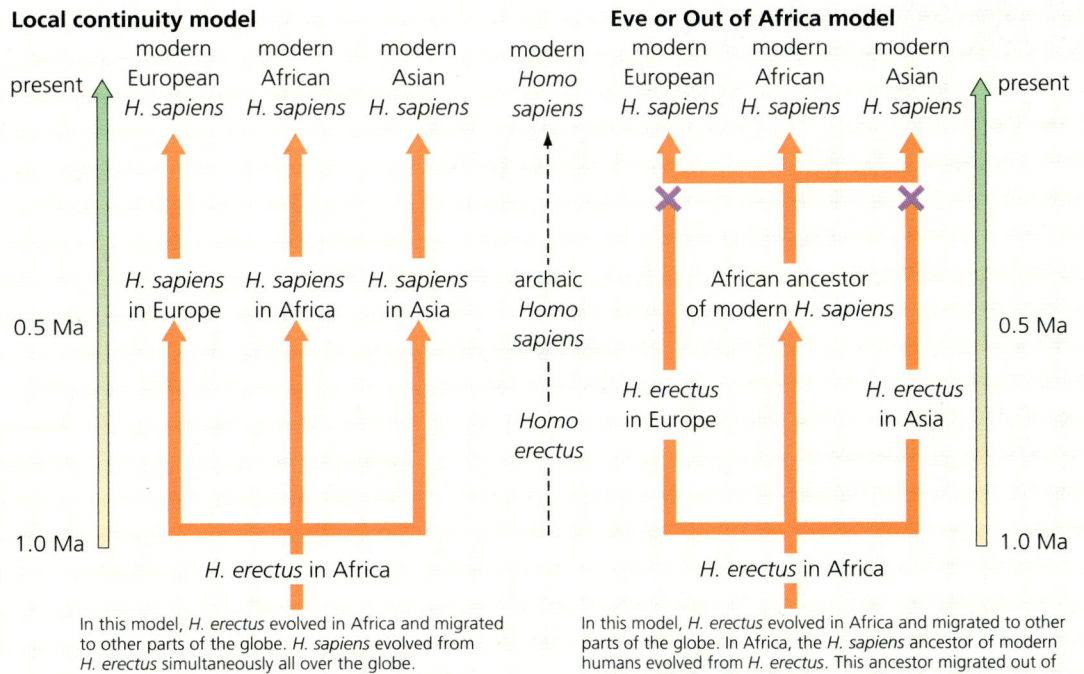

Local continuity model

present — modern European H. sapiens, modern African H. sapiens, modern Asian H. sapiens, modern Homo sapiens

0.5 Ma — H. sapiens in Europe, H. sapiens in Africa, H. sapiens in Asia, archaic Homo sapiens

1.0 Ma — H. erectus in Africa — Homo erectus

In this model, H. erectus evolved in Africa and migrated to other parts of the globe. H. sapiens evolved from H. erectus simultaneously all over the globe.

Eve or Out of Africa model

modern European H. sapiens, modern African H. sapiens, modern Asian H. sapiens — present

African ancestor of modern H. sapiens — 0.5 Ma

H. erectus in Europe, H. erectus in Asia

1.0 Ma — H. erectus in Africa

In this model, H. erectus evolved in Africa and migrated to other parts of the globe. In Africa, the H. sapiens ancestor of modern humans evolved from H. erectus. This ancestor migrated out of Africa and displaced H. erectus.

Source: Lewin, *Human Evolution, an Illustrated Introduction*, 2nd edn, Blackwell Scientific, 1989

2 Life expectancy

Average life expectancy	London	Liverpool	Manchester	Surrey	Glasgow
Table 1 Life expectancy changes with time and conditions					
1841					
• males	35	25		44	
• females	38	27		46	
1881					
• males			29	51	35
• females			33	54	44

Continent	Region	Average life expectancy for the region/years	Range of average life expectancy/years*
Table 2 Average life expectancy around the world in 1996			
Africa	Northern Africa	64	46–68
	Western Africa	53	44–65
	Eastern Africa	50	46–73
	Middle Africa	49	46–63
	Southern Africa	65	55–66
North America		76	76–78
Latin America	Central America	71	65–76
	Caribbean	69	57–78
	South America	68	60–73
Asia	Western Asia	67	52–77
	South Central Asia	59	43–73
	South East Asia	64	49–76
	East Asia	71	64–80
Europe	Northern Europe	76	67–79
	Western Europe	77	72–78
	Eastern Europe	68	65–73
	Southern Europe	76	71–77
Australasia		73	56–78

* Average life expectancy varies from country to country within each region

3 Imbalances in the brain

Over 40 substances are known or strongly suspected to be neurotransmitters – transmitters in the brain – and more are likely to be found. Discovering what each does is not easy because nerve cells are tightly packed in the brain and only 500–1000 molecules of transmitter are released from a synaptic terminal at a time. Some clinical conditions are thought to be caused by an imbalance in brain transmitters – this is the focus of much recent research (Table 3).

Table 3 Neurotransmitters and drugs that act at their synapses

Transmitter	Function	Clinical effect of imbalance in transmitter levels	Drugs acting at the same synapses as the transmitter	Use of the drug
Amino acids				
• GABA	• widespread inhibitory effect	• possibly linked to anxiety	• benzodiazepines (e.g. Temazepam)	• tranquillisers
			• barbiturates (e.g. Phenobarbitone)	• antiepileptics and anaesthetics
• glycine	• widespread inhibitory effect		• strychnine and tetanus toxin	• non-deadly poisons
• glutamate	• widespread excitatory effect		• ketamine and phencyclidine	• general anaesthetics
Amines				
• acetylcholine	• affects mood and memory	• low levels possibly involved in Alzheimer's disease	• Tacrine (acetylcholinesterase blocker – under test)	• delays progress of Alzheimer's disease
• noradrenaline	• affects mood, pleasure, blood pressure, alertness	• possibly linked to depression	• amphetamines and cocaine	• stimulants that boost levels of noradrenaline (addictive)
• dopamine	• affects emotions and skeletal muscle movement	• low levels possibly associated with Parkinson's disease	• L-DOPA	• raises dopamine levels (transplants of fetal brain tissue rich in dopamine are under trial)
		• high levels possibly linked with schizophrenia	• clozapine	• antischizophrenic
• serotonin	• affects sleep, perception of pain, mood, body temperature, feeding	• possibly associated with depression	• selective serotonin re-uptake inhibitors (SSRIs, e.g. Prozac)	• treat low self-esteem and negative feelings
			• LSD	• hallucinogen
			• dexfenfluramine	• appetite suppressant
Neuropeptides				
• endorphins (natural opiates)	• inhibit pain	• possibly associated with schizophrenia	• opiates (e.g. morphine)	• pain and nausea relief
• enkephalins	• help memory and learning; cause pleasure; control sexual drive		• heroin	• drug of abuse
				• control of pain by acupuncture might work buy the release of natural endorphins and enkephalins

Depression

About 100 000 people in Britain poison themselves each year (not always fatally) because they are suffering from depression. One in ten people will suffer from depression at some point in their lives. Sometimes, depression is a temporary response to a disappointment, an illness or a bereavement. But some people are permanently and deeply unhappy without an apparent cause.

Clinically depressed people need medical help. In this condition, sleeping and eating patterns are often disturbed, everything seems an effort and body pains might be experienced. Interests, ambitions, pride in personal appearance and sense of self-worth are lost.

People with manic depression swing suddenly from gloom and despair to excitement and elation, then back again to despair. This condition is an extreme version of the normal mood changes that happen in every day life.

Depression can be treated with drugs or psychotherapy or both. Modern anti-depressive drugs work by altering the activity of monoamine transmitters, such as serotonin, which are known to control neurones in parts of the brain that influence mood. Research on vervet monkeys shows that dominant males have higher serotonin levels than subordinate males. So, it has been suggested that low social status, whether real or perceived, can lead to depression in humans through a lowering of serotonin levels. Drugs such as Prozac work by raising the serotonin level, but they are not successful in all cases.

Psychotherapy often aims to guide the patient to explore past events that might have led to the development of depression. Alternatively, some therapists try to help patients to examine situations in their present lives that lead to difficulties, and then help the patients to adjust their behaviour so as to relieve the problem.

Anxiety states are not the same as depression but the two can occur together. Drug treatments for anxiety and stress include tranquillisers such as benzodiazepines.

Alzheimer's disease

In Alzheimer's disease (AD) there is a progressive and irreversible loss of memory. The disease affects 11% of the population over 65 years of age. Patients start to have difficulty remembering recent events. Gradually, they become more confused and forgetful and can get lost in once familiar places. As the deterioration continues, they eventually lose the ability to read, write, talk, eat or walk. It can be difficult to distinguish AD from other degenerative disorders, but post mortem examination (autopsy) confirms the diagnosis. The brains of patients who have died from AD have three types of abnormality:

- loss of neurones in the hippocampus and cerebral cortex (affects memory);
- amyloid plaques are present (these consist of broken-down neurones containing an abnormal protein called β amyloid);
- abnormal tangles of protein fibrils in the neurones.

There are several different kinds of AD and a number of different causes have been suggested:

- a virus;
- a toxin such as aluminium;
- a genetic factor (some families have a history of the disease).

An enzyme involved in the production of acetylcholine is much reduced in patients with AD, and drugs that raise acetylcholine levels in the brain are under trial. It is estimated that the number of AD patients would be halved if the onset of the disease could be delayed by 5 years.

Parkinson's disease

Parkinson's disease (PD) usually affects people around the age of 60, but its cause is unknown. Only 5% of patients have a family history of PD. Neurones in the part of the cerebrum that controls skeletal muscle contraction begin to degenerate, particularly neurones that produce the transmitter, dopamine. Muscles can either shake or become rigid. Hand movements become slow, and walking becomes a shuffle. Speech might eventually be affected, but there is little mental degeneration. The disease progresses over many years, and can be relieved to some extent by dopamine-boosting drugs. There have been trials of fetal brain tissue transplants. This technique has had limited success, but raises many ethical problems.

Schizophrenia

This is a serious disease affecting the processes of thought, emotion and perception. It affects about 1% of people and often begins in young adults. Many schizophrenics can not filter out unnecessary sights, sounds and other stimuli that would be ignored by others. In severe cases, patients might develop strange abnormal beliefs, have hallucinations or hear voices. Thought processes become highly disturbed. The disease tends to run in families, and a gene has been identified that seems to increase the risk of developing the disease.

4 A classic study revisited

The reproductive behaviour of sticklebacks is the subject of one of the classic studies in ethology (the study of the behaviour of animals in their natural habitats). Niko Tinbergen first described the behaviour of these fish in 1937, and subsequently won the Nobel prize for his research. Stickleback behaviour seems to be a clear example of the way in which animals respond to stimuli. However, recent efforts to repeat some of Tinbergen's observations have not been successful.

Tinbergen noticed that male sticklebacks in his aquarium became aggressive when a red post van drove past a nearby window. This inspired him to design a series of experiments using models. Some of the models were close imitations of sticklebacks, others were not shaped like fish but had the colours of a male stickleback in spring: a blue–grey back with a contrasting red belly. Tinbergen presented the models to male sticklebacks and counted the number of times each model was attacked. He found that the realistically shaped but non-red models provoked little interest, but even simple shapes painted red on their lower surfaces provoked strong aggression. It seemed that it was the stimulus of the red belly rather than the shape of the fish that released the aggressive behaviour of the male stickleback.

In the 1980s, W. J. Rowland carefully repeated Tinbergen's experiments. He found that a red model was *less* likely to be attacked by a male stickleback than a grey model, the exact opposite of Tinbergen's findings. Rowland tested 14 males by offering them grey and red models simultaneously. Twelve of the males attacked the grey model more than the red model, two males attacked the red model slightly more. In a second experiment with 13 males, the grey and red model were presented one after the other, alternating which was shown first. Ten of the males attacked the grey model more, three males attacked the red model more. In his review of these findings, Michael Reiss said 'such an emphatic contradiction of perhaps the most famous of all ethological experiments seems remarkable, but was statistically significant'.

It might be that Tinbergen's experimental conditions were different in some way from Rowland's, or that not all male sticklebacks behave in the same way. The males used by Rowland were quite variable in their behaviour. What we call scientific knowledge is always just a working hypothesis. Scientists like Tinbergen, who inspire others with a new way of looking at the world, deserve credit even when their results turn out not to be true in all circumstances at all times.

(summarised from Michael J Reiss 'Courtship and reproduction in the three-spined stickleback' *Journal of Biological Education* **18**, 1984)

Answers to questions

Chapter 1

1 a A nerve is a bundle of axons belonging to several neurones, surrounded by connective tissue. A neurone is a single nerve cell.

 b Damage to a sensory nerve prevents impulses from sense organs reaching the brain.

2 The resting potential is maintained by the Na^+/K^+ pump, which uses active transport. Active transport needs ATP energy.

3 Na^+ ions are about ten times more concentrated outside an axon than inside it. During an action potential, Na^+ ions enter the axon by diffusion down their concentration gradient. When Na^+ ions leave an axon, they are moving against their concentration gradient, so they require active transport.

4 Na^+ ion permeability increases rapidly to the peak of the action potential, then rapidly decreases again as the potential returns towards the normal resting potential.

5 As the intensity of a stimulus increases, the frequency of action potentials in the sensory neurones increases. The brain interprets the increased frequency as stronger pressure.

6 a The source of energy for the electronic grip is electric batteries. In the natural system, energy comes from the resting potential, which is maintained by the Na^+ pump using ATP energy from respiration.

 b In the electronic grip, information travels to the muscles as electrons, through metal electrodes. The natural system uses nerve impulses based on the movements of ions across the axon membrane.

7 a Eserine prevents the breakdown of acetylcholine by acetylcholinesterase. So, at a cholinergic synapse, acetylcholine remains in place, and the neurone continues firing.

 b Eserine has no effect at an adrenergic synapse because the transmitter is noradrenaline, not acetylcholine.

8 ATP is needed to resynthesise acetylcholine from choline and acetate. If there is a shortage of ATP, the transmitter is released from the synaptic vesicles faster than it can be made, so it runs out.

9 Transmitters can affect learning by increasing the number of ion channels or receptor proteins (for the transmitter) at a synapse.

10 To understand the brain we need the work of neurophysiologists studying individual cells, and also anatomists, psychologists, ethologists (studying behaviour), biochemists, physicists and computer scientists.

11 a Because of its shape, caffeine binds to adenosine receptors at synapses and excites nerve cells.

 b Too much stimulation by caffeine could cause the brain to reduce the number of adenosine receptors. The person would then feel lethargic without caffeine and would have an urge to take some in order to feel normal again.

12 A large dose of digitalin could stimulate the heart too much and cause a heart attack.

13 Natural selection has favoured poisonous plants because they do not get eaten by animals.

14 Similarities:
 • use of acetylcholine;
 • the transmitter bonds to receptors and depolarises the post-synaptic membrane or sarcolemma and triggers action potentials.
Dissimilarities:
 • no summation;
 • acetylcholine is the only transmitter at neuromuscular junctions in skeletal muscle.

Chapter 2

1 a For near vision the ciliary muscles are contracted. This means they can become tired.

 b The ciliary muscles are relaxed, so the internal pressure opens the ciliary ring and creates tension in the suspensory ligaments. The ligaments pull the lens into a thin shape for distant vision.

2 a A cataract is cloudiness of the lens. When the lens is opaque, not enough light falls on the retina for good vision.

 b An artificial lens can not carry out accommodation, so glasses will be needed. Because an artificial lens does not have the slight yellow coloration of a natural lens, it does not screen out blue light in the same way, so colour vision is not restored to its original state. Also, the focus is not so sharp.

3 a The cross disappears when the image falls on the blind spot. When you look with your right eye, the image is does not disappear because it always falls on the retina; the blind spot is on the opposite side of the right eye.

 b We are not usually aware of the blind spot because any point is always seen by at least one of the two eyes.

4 The curvature of the cornea is important to its role as a convex lens. It does most of the necessary light refraction. Distortion of the cornea would cause distortion of the image on the retina.

5 a Colours are perceived by cones – and cones only work in bright light.

 b Several rods converge on one bipolar cell, so information from the rods is pooled to send a nerve impulse to the brain.

 c At the fovea, each cone is linked to a single bipolar cell, so each cone sends separate inputs to the brain.

6 It takes time for dark adaptation – the remaking of rhodopsin from retinal and opsin.

7 We can see colours right to the edge of our field of vision, not just at the centre, so there must be cones in all parts of the retina.

8 a The green cones (a little) and the red cones (a lot).

 b The brain computes the colour from the extent to which the three types of cone are stimulated.

 c The white paper stimulates all three cone types equally; black letters do not stimulate any of the cones.

9 The genes for the cone pigments are on the X chromosome. As women have two X chromosomes and men have only one, a defective gene is more likely to be expressed in men than in women.

10 Looking fixedly at the red shape tires the red cones. So, when you switch to looking at a white surface, the red cones in the area that originally responded to the red shape do not send enough impulses to the brain. Only the green and blue cones in that area respond fully, so instead of white, a patch of the paper is seen as greeny blue.

11 The brain uses its past experience to interpret images. The two converging lines are interpreted as parallel lines receding into the distance (like two railway lines). With this perspective, the upper horizontal line is interpreted as being farther away than the lower one. The upper line therefore appears to be the same length *and* farther away, so it is interpreted as being the longer of the two lines.

Chapter 3

1 Spring flowering plants are short-day (long-night) plants. When long nights return in the autumn, flowering can be initiated again, and the plant will then come into flower if weather conditions are suitable.

2 Light shining during the night can prevent short-day (long-night) plants from coming into flower. This is the night break effect.

3 a No. A night break would not occur because far-red light is the last to be shone on the plant and far-red light prevents the night break effect of the red light.

 b In this sequence, the far-red light does not prevent the night break effect of the red light because over 30 minutes have passed since the red light was shone on the plant. So, flowering does not occur.

4 Pr changes to Pfr by absorbing red light or white light and changing the three-dimensional shape of the molecule.

5 Flowering in short-day plants can be prevented by shining light on the plants in the middle of the night, or by extending daylight using artificial lighting. This reduces the amount of Pr phytochrome. The Pr/Pfr ratio must be low to prevent flowering in short-day plants.

6 Florigen causes plants to make flower buds, so it could be used to bring crops or ornamental plants into flower at any time of the year without changing the photoperiod.

7 The Pr/Pfr ratio changes quite quickly. A more gradual change would be needed to measure time periods of several hours. Also, the rate of change is affected by temperature, but biological clocks are independent of temperature.

8 a Plants are highly variable and coleoptiles bend in a variety of directions. In order to see an overall pattern in the direction of bending, a large sample is needed.

 b Germinating seeds from plants with coleoptiles can be grown in the dark in a horizontal position. They soon grow upwards, away from gravity. The seeds can also be planted on a vertical board attached to a slowly rotating drum called a klinostat. The rotation of the board means that the direction of gravity is constantly changing, so the plants grow randomly in all directions. Seeds grown in the dark in zero gravity in space, show the same random growth patterns.

9 Coleoptiles have been used for the majority of experiments on tropisms. They are unlikely to reflect what happens in plants such as beans and sunflowers, which do not have coleoptiles.

10 Figure 7 shows that growth decreased on the illuminated side of the coleoptile and increased on the dark side. If auxin was destroyed on the illuminated side, growth would still decrease on that side, but it would not increase on the dark side.

11 Auxin has a higher concentration on the dark side than on the illuminated side of the coleoptile tip. Auxin is transported down the coleoptile, mostly on the dark side. Auxin causes cells to expand more on the dark side than the illuminated side. This cell expansion causes the coleoptile to bend towards the light.

12 a IAA is indoleacetic acid, an auxin.
   b Plant growth regulators are chemicals that mimic the action of plant growth hormones.
13 a Some artificial auxins can be used to stimulate root growth in cuttings; some act as weedkillers on broad-leaved plants; certain artificial auxins prevent plants from dropping ripe fruit; the development of seedless fruit can be enhanced by the use of artificial auxins.
   b Anti-gibberellins are used to stop gibberellin growth hormones from causing stem growth.

Chapter 4

1 When the gastrocnemius contracts, the tibialis anterior relaxes and the heel is raised. When the gastrocnemius relaxes, the tibialis anterior contracts and pulls up the top of the foot – this raises the toes and the instep as the foot is lifted and swung forward.
2 When elastin is damaged, it is harder to move a joint in its socket because the ligaments are not so easily stretched.
3 Myosin.
4 a No, both actin and myosin filaments stay the same length when a sarcomere shortens.
   b When a sarcomere contracts, the actin filaments slide further between the myosin filaments. This reduces the width of the H-zone.
5 ATP is used to detach the myosin cross-bridges and bend them forward ready for the active stroke.
6 Nerve impulse → acetylcholine released at synapse → action potentials in sarcolemma → action potentials transmitted down T-tubules → Ca^{2+} ions released from sarcoplasmic reticulum → Ca^{2+} ions bind to switch proteins and expose actin binding sites → myosin binds to actin and cross-bridges move → fibre contracts.

7 ATP is needed to release the myosin cross-bridges from the actin binding sites. Without ATP, the myofibril filaments are locked together and the muscles remain contracted.
8 A motor unit is a group of muscle fibres served by branches from a single axon. All the fibres in a motor unit contract together.
9 Movement is controlled by the brain controlling how strongly a muscle contracts. The strength of contraction made by a muscle can be increased by:
   • increasing how much each fibre in a motor unit contracts;
   • increasing how many motor units contract.
10 Muscle fibres become thicker when they are exercised vigorously. In weightless conditions, the muscles have very little work to do and the fibres shrink. This causes loss of muscle strength.
11 a Sprinters use mainly white fibres because these fibres contract rapidly (but tire quickly).
   b In an averagely fit individual, the numbers of red and white fibres are about equal.
12 Reflex actions use only three neurones and two synapses. There is very little delay as impulses travel around the reflex arc to bring about a response.
13 Reflexes are built into the nervous system, but babies need to learn how to link the appropriate reflexes together in the ordered sequence needed for walking. They also need to learn how to balance as they walk.
14 The cerebral cortex is the folded outer surface of the cerebrum. It is about 3 mm thick.
15 Sensory areas of the cortex receive and process nerve impulses from sense organs. The association areas receive processed information from the sensory areas about changes in the environment. Using this information, and relating it to past

experience stored in the memory, the association areas send instructions to the motor areas. The motor areas send nerve impulses to the muscles to make appropriate movements.
16 The auditory sensory areas would be receiving signals from the ears, the auditory association areas would be receiving information from the auditory sensory areas and sending instructions to the hands via the motor areas. There would also be visual input from the eyes to the visual sensory areas as part of the mechanism controlling note-taking.
17 a When you sit an examination, you need working memory to keep track of the sentences you are reading or writing, and long-term memory to recall your knowledge and understanding of the subject.
   b To answer this question you need working memory to remember the question, and long-term memory to remember what you have read earlier in the chapter.
18 Accurate speech needs feedback from hearing what we say. Any imperfections in the way we speak can then be put right. People born deaf can not learn in this way.
19 To maintain balance, we need feedback from our vision of the horizon and objects around us. In a white-out there is no such feedback.
20 a Cardiac muscle and smooth muscle.
   b The medulla controls the autonomic nervous system. This part of the nervous system controls the actions of the internal organs of the body. It therefore controls heart beat and breathing rate.
21 a A ganglion is a small area of grey matter outside the central nervous system.
   b Parasympathetic ganglia are found in organs controlled by the parasympathetic nervous system.

22 Postganglionic neurones of the sympathetic nervous system secrete noradrenaline. The effects of stimulating the sympathetic nervous system are similar to the longer-term effects of the hormone adrenaline, but the sympathetic nervous system acts first. Adrenaline maintains the ready-for-danger state brought about by the sympathetic nervous system.

23 Mike Stroud lost more water in the Sahara because he sweated much more than he did in the Antarctic.

24 The osmoregulatory system responds by stopping the release of antidiuretic hormone from the hypothalamus. The kidneys then make a more dilute urine.

Chapter 5

1 Learned behaviour is not inborn. Learning happens through changes in the structure of neurones and synapses. These changes do not alter the DNA in the nuclei of sperms or eggs, so learned behaviour can not be inherited.

2 The migration of blackcaps is innate.

3 Young geese learn the location of good winter feeding areas by following their parents. This allows for changes in migration routes if feeding grounds change.

4 Only reflex actions (for example, moving the hand from a hot or sharp object, the stretch reflexes in muscles) are innate in humans. See Chapter 4 for more information on reflex actions.

5 A taxis is orienting behaviour in which an animal keeps turning towards the side where a stimulus is more intense (positive taxis) or less intense (negative taxis).

6 a A taxis is an orienting, directional response towards or away from a stimulus. A kinesis is a change in the rate of movement or turning; the movement is not directionally related to the stimulus.

b When flatworms encounter a chemical gradient, they start to turn at an increasing rate until they touch the food. This is a kinesis because the direction of movement is not related to the direction from which the stimulus is coming.

7 A robin would defend territory from another robin but would not attack a male of another species such as a sparrow.

8 Formal displays and conventions resolve conflicts without violent attack and possible injury.

9 a The features likely to be sign stimuli are: large forward-facing eyes, a hooked beak, streaky brown feathers, and a large head.

b Each model should be tested for the same length of time at the same time of day in an aviary, and mobbing incidents counted and compared. The same number of small birds should be used each time, but not the same individuals.

10 Habituation leads to a change in behaviour (ceasing to respond to a stimulus) that lasts for at least several hours.

11 There must be a repeating pattern of exactly the same stimulus.

12 Imprinting changes the behaviour of chicks and ducklings. They learn to follow a particular object to which they did not respond before imprinting.

13 The glove puppets ensure that the birds imprint on and learn to recognise their own species, rather than imprinting on humans.

14 a Pavlov's dogs learned to associate the sound of the bell with the delivery of food. This is a conditioned reflex (also called classical conditioning).

b Reinforcement develops operant conditioning.

15 The Speedo advertisement suggests that by buying this brand of swimwear you will become as proficient and at home in the water as sea creatures themselves. The advertisement for Häagen-Dazs suggests that eating this ice cream leads to romance.

16 Children can be taught to associate poor behaviour with negative reinforcement and good behaviour with positive reinforcement. For example, small children could be ignored or discouraged when behaving badly, and given attention and encouragement when behaving well.

Chapter 6

1 Pyramids of biomass show that in any habitat there is less food available at higher trophic levels. So, if they are to find enough food, birds of prey need larger territories than herbivorous birds.

2 Breeding success was higher at the better quality nest sites.

3 Shags can not defend a feeding area because they feed out to sea on shoals of fish, which are continuously moving.

4 a Houses and gardens are like territories. They have fences and walls and lockable doors, and are defended from other people. Humans also have home ranges, such as town centres, that are shared with many others.

b National boundaries are defended from other nations, so they are like territorial boundaries. However, humans can recognise friendly or hostile intent, and can choose to allow others into their houses or countries.

5 Individual space prevents fighting over food and also reduces the chances of catching a disease or parasites.

6 In temperate latitudes, increasing day length in spring triggers the breeding cycle through its effect on sex hormone secretion.

7 In polygynous species, one male commonly attracts several females. A male with a slight advantage in attractiveness to females will have a great reproductive advantage.

8 Parental investment is the term for the resources of energy and time that are put into producing and rearing offspring.

9 The chart should have a column for each of the separate courtship stages and numbers down the side to show the action sequence in which the stages were observed.

Action no.	Se	A	F	P	L	S	CA
1	✓						
2		✓					
3	✓						
4		✓					
5			✓				
6		✓					
7			✓				
8				✓			
9					✓		
10						✓	
11							✓
12							

In this chart, the sequence of events was: Se → A → Se → A → F → A → F → P → L → S → CA. After many tick charts have been completed, an ethogram can be drawn with the thickness of the lines showing the average frequency of each step in the sequence.

10 Skylark: song, hopping and bowing display, ruffling neck and crest feathers.
Swallow: song, tail streamers.
Bird of paradise: tail display.
Cuttlefish: rippling colour changes of skin made by chromatophores.
Fiddler crab: pattern of claw waving.
Crickets and grasshoppers: pattern and frequency of sounds.

11 a Pheromones are chemicals used for communication by smell between members of a species.

b Pheromones can move round objects and are energy efficient.

c Visual communication is effective over long distances, is not affected by wind, and gives a precise location for the sender of the message.

12 The caterpillars could become resistant to pesticides. Pesticides might also kill other insects that are useful. The pesticides could be washed into lakes and rivers and harm other animals.

13 Pheromones cannot travel up wind, they cannot travel as far as visual or sound signals, and they do not show the exact location of the sender.

14 a Farmers keep a bull to impregnate their dairy cows at the appropriate time so that they continue to produce milk.

b Artificial insemination.

15 Young swans and geese stay together during the winter, so they need to recognise each other. Larger family groups have an advantage in competing for food in the flock.

16 Primates have complex social lives. They need to learn about the behaviour of other members of their group in order to breed and rear their young successfully.

17 a Play probably helps children to learn how to socialise successfully, being neither too aggressive nor too submissive. (Among adults it may help to reduce social conflict.)

b Humans can signal the desire to play by smiling, explaining through speech, and demonstrating – for example, by throwing a ball.

c The ability to predict behaviour can be helpful in finding food or a mate.

18 Each pair of bitterns needs about 1 km² (100 hectares) of reedbed. 100 pairs would need 100 km² (10 000 hectares).

19 Many people think we have a responsibility for the conservation of all our native species, including wolves. The argument is that if we do not conserve our own wildlife, we cannot justify trying to persuade other countries to conserve theirs. However, there are practical problems in reintroducing large mammals such as wolves and bears to a small crowded country like Britain. On the other hand, the reintroduction of smaller animals like beavers may be possible in the near future, and the reintroduction of otters to parts of England has already been very successful.

Chapter 7

1 Primate features related to life in trees are: stereoscopic vision, dextrous ability, and some reproductive characteristics.

2 a They all display a range of hand angles (a normal distribution). The most common class is 40–50°.

b The primate hand is a grasping hand with opposability. This enables animals to hold onto branches or, in the case of the young, to hold onto the mother's fur.

3 Single births allow for more effective rearing and care of the young in the treetop environment. A firm grasp is useful as the young can be carried by holding onto an adult's fur.
Pectorally mounted mammary glands make it easy for a female to feed her young while sitting on a branch and cradling the young animal.

4 a Gestation is about 1.4% of squirrel monkey life span.

b The juvenile phase is about 13% of squirrel monkey life span.

c Adulthood is about 86% of squirrel monkey life span.

5 a Gestation is about 1% of human life span.

b The juvenile phase is about 20% of human life span.

c Adulthood is about 78% of human life span.

6 The mother always has less sleep than her offspring, but the patterns are similar. The general trend for both is a decrease in amount of time spent sleeping. The patterns are similar because a newborn infant spends a lot of time asleep, and the mother mimics the young monkey's activity patterns as she cares for it.

7 Extended dependency increases the young animal's chance of survival. It also allows time for social and physical skills to be learned by the young.

8 The larger the cortex size, the larger the group size (a positive correlation).

9 Modern lemurs are found in the wild only in Madagascar (an island off the east coast of Africa).

10 Lemurs communicate by scent marking and by sound. Some species also use urine as an additional scent marker.

11 Acting like a submissive gorilla gives the observer the best chance of not being attacked. The gorillas tend to ignore the observer, so he or she is able to record natural gorilla behaviour.

12 a The silverback male is called Beethoven.
   b Pantsy – the offspring of the match is called Banjo.

13 It is a social activity. It reinforces the bonds between gorillas and helps to maintain the hierarchy of the gorilla group.

14 The adult dominant male (the silverback) associates closely with the females. Immature males keep their distance from the silverback – their presence could be seen as a threat to his dominance.
The adult females spend a lot of time with each other and their young.
Immature males keep their distance from females – only the silverback can mate with the females and he might see immature males as a

threat. As the young gorillas are with the females, the immature males do not get closer to the juveniles than about 5 metres.
The juveniles associate with each other a lot, probably in play.

15 Bonobos live in groups that include mature animals of both sexes. Gorillas have only one mature male in a group.
Large silverback male gorillas exert their dominance by strength, whereas bonobo males and females are of a similar size and do not have a hierarchy based on strength.
Only the silverback gorilla can mate with the mature female gorillas, whereas all mature male and female bonobos in a group can mate freely. Female bonobos show they are receptive to males by sexual swellings present for most of their 35-day cycle. Gorilla females are only receptive to the silverback for 5 days each month.
Bonobos associate closely with all members of the group, often with sexual behaviour. Gorillas have a strict hierarchy within the group.

16 Primate species are adapted to live in the forests. So, they are threatened by human expansion into forested areas. Such expansion includes expanding farming settlements, felling trees for timber, and hunting.

17 There are many issues to be considered in the debate about whether we should protect species in the wild or keep them in zoos. Protection in the wild ensures that the habitat is protected as well as the species, but there are often problems of animal welfare and protection from poachers. On the other hand, responsible modern zoos guarantee the welfare and the provision of food for endangered species. However, in the artificial conditions of a zoo, many aspects of animal behaviour can be affected and the animals might become unable to be returned to the wild. At the same

time, habitats might disappear before animals can be returned to them.

Chapter 8

1 Hunter–gatherer communities have a way of life in which plant food is gathered and animals are hunted from the natural environment.

2 The limiting factors are a long period between births (enhanced by extended sucking of infants) and short life expectancy in the harsh bush environment.

3 Humans are now believed to be more closely related to the African apes (chimpanzee and gorilla) than the Asian apes (for example, the orang-utan).

4 If it were true, the calculation that humans spread around the world in about 200 000 years would be unreliable. In that case, modern humans might not be descended from a common African human ancestor, but might have evolved from earlier migrations of *Homo erectus*. (See also p.189, Fig. 6)

5 The pelvis and a whole femur would be most useful. An S-shaped spinal column and a skull with a central foramen magnum would be further evidence of bipedalism.

6 First, fill the upturned skull with sand through the foramen magnum. Then, empty the sand into a measuring cylinder. As most skulls are assembled from incomplete fragments, this is difficult to do accurately with an actual skull; it can be done using a cast of the skull.

7 Chimpanzees have square jaws with projecting incisors, prominent canines, and a diastema. Humans have rounded jaws with vertical teeth; the canines are in line with the other teeth, and there is no diastema. Humans have a chin and a short muzzle or snout, whereas chimpanzees have a long muzzle.

8 Fossil bones and skulls can indicate how the animal moved (locomotion), what its diet was, and how big its brain was.

9 a Potassium dating would be appropriate for a specimen over 4 million years old.

   b For a specimen less than 40 000 years old, carbon dating is the most suitable method.

10 The trends from *A. afarensis* through *A. africanus* to *A. robustus* are:

   a The profile of the face becomes flatter with less of a snout or muzzle.

   b The cranium becomes larger and more domed.

   c The teeth become more vertical and the prominent canine disappears. The jaws become deeper and more robust (heavily built).

11 a The profile of the face becomes flatter with less of a snout.

   b The cranium becomes larger and more domed.

   c The teeth become more vertical and smaller. The jaws become shallower and more gracile (lightly built).

12 a *A. afarensis* has a shallower jaw and a more gently curving zygomatic arch than *A. africanus*. In these features, *A. afarensis* is the more like *H. habilis*.

   b *A. africanus* has a larger brain and smaller brow ridges than *A. afarensis*. In these features, *A. africanus* is the more like *H. habilis*.

13 The features that do not seem to fit are: the thickness of tooth enamel; the size of the molars; the robustness of the jaws. The structure of teeth and jaws might reflect local dietary adaptations rather than an overall evolutionary progression.

14 a Drawing (i) is an Acheulian hand axe; drawing (ii) is an Oldowan chopping tool; drawing (iii) shows a Mousterian scapper and hand axe.

   b Drawing (iv) is a pointed drill; drawing (v) is a barbed blade; drawing (vi) is an arrow head; drawing (vii) is a scraper.

15 Fire can provide warmth, light and protection against predators. It can also be used for cooking and hardening wooden sticks so they can be used as spears or for digging. We can conjecture that fire might have provided a social focus for our ancestors while eating and tool making.

16 No. Kanzi can understand some of the human language but does not have the vocal apparatus to produce speech.

17 Electronic transfer of information has increased the speed at which information is transmitted around the world. Interaction via keyboards, video and other media has added another dimension to the ways in which humans communicate. A new vocabulary associated with information technology has come into being. However, whether or not these additions will fundamentally alter social evolution, it is not possible to say.

Chapter 9

1 a Features showing discontinuous variation have two (or more) distinct categories. Features showing continuous variation have a range of values.

   b Human ear length is a feature showing continuous variation; Darwin's point is a feature showing discontinuous variation. Table 1 on page 126 gives other examples.

2 Cosmetic change affects the appearance (or phenotype) but does not change the genetic material.

3 a Dizygotic twins grow from two ova fertilised by two separate sperm.

   b Identical twins begin life as one fertilised ovum. As they have the same genetic material, they are genetically identical and must be the same sex.

4 a Table 3 suggests that the more closely related two individuals are, the more similar their intelligence test results. This suggests that genetic factors are important.

   b Identical twins reared apart show less of a correlation than identical twins reared together. This suggests that the environment affects intelligence scores.

5 a The scattergram in Figure 2 shows a positive correlation between the average ridge count of the parents and the ridge count of their offspring. This suggests that genetic factors are important. Table 4 also supports the genetic inheritance of fingerprint ridge counts. For example, a child inherits 50% of its genes from its mother. This is very close to the figure for the mother–child correlation.

   b The scattergram in Figure 2 shows that the correlation is not perfect between the average ridge count of the parents and the ridge counts of offspring. The observed correlations in Table 4 do not exactly match the theoretical correlation. These pieces of evidence suggest that environmental factors make a small contribution to ridge counts in individuals.

   c If a pregnant woman is infected with the rubella virus, the sight and hearing of her unborn baby can be affected. Rickets can lead to deformed bones. Other environmental factors include exposure to the sun (which affects the condition of the skin) and diet (an excess of nutrients leads to weight gain and obesity, a lack of nutrients can lead to impaired growth).

6 Vitamin D is important in bone growth and development. A lack of vitamin D causes rickets – a disease in which the bones, while still relatively soft, deform under the weight of the body.

7 Short wavelength ultraviolet radiation, UVC, is very efficient at causing sunburn but natural UVC is absorbed by ozone in the upper atmosphere and does not not reach Earth. The mid-wavelength ultraviolet radiation, UVB, is slightly less efficient at causing sunburn, but it does reach sea level and is the natural cause of sunburn. The long wavelength ultraviolet radiation, UVA, does not cause sunburn.

8 a Melanocytes produce the pigment melanin (in melanosomes) and thereby protect the nuclei of skin cells from the harmful wavelengths of ultraviolet radiation.

  b UVB.

9 Pigmentation is darkest where the ultraviolet light intensity is greatest. The lightest levels of pigmentation are found in regions where ultraviolet light intensity is lowest.

10 Skin pigmentation varies. Lighter-skinned individuals would have an advantage in regions of lower light intensity because a paler skin produces more vitamin D. Over many generations, natural selection would favour pale skins in these regions since individuals with pale skin would not suffer from vitamin D deficiency.

11 Gene pool is the term for the total number of alleles in a population.

12 Pale skins are a disadvantage in sunnier climates. The sunnier the location, the greater the risk of skin cancer for white-skinned people.

13 a Of the 100 children with malaria, 96% had the genotype Hb^AHb^A whereas only 4% had the genotype Hb^AHb^S.

  b The genotype Hb^SHb^S is rare and the sample was of only 100 children. Without medical care, few individuals with this genotype survive childhood.

14 In general, the Hb^S allele is found in areas where malaria is common. However, there is a widespread distribution of malaria in areas without the Hb^S allele. There are also some areas where the Hb^S allele is found but there is little or no incidence of malaria. It is thought that this might be due to immigration of people carrying the Hb^S allele.

15 The advantage of having the genotype Hb^AHb^S is that the chance of developing malaria is reduced. The disadvantage is that some sickling of the red cells occurs in these individuals.

16 The cystic fibrosis gene transfer inhaler is used to provide a liposome–DNA complex that can overcome the faulty genes in lung cells. The lipid in the liposome–DNA complex fuses with the cell membrane and releases DNA into the cell. The DNA enters the cell nucleus and enables the synthesis of normal CFTR protein. This protein allows the transport of chloride ions that is blocked by faulty CFTR protein.

17 Gene therapy might help sufferers of the disease but it will have no direct effect on the incidence of genetic disease. The number of patients with such diseases might increase if therapy can overcome the effects of the disease.

Chapter 10

1 The Neolithic revolution is the change from hunter–gatherer to farmer. It began about 10 000 years ago in the Middle East.

2 Cultivated wheat has smaller and less tightly fitting glumes (naked grains) and a stronger rachis.

3 Modern cultivated wheat is polyploid (has more than two sets of chromosomes) whereas the wild ancestors are diploid (have two sets of chromosomes).

4 Short straw: less energy goes into stalk growth so more is available for the grain.
Standing power: more grain is collected because there is less collapse from weak stalks.
Resistance to shedding: less grain is lost before and during the harvest.
Resistance to fungi: there is less damage to leaves, so more efficient photosynthesis is possible.

5 a The pig is an omnivore whereas most domesticated farm animals are herbivores.

  b Trampling and eating of the vegetation by animals exposes the soil to erosion by wind and rain.

6 a Cattle are a source of food (meat and milk), clothing (skins) and material for tools and utensils (bone and horn). They can also provide transport and be used for ploughing.

  b Selecting useful characteristics and features suitable for local conditions eventually leads to many localised types of cattle.

7 a Modern farming is highly competitive and uses only a limited number of specialised and profitable breeds. This reduces genetic diversity.

  b As uneconomic breeds are lost from farming, any characteristics they possess that might be useful in the future are lost.

8 Grain: after ripening on the plant, grain can be harvested and stored in a cool, dry, well ventilated place such as a roofed building.
Meat: after removal from the carcass, meat can be either dried over a fire or salted. The dried or salted meat can be stored in a cool, dry, well ventilated place.

9 During the transition from hunting and gathering to farming, communities became settled (sedentary) rather than nomadic; gained specialist knowledge such as how to cultivate plants and tend animals; increased in size; started to trade with other communities.

10 By 7000 BC, the diet appears to rely heavily on domesticated species such as sheep, goats and cattle. There seems to be less dependence on gazelle and fox for food.

11 Many butchering tools found in or near to the home site suggest systematic killing and butchering of domesticated animals; large numbers of bones recognisable as those of domesticated species indicate animal husbandry; a high proportion of bones from young animals indicates the maintenance of a breeding herd.

12 A picture of our agricultural past has been assembled from evidence of increasing settlement size; the shape and size of buildings; evidence of enclosures for animals; evidence of irrigation systems; remains of farming tools; animal remains; plant remains; comparisons of cultivated and wild wheats.

13 Planting immediately after clearance, replanting after harvest, and maintaining the vegetation cover all help to protect soil from erosion by rainwater because plant roots bind the soil together.
Using windbreaks of native species and conserving areas of natural forest protects crops and reduces erosion by wind and rain.
Contour ploughing and installing drainage systems prevent soil being washed away by streams that form and run down the hillside.
Improving soil fertility with vegetable waste and manures improves soil structure and promotes better root growth which then holds soil together.

14 a A human must work with the oxen during ploughing. The human uses energy.
   b It is much quicker to use oxen (about 6 times faster than using human power alone).

15 Cash crops are crops that are grown for export to bring in foreign currency rather than grown to feed the farmers and their families.

16 a One organic rotation takes 7 years; a convention rotation takes 5 years.
   b The organic system takes longer and has fewer cereal crops in a rotation (two instead of three). The organic system has four consecutive years of grass/clover lea, which doesn't feature in the conventional system.
   c Peas (and beans) are legumes – they fix nitrogen. This incorporates nitrate into the soil. The plant wastes contribute to soil structure and fertility.

17 Silage requires the use of artificial fertilisers to get good grass growth in the spring. Highly specialised machinery with a high energy consumption is required. This machinery is expensive, cannot be put to other uses, and requires fossil fuel.

18 Even with crop protection products, total estimated losses are about 44%.

19 Diseases are more readily transmitted.

20

Intensive farming	Subsistence farming
• highly mechanised	• little, if any, mechanisation
• high use of crop protection products	• few, if any crop protection products used
• high energy consumption of fossil fuels	• low energy consumption; reliance on human and animal power
• high yields	• low yields
• highly specialised (little diversity)	• wide range of crops

21 Fossil fuels are not renewable and will be used up.

22 a Biomass is plant material produced by photosynthesis.
   b Ethanol and methane (biogas) are biological fuels.

Chapter 11

1 Fertilisation occurs in the oviduct.

2 If the oviduct is blocked, the sperm and ovum cannot meet. So, no fertilisation occurs.

3 a The first day of a new menstruation is taken as the first day of a new reproductive cycle.
   b Ovulation in humans occurs at about day 14 of the cycle.
   c Ovulation occurs about every 28 days in women.

4 a FSH (follicle stimulating hormone) and LH (luteinising hormone) are produced by the pituitary gland.
   b Oestrogen and progesterone.

5 a Oestrogen and progesterone levels are controlled by negative feedback. An increasing level of oestrogen inhibits FSH; without FSH to stimulate oestrogen production, the level of oestrogen falls. This is negative feedback. Similarly, an increasing level of progesterone inhibits LH and leads to a reduction in progesterone production.
   b If fertilisation occurs, the corpus luteum does not degenerate and continues to make progesterone. The endometrium does not break down. The reproductive cycle does not start again until the pregnancy is over.

6 The artificially high level of oestrogen inhibits the production of FSH. So, no ovum develops, ovulation does not occur and pregnancy is not possible.

7 Without contraception, there is a 50% chance of conception.
Natural methods avoid intercourse at fertile times of the cycle and reduce the chances of pregnancy by half. Spermicides used alone are not very effective as not all sperms are killed.

Physical barriers such as the condom and diaphragm are quite effective because they prevent sperm and ova from meeting.

The IUD prevents implantation and is more effective than physical barriers. The pill prevents ovulation and is very effective.

Sterilisation is almost 100% effective. Sperms are not released following vasectomy; ova cannot pass down the oviduct after tubal ligation.

8 The head and brain.

9 **a** The boy was born 4 weeks early and was heavier than normal.

   b During the first 10 weeks of life growth, was slow (only 1 cm gain). In the next 10 weeks, growth rate was normal but slightly below the mean. During weeks 20–50, the growth rate was normal but slightly above the mean.

10 The tight grasp of a newborn baby is an innate reflex. Innate reflexes are present from birth but soon disappear. Learned behaviour is acquired by experience. Examples are: learning to walk, sharing, responding appropriately to a variety of social situations.

11 **a** In males, the growth rate is greatest during gestation and in adolescence at about age 14 years.

   b In females, the growth rate is greatest during gestation and in adolescence at about age 12 years.

12 As humans grow, the head becomes relatively smaller, the body becomes relatively slightly smaller, and the legs become relatively longer.

13 Growth is controlled by PGH (pituitary growth hormone) acting on the long bones, the soft tissues and the thyroid. The thyroid produces thyroxine which controls growth and metabolism.

14 **a** The pituitary stimulates the testes to produce testosterone. This causes the adolescent growth spurt.

   b On average, females go through puberty before males but the growth rate is greater in males than females during puberty.

15 Sex differences are the physical differences between males and females. Gender appropriate behaviour is the term to describe the activities seen as appropriate by society for males and females.

16 Ageing includes all physical changes and begins at birth.

17 **a** The maximum decline in efficiency is in breathing capacity.

   b The effect of the decline in breathing capacity is a reduction in gaseous exchange and reduced efficiency of breathing.

18 Physical changes that are sex-related: breasts loose firmness with age; ovaries cease ovulation and oestrogen release; testosterone release from the testes declines; the prostate gland enlarges.

19 Longitudinal studies take a long time to get results. In cross-sectional studies, other factors such as diet can influence the apparent effects of ageing in each age group studied. Cross-sectional studies do not take account of the fact that people of the same numerical age do not necessarily age at the same rate.

20 **a** Increasing somatic mutations lead to changes in cell function, and hence ageing. Environmental factors could cause or accelerate somatic mutation.

   b Incorrect repair and 'wear and tear' lead to degeneration of tissues. Degeneration might be accelerated by pollutants.

   c Inefficiency of the immune system can be caused by genetic or environmental factors. The immune system might also develop autoimmunity.

Chapter 12

1 Pathogens use host nutrients to grow. They produce waste products and toxins. Increasing numbers of pathogens disrupt host cell functions and can damage cells.

2 Physical barriers: skin, mucus and cilia. Chemical barriers: stomach acid, lysozyme, salt in sweat.

3 Pathogens have antigens that are different from the body's own. Foreign antigens are recognised and specific antibodies produced to combine with the antigens and destroy or inactivate the pathogens.

4 Artificial immunity can be provided by live vaccines, dead vaccines, antisera, and toxoids.

5 Immunity to poliomyelitis is provided by a live vaccine, whereas immunity to cholera is provided by a dead vaccine. Live vaccines produce a longer-lasting response than dead vaccines.

6 Washing hands prevents the transfer of pathogens from hands to food. Fully cooking food kills any pathogens present in the food. Roadside stalls might not have hygienic conditions for the handling and preparation of food.

Using bottled or purified water ensures that drinking water is not contaminated.

You should avoid using ice in drinks because freezing does not kill pathogens in contaminated water. Peeling uncooked fruit and vegetables removes surface contamination.

Taking your own sterile syringe and needle guarantees sterility should you need any injections.

7 Percentage growth rate in a given period is calculated from the population numbers at the beginning and end of the period. It can also be calculated from the numbers of births, deaths, immigrants and emigrants during a time interval.

8 **a** and **b**

Period	Population increase thousands	growth rate %
1950–55	684	11.2
1955–60	848	12.5
1960–65	965	12.7
1965–70	925	10.8
1970–75	876	9.2
1975–80	765	7.4
1980–85	977	8.8
1985–90	1051	8.7
1990–95	1064	8.1

c The percentage growth rate was increasing from the period 1950–55 up to the period 1960–65. Since then, the general trend has been for the figure to decrease (despite an increase for the period 1980–85).

9 In a developing country the average life expectancy is about 54 years; in a developed country it is about 74 years.

10 a

Population pyramid for 1950

males   females

age 65

age 15

8 6 4 2 0 2 4 6 8
percentage of total population

Population pyramid for 2000 (projected)

males   females

age 65

age 15

8 6 4 2 0 2 4 6 8
percentage of total population

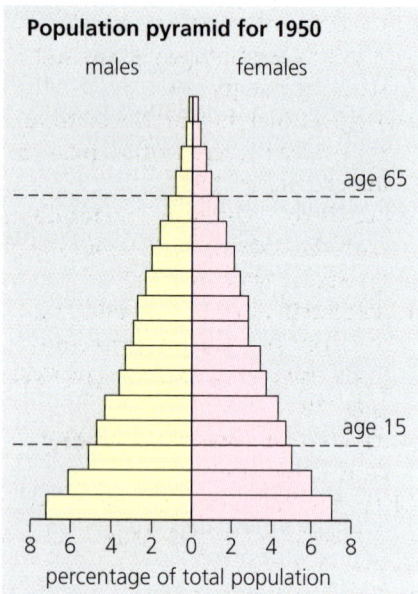

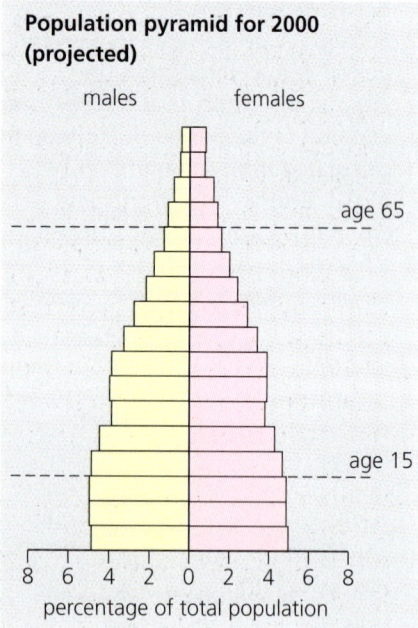

b The pyramid for 2000 has a narrower base than the one for 1950 because there is a lower proportion of young children in the population; it has steeper sides because people are living longer; it is wider at the top because there are more older people.

11 The shorter the doubling time, the greater the percentage growth rate per year.

12 Stage 1: high birth rate and high (fluctuating) death rate; low (but stable) total population.
Stage 2: high birth rate and declining death rate; increasing total population.
Stage 3: declining birth rate and low death rate; increasing population.
Stage 4: low (fluctuating) birth rate and low death rate; high (but stable) total population.

13 Pyramid A represents stage 3; pyramid B represents stage 2; pyramid C represents stage 4; pyramid D represents stage 1.

14 a Sweden: the general trend is a decline in both birth rate and death rate since 1900.
Sri Lanka: from 1900 to 1930 the birth rate increased; from 1930 to 1960 the birth rate was fluctuating; from 1960 onward the birth rate has declined. The death rate increased to 1920, and has generally declined since. The difference between the birth and death rates in Sri Lanka is very much greater than in Sweden; the population of Sri Lanka is growing more rapidly.
b Sweden is in stage 4 of the demographic transition model; Sri Lanka is in stage 3.

15 a An increase in food supply leads to increased fertility rate and birth rate. With better nutrition, there is a decrease in the infant mortality rate and an increase in life expectancy. The result is an increase in population growth.

b A decrease in food supply leads to decreased fertility rate and birth rate. There is an increased incidence of disease, high death rate and poor life expectancy. The result is a decrease in population growth.

16 If couples marry later, the fertility rate tends to decrease.

17 a Cholera spreads rapidly in conditions where the population is overcrowded, water supplies are shared, and there is inadequate sewage treatment. These conditions tend to be found among poor urban populations.
b The main factors in eradicating cholera from the UK were improvements in housing conditions leading to a reduction in overcrowding, and the establishment of efficient water purification and sewage treatment.

18 a With a growth rate of 1.1% per year, the total population each year is:
1996 $\left(\dfrac{1.218}{100} \times 1.1\right) + 1.218$
      = 1.231 billion
1997   1.245 billion
1998   1.259 billion
1999   1.273 billion
2000   1.287 billion
b Radical birth control policies to substantially reduce the birth rate are the only way to achieve the target population of 1.2 billion rather than the projected population of 1.287 billion.

19 Poor rural populations depend on a large labour force for agriculture. Because of this, people in rural areas are more likely to resist the one-child policy. It is more difficult to impose family planning controls in scattered rural communities.

20 The pill reduces blood loss during menstruation. So loss of iron, which is contained in the red blood cells, is reduced.

Glossary

absolute date the definite age of a fossil or rock 114

absolute dating a way of finding the precise age of a fossil or rock 114

accommodation changes in the thickness of the lens of the eye in order to adjust the focal length of the eye for near and distant vision 25

acetylcholine a transmitter at some synapses and neuromuscular junctions 14

acetylcholinesterase the enzyme that breaks down acetylcholine into choline and acetate at a synapse 15

actin the protein that forms the thin filaments in myofibrils 50

action potential the brief reversal of electrical potential between the inside and the outside of an axon or muscle fibre membrane 10

active immunity immunity in which the body produces its own antibodies 173

acuity the ability of the eye to see detail 28

adaptation the loss of response caused by a shortage of transmitter when a synapse is stimulated very frequently 17

adolescence the period of physical growth and sexual development that divides childhood from adulthood 164

adrenergic an adrenergic synapse uses the transmitter noradrenaline 14

after-image an image seen after the eye has looked away from an object 32

all-or-nothing law the size of an action potential is not affected by the strength of the stimulus – all action potentials are the same size but they can vary in frequency 12

allometric growth the growth of organs and body parts at different rates 161

amacrine cell a cell in the retina that links bipolar and ganglion cells 33

antagonistic pair a pair of muscles controlling a joint (when one contracts, the other relaxes) 48

anthropoid an advanced primate (for example, a monkey or ape) 95

anti-gibberellin a substance that prevents the stimulation of growth caused by gibberellin growth hormones 46

antibody a protein produced by the immune system in response to an antigen 172

antigen a substance, usually on the surface of a cell, that allows the body to determine self from non-self 172

antiserum (plural **antisera**) a preparation that contains ready-made antibodies 173

aqueous humour the watery fluid between the cornea and lens of the eye 25

artificial auxin a synthetic substance that has an effect on plant growth and development similar to that of a natural auxin 45

artificial immunity immunity produced without infection by a pathogenic organism 173

associate to link one stimulus with another in the process of learning 75

association area a region of the brain receiving information from sensory areas and sending appropriate instructions to motor areas 59

attenuated vaccine (live vaccine) a preparation that contains live but weakened pathogens 173

autoimmunity the condition in which the immune system fails to recognise body cells as self and attacks them 170

autonomic nervous system the part of the nervous system that controls smooth muscles (muscles not attached to bones) and glands 14

auxin one of a group of plant hormones that regulate plant growth and development (auxins are produced in regions of active cell division) 43

auxin inhibitor a substance that inhibits the effects of auxins on plant growth and development 43

average life expectancy the age at which 50% of a population have died 177

benign the term to describe a cancerous tumour that remains a harmless lump 133

biogas methane produced by the fermentation of organic material 153

biological clock a mechanism in cells that controls such cycles and rhythms as photoperiodism and sleeping/waking 40

biological fuel fuel produced from renewable sources such as plant or animal wastes 153

bipedalism walking on the hind legs with the body upright (also called bipedal movement) 112

bipolar cell a cell in the retina that links one or more photoreceptors (rods and cones) to optic ganglion cells 28

birth control ways of preventing conception 159

birth rate the number of births per thousand of the population per year 176

blind spot (optic disc) the point on the retina at which sensory nerves converge to form the optic nerve (so there are no photoreceptors and no sensitivity to light) 26

bombycol a pheromone made by female silk moths 86

calcium switch the activation by calcium of the protein mechanism in myofibrils (the presence of Ca^{2+} ions allows myosin to bind to actin) 51

cardiac muscle heart muscle 61

cardiovascular centre the area of the medulla that controls heart rate 61

carnivorous feeding only on other animals 113

cascade amplification a sequence of enzyme reactions in which an activated enzyme activates the next in the sequence which in turn amplifies the initial response 30

centile chart a type of graph to show the expected growth pattern of a child's height or mass with age 162

central nervous system (CNS) the brain and spinal cord 7

cerebellum the part of the hindbrain that coordinates balance and fine movement 60

cerebral cortex the layer of grey matter, rich in synapses, that covers the surface of the cerebral hemispheres 58

cerebral hemisphere one of the two hemispherical halves of the cerebrum 58

cerebrum the main part of the forebrain (it is greatly enlarged in mammals, particularly in humans) 57

cholinergic a cholinergic synapse uses the transmitter acetylcholine 14

choroid layer the layer behind the retina that contains black pigment and blood vessels 27

classical conditioning the conditioning of a reflex action so that an animal performs the reflex in response to a new stimulus 75

coleoptile the thin layer of cells covering and protecting the germinating shoot of grasses and corn 42

complex cell a cell in the visual cortex sensitive to sloping edges or shadows 33

conditioned reflex a reflex that has been modified by classical conditioning 75

conditioning associative learning (conditioning is divided into classical conditioning and operant conditioning) 75

constraint on learning a restriction on the ability of animals to learn (the restriction is determined by the structure of the nervous system) 66

continuous variation the type of variation that displays a range of values (for example, height) 126

contraception methods of birth control 159

convention (in behaviour studies) behaviour that determines the outcome of competition for territory without the need for fighting 71

core area the central part of a home range (it is not shared with any other animal or group of animals) 82

cornea the transparent convex window in the front of the eye that carries out most of the refraction 25

corpus callosum the band of nerve fibres connecting the two cerebral hemispheres 58

corpus luteum the structure that develops from the empty follicle after ovulation 157

correlation a link between two measured variables 128

courtship display advertising behaviour that attracts a mate 84

crop protection product a chemical for pest control or for use as fertiliser 151

cross-bridges the heads of myosin filaments that link to actin filaments and swing forwards to make a myofibril contract 51

cross-sectional study the simultaneous study of a sample of people of different ages 168

cultural tradition behaviour passed on by learning from one generation to the next 91

cyclic GMP (cGMP) cyclic guanosine monophosphate, one of a family of secondary messengers that are produced when a stimulus or a hormone reaches a cell 30

dark adaptation the gradual change of retinal and opsin to rhodopsin (dark adaptation gives gradually improving sensitivity when entering a poorly lit place) 29

dark band a band in skeletal muscle fibres formed by myosin filaments and including the area in which myosin and actin filaments overlap (dark and light bands are lined up across many myofibrils to give a banded pattern to the whole fibre) 50

dead vaccine a preparation that contains dead pathogens (killed by means of heat or chemical treatment) 173

death rate the number of deaths per thousand of the population per year 176

demographic transition model a model to explain the changes in birth rate, death rate, and total population growth in response to social and economic factors 179

dentition the type, position and number of teeth in an animal 113

depolarised the term to describe the membrane of a neurone or muscle fibre with a reduced potential difference across it 9

dextrous ability the ability to manipulate things with the hands 96

diastema the gap between the canine and incisor teeth (typical of herbivore dentition) 113

diploid having two sets of chromosomes in the nucleus of each cell 141

discontinuous variation the type of variation in which distinct forms can be recognised (for example, having dimples or not) 126

dizygotic twins twins conceived from two separate ova fertilised by two different sperm cells 127

effector a cell or organ that responds to a stimulus (refers especially to muscles and glands controlled by motor nerves) 8

endogenous controlled from within a living organism (not by changes in the external environment) 40

endometrium the lining of the uterus 156

epidemic a sudden and widespread outbreak of infectious disease 180

epidermis the outer layer of the skin 131

ethogram a diagram showing how frequently one behaviour pattern is followed by another 85

excitatory causing an increased response 14

extended dependency the relatively long period in which young primates remain with the adults 98

fertility rate the average number of children born to a woman of child-bearing age in a population 176

fetus the embryo of a mammal from the time when it has formed the main organ systems of the adult body (from about 2 months in humans) 156

filament a protein strand in myofibrils (thin filaments of actin overlap with thick filaments of myosin) 50

flavoprotein a compound that contains the pigment flavin joined to a protein 43

florigen a plant hormone (not yet identified) that causes the initiation of flower buds 40

foramen magnum the hole in skull through which the spinal column passes 112

fovea the place on the retina where there is a high density of cones but no rods (the fovea is on the central axis of the eye) 26

ganglion a region of grey matter (nervous tissue containing synapses) outside the central nervous system 61

gender appropriate behaviour behaviour that is generally accepted by members of a society as appropriate for the men and women in that society 167

gender role the type of activity generally undertaken by each sex in a society 167

gene pool the total number of alleles in a population 132

gene therapy the use of normal genes to override or replace defective ones (for example, trials of the CFTR inhaler to relieve the symptoms of cystic fibrosis) 136

genetic counselling professional help for prospective parents about the chances of their baby inheriting particular combinations of genes 137

genotype the genetic constitution of an organism 134

gestation the time from conception to birth 161

glume one of a pair of structures that enclose the separate grains of wheat 140

gonadotrophic hormone a hormone produced by the pituitary to control the ovaries or testes 157

gracile lightly built (often used to describe a hominid with slender body features) 116

grooming a social activity among primates in which flakes of skin, parasites and plant debris are removed from one animal's fur by another 104

growth (of population) the change in the number of people in the population 176

growth factor a substance produced in the body to control the growth, division, survival or differentiation of cells (nerve growth factor increases axon growth and the survival of neurones) 21

growth rate (of an individual) growth per unit time (for example, increase in height in cm per year) 164

H-zone a paler area in the centre of each dark band where there is no overlap of myosin and actin filaments 50

habituation ceasing to respond to a frequently repeated stimulus (this is a simple type of learning) 73

hand angle a method for comparing the shape of hands and the arrangement of the fingers 97

herd immunity effect the effect of immunising a sufficiently large number of people to protect a whole population from the spread of particular diseases 174

hexaploid having six sets of chromosomes in the nucleus of each cell 141

hierarchy the structure and dominance relationships within a group of animals 103

hippocampus a part of the cerebral cortex used in the storage and retrieval of memory 60

home range an area occupied by more than one animal or group of animals but not actively defended 82

horizontal cell a cell in the retina linking rods, cones and bipolar cells 33

hormone (endocrine) system the system that secretes small quantities of organic substances (hormones) from ductless (endocrine) glands into the blood stream (hormones cause responses in specific target organs) 62

hypercomplex cell a cell in the visual cortex that receives impulses from complex cells and responds to objects that are moving 33

hypothalamus the region of the forebrain that controls activities such as temperature regulation and osmoregulation (it is connected to the pituitary gland and thereby links the nervous system to the endocrine system) 62

immune system the system that resists the effects of pathogenic organisms that have entered the body 172

immunisation the process of providing artificial immunity 173

immunity the ability of an organism to resist infection 173

imprinting a type of learning in which birds that leave the nest shortly after hatching follow the first moving object they see (the young birds then continue to follow that object and no other) 73

individual space the space maintained between individuals of group-living species 82

infant mortality rate the number of deaths of children (up to age 1 year) per 1000 births per year 176

infectious disease a disease that can be passed from an infected individual to another person 174

inhibitory causing a decreased response 14

innate inborn, controlled by genes and not influenced by learning 65

innate releaser mechanism a mechanism in an animal's brain that causes the animal to respond in an automatic way to a sign stimulus 70

interbirth interval the time between births 104

kinesis (plural **kineses**) orienting behaviour in which an animal reduces its rate of movement or increases its rate of turning as the intensity of the stimulus increases 67

lactation the production of milk in mammals 144

larynx voice box 122

learned behaviour a long-lasting change in behaviour brought about by experience 65

lens the transparent convex structure behind the iris that changes shape to adjust the focal length of the eye (accommodation) 25

light band a band in skeletal muscle fibres formed by actin filaments attached to a central Z disc but excluding the region of overlap with myosin filaments (light and dark bands are lined up across many myofibrils to give a banded pattern to the whole fibre) 50

live vaccine (attenuated vaccine) a preparation that contains live but weakened pathogens 173

localisation of function the arrangement of brain function so that the muscles or sensory receptors are mapped out on the cerebral cortex (particular regions of the cortex serve specific regions of the body) 59

long-term memory memory that lasts for longer than a few minutes 60

longitudinal study a study of the same sample of people over a period of time 168

malignant the term to describe a cancerous tumour that will grow, spread and lead to death 133

maturation physical changes in the body with age 162

medulla (medulla oblongata) part of the hindbrain at the top of the spinal cord that helps to control the autonomic nervous system 61

melanin the pigment found in varying degrees in skin 131

melanocyte a cell in the epidermis that produces the pigment melanin 131

melanosome a structure that is found in the cytoplasm of a melanocyte and contains melanin 131

memory cell a cell that retains the ability to produce specific antibodies after infection 173

menstruation the breakdown and loss of the endometrium in mammals 156

mobbing behaviour shown by birds to some predators such as owls or cats (the predator is approached to within a safe distance and subjected to scolding calls and 'dive-bombing') 72

molecular evidence the comparison of chemicals in cells to find out how closely related certain species are 110

monoculture the growth of the same crop in a field for many years 152

monozygotic twins identical twins who start life as a single fertilised egg 127

morula a ball of cells produced by cell division of the zygote 156

motivational state the state of the nervous system or brain that makes an animal more or less likely to respond to a given stimulus 70

motor area (motor cortex) an area of the cerebral cortex controlling the movement of skeletal muscles 59

motor unit a group of muscle fibres that are controlled by a single axon and contract in unison 53

muscle spindle the sense organ inside a muscle that sends sensory information to the brain about the length of the muscle 56

muscle twitch a brief shortening of a muscle fibre 21

myelin sheath the fatty covering round an axon that speeds up the rate of conduction of action potentials and insulates the axon 8

myofibril one of many units that make up a muscle fibre (also called a muscle fibril) 50

myoglobin the protein in red muscle fibres that stores and releases oxygen 54

myosin the protein that forms the thick filaments in myofibrils 50

Na$^+$/K$^+$ pump an ion pump in axon membranes that pumps Na$^+$ ions out of the cell and K$^+$ ions into the cell using ATP energy from respiration 9

natural selection the maintenance in a population of characteristics that give an organism an advantage in a particular environment 132

negative feedback the method of regulation in which increasing the level of a substance brings about changes that reduce its level 158

negative geotropism growth of a plant away from the stimulus of gravity (also called negative gravitropism) 42

negative phototaxis movement, using a taxis, away from the stimulus of light 68

Neolithic revolution the transition from a hunter–gatherer way of life to farming 139

nerve a group of axons leading to or from the central nervous system and enclosed in connective tissue 7

neuromuscular junction a junction between a motor neurone and a muscle fibre 14

neurone a nerve cell 7

New World monkey any species of monkey native to the Americas 95

nomadic pastoralism the way of life in which humans move with their herds of grazing animals 142

noradrenaline transmitter at adrenergic synapses in the sympathetic nervous system 14

normal distribution curve the symmetrical curve produced when a random population is sampled for a feature showing continuous variation 126

oestrus behaviour the behaviour accompanying the stage of a female mammal's reproductive cycle when ovulation occurs 88

Old World monkey any species of monkey native to Africa or Asia 95

olfactory to do with smell 96

omnivorous feeding on a mixed diet of animal and plant material 113

operant conditioning conditioning in which a normal behaviour pattern is reinforced so that the behaviour becomes more frequent 75

opposability the ability to grip objects between the finger tips and tip of the thumb 97

opposable thumb a thumb that can be folded over the palm 97

opsin a protein that combines reversibly with retinal to form rhodopsin 29

optic disc (blind spot) the point on the retina at which sensory nerves converge to form the optic nerve (so there are no photoreceptors and no sensitivity to light) 26

optic nerve the sensory nerve leading from each eye to the brain 26

oral contraception hormones in tablet form that are taken by mouth to prevent conception 159

osmoregulation the regulation of solute concentration in the blood plasma 62

ovulation the release of a mature ovum from an ovary 156

ovum (plural **ova**) an egg cell 155

pair bond a behavioural attachment between male and female birds or mammals that is necessary for coordination of breeding 83

pandemic an epidemic that spreads across international boundaries 180

parasympathetic nervous system the part of the autonomic nervous system that (usually) reduces the activity of the internal organs 61

parching heating grains of wheat for easier removal of the glumes 141

parental investment the time and energy put into producing and rearing offspring 84

passive immunity immunity resulting from the injection of ready-made antibodies 173

pathogen an organism that causes disease 172

percentage growth rate (of population) the change in the number of people per unit time expressed as a percentage 176

Pfr one of two interchangeable forms of phytochrome (when the Pfr form absorbs far-red light it changes into the Pr form) 39

pharynx the area behind the mouth 123

phenotype the physical characteristics of an organism 126

pheromone a chemical substance that affects the behaviour of other members of the same animal species 86

photochemical change a chemical reaction triggered by light energy 30

photoperiod the length of light and darkness during one day 37

photoperiodism the response of living organisms to changes in photoperiod (for example, by coming into flower) 37

phytochrome the pigment in plant leaves that responds to the photoperiod by changing from one form to another and thereby triggers changes in growth or development 39

pituitary gland the small outgrowth at the base of the brain that controls the endocrine system (the pituitary gland is under the influence of the hypothalamus) 62

pituitary growth hormone (PGH) a hormone produced by the pituitary that stimulates growth of the skeleton and muscles (PGH also stimulates the thyroid to produce thyroxine) 165

placenta the disc of tissue that allows exchange of materials between the separate blood systems of the mother and developing fetus in mammals 156

plant hormone an organic chemical substance that controls plant growth and development (also called a plant growth substance, examples are auxins and gibberellins) 43

play behaviour mock fighting and chasing by (usually) young animals 91

polarised the term to describe the membrane of a neurone or muscle fibre with a potential difference across it 9

polyandrous the term to describe a female with more than one mate 82

polygamous having more than one mate 82

polygynous the term to describe a male with more than one mate 82

polyploid having more than two sets of chromosomes in the nucleus of each cell 141

population pyramid a visual representation of the age structure of a population 176

positive geotropism the growth of a plant towards the stimulus of gravity (also called positive gravitropism) 42

positive phototaxis movement, using a taxis, towards the stimulus of light 68

positive phototropism the growth of a plant towards the stimulus of light 42

positron emission tomography (PET) a method of scanning the brain to show which areas are the most active when carrying out different thought processes 59

positron a subatomic particle with a positive charge (anti-particle of an electron) 59

Pr one of two interchangeable forms of phytochrome (when the Pr form absorbs red light or white light it changes into the Pfr form) 39

pre-pubertal (stage) the period of time from birth to puberty 164

prehensile tail a tail that can wrap around and grasp objects 96

prehensility the ability to grasp 97

primary sexual characteristics the sex organs – either male or female – that we are born with 166

primates the order of animals that includes the anthropoids (monkeys and apes) and the prosimians (primitive primates such as the lemurs and the tarsier) 90

proliferation an increase in the number of cells in the endometrium of the uterus as it thickens and develops a rich blood supply 156

prosimian a primitive primate (for example, lemurs and the tarsier) 95

puberty the change from child to adult during which the child develops the secondary sexual characteristics found in adults 164

pupil the adjustable hole in the centre of the iris through which light enters the eye 26

queen substance a pheromone made by queen bees that controls the behaviour of other bees in the colony 87

rachis the flowering stem on which grains of wheat are found (in most grasses it shatters easily to aid seed dispersal) 140

ratchet mechanism the means of shortening of a myofibril by the movement of myosin cross-bridges that attach, swing and detach from actin filaments many times a second 51

receptor a neurone that responds to a stimulus (by a change in the resting potential of the membrane) 8

receptor potential depolarisation in a sensory receptor 12

reflex action a rapid, automatic, innate response to a stimulus 55

reflex arc a chain of two or three neurones linking a receptor to an effector via the central nervous system and controlling a reflex action 56

reflex escape response a rapid, automatic, innate fleeing movement made by an animal in response to danger, often controlled by fast-conducting axons 67

reinforcement a reward (positive reinforcement) or punishment (negative reinforcement) used in conditioning 75

relative dating judging the relative date of a fossil from the rock layer in which it was found 114

repair (in reproduction) the stage of the ovarian cycle in which the womb develops a new lining 156

respiratory centre the area of the medulla that controls breathing rate 61

resting potential the potential difference across a neurone or muscle fibre membrane when it is not conducting an action potential 9

retina the layer of photoreceptors (rods and cones) on which an image is focused 25

retinal a substance that combines reversibly with opsin to form rhodopsin (retinal is also called retinene and is made in the body from vitamin A) 29

rhodopsin the visual pigment in rod cells 29

robust heavily built (often used to describe a hominid with a large bone structure) 116

sarcolemma the membrane of a muscle fibre 52

sarcomere the area between two Z discs – it is the repeating unit of a myofibril 50

sarcoplasmic reticulum the membrane sacs that store Ca^{2+} ions in a muscle fibre 52

Schwann cell a cell that produces a section of the myelin sheath surrounding an axon 8

secondary messenger a signal inside a cell that is triggered by the binding of a hormone or transmitter to receptors on the cell surface membrane 17

secondary response the rapid production of antibody following a second infection by a pathogen 173

secondary sexual characteristics the characteristics that develop at puberty in males or females 166

senescence the deterioration of body function or appearance in old age 168

sensitive period a time in the life cycle of an animal when it is especially likely to learn 74

sensitivity the ability of the eye to see objects at very low light intensity 29

sensory area a region of the brain receiving and processing information from sense organs and passing on nerve impulses to association areas 59

set-aside farmland land taken out of crop production so as to reduce the overproduction of crops (supported by EC subsidy) 80

sex hormone a hormone that controls the reproductive cycle and reproductive behaviour 83

sexual selection the selection of a mate based on attractive features of the opposite sex (leads to the evolution of colours and structures used in courtship) 83

sign stimulus a shape or colour that triggers species specific behaviour 70

simple cell a cell in the visual cortex that responds to information from a group of rods and cones 33

skeletal muscle a muscle that is attached to a bone 48

Skinner box a conditioning apparatus in which an animal such as a rat presses a lever to gain a reward 75

sliding filament hypothesis the idea that muscles contract by actin and myosin filaments sliding between each other 51

smooth muscle muscle that surrounds internal organs such as blood vessels and the gut 61

social organisation living in extended family groups 96

somatic mutation a mutation in body cells 169

spatial distribution (of population) the areas in which a population is found 175

spatial summation the adding together of the effects of two or more axons at a synapse 16

species specific behaviour a response that always occurs in the same way to the same stimulus and is common to all members of the species (also called a fixed action pattern) 70

stereoscopic vision the ability to judge distance 96

stimulus a change in an animal's environment 8

stratigraphy the study of rock layers (rocks are classified by how old they are) 114

subsistence farming production of food for personal consumption by the farmer's family 148

survival curve a graph that shows the number of survivors of a group of 10 000 people plotted against time 177

sympathetic nervous system the part of the autonomic nervous system that (usually) increases the activity of the internal organs 61

synapse a junction between two neurones or between a motor neurone and a muscle fibre where the message is carried by a chemical transmitter 8

synaptic vesicle one of many membrane sacs found in synaptic knobs and containing a transmitter chemical 14

T-tubule a finger-shaped infold of the sarcolemma that conducts action potentials to the interior of a muscle fibre 52

taxis (plural **taxes**) orienting behaviour in which an animal turns towards or away from a stimulus (such as light) 67

temporal summation the adding together of the effects of a series of action potentials arriving at a synapse in quick succession 16

territory area defended from members of the same species (area used for feeding or reproduction) 79

tetanus a sustained contraction of a muscle fibre or whole muscle 53

tetraploid having four sets of chromosomes in the nucleus of each cell 141

thyroxine a hormone produced by the thyroid gland that controls growth and metabolic rate 165

tool culture any of a number of distinctive styles of early tool making 119

toxoid a harmless form of a toxin 174

transmissible disease a disease that can be passed from person to person 174

transmitted (of disease) passed from person to person 174

transmitter a chemical able to carry an impulse across the gap between a synaptic knob and the adjacent neurone 14

trichromatic theory the theory that colour vision occurs through the combined effects of cone cells sensitive to the three primary colours (red, green and blue) 31

tropism the growth of plants towards or away from a stimulus 42

tubal ligation a method of sterilisation for women by putting clips on the oviducts 159

twitch a brief shortening of a muscle fibre in response to one action potential 53

ultraviolet radiation (UVR) a short wavelength section of the electromagnetic spectrum that is particularly harmful to living organisms 130

urbanisation the process by which a previously rural population becomes town-based 182

vaccine a preparation that contains pathogens 173

vas deferens the tube that carries semen from the testis (also called the sperm duct) 155

vasectomy a method of sterilisation for men by removing a section of the vas deferens 159

vitreous humour the transparent jelly that fills the eye behind the lens and keeps the eye in shape 25

vocalisation sounds made by animals to communicate with each other 102

voltage-gated channel an ion diffusion channel that changes its permeability to ions when the potential difference across the membrane changes 11

warning coloration the bright and conspicuous coloration found in many poisonous or distasteful animal species 76

wave of depolarisation the passage of an action potential along an axon 11

working memory memory that lasts for a few seconds or minutes (also called short-term memory) 60

yellow spot the area around the fovea that contains yellow pigment (also called the macula lutea, it improves the sharpness of the image by absorbing blue light) 27

Z disc the boundary between two sarcomeres to which actin filaments are attached (also called the Z line) 50

zygote the first cell produced when a sperm fertilises an ovum 127

Index